Springer Undergraduate Mathematics Series

The Springer Undergraduate Mathematics Series (SUMS) is a series designed for undergraduates in mathematics and the sciences worldwide. From core foundational material to final year topics, SUMS books take a fresh and modern approach. Textual explanations are supported by a wealth of examples, problems and fully-worked solutions, with particular attention paid to universal areas of difficulty. These practical and concise texts are designed for a one- or two-semester course but the self-study approach makes them ideal for independent use.

More information about this series at http://www.springer.com/series/3423

Arkadii Slinko

Algebra for Applications

Cryptography, Secret Sharing,
Error-Correcting, Fingerprinting,
Compression

Second Edition

 Springer

Arkadii Slinko
Department of Mathematics
The University of Auckland
Auckland, New Zealand

ISSN 1615-2085 ISSN 2197-4144 (electronic)
Springer Undergraduate Mathematics Series
ISBN 978-3-030-44073-2 ISBN 978-3-030-44074-9 (eBook)
https://doi.org/10.1007/978-3-030-44074-9

Mathematics Subject Classification (2020): 11A05, 11A07, 11T71, 11Y05, 11Y11, 68P25, 68P30

This Springer imprint is published by the registered company Springer Nature Switzerland AG
The registered company address is: Gewerbestrasse 11, 6330 Cham, Switzerland

*To my parents Michael and Zinaida,
my wife Lilia,
my children Irina and Michael, and
my grandchildren Erik and Yuri.*

Preface to the Second Edition

In our work, we are always between Scylla and Charybdis; we may fail to abstract enough, and miss important physics, or we may abstract too much and end up with fictitious objects in our models turning into real monsters that devour us.

Murray Gell-Mann (*Nobel Prize in Physics in 1969*)

My goals for this edition remain the same. I would like this book to be a basis for a one-semester undergraduate course in applied algebra. I want it to be mathematically rigorous and self-contained, and at the same time to provide a glimpse into the exciting world of applications. The challenge for such a course is to avoid getting overexcited about proving theorems and, on the other hand, not to get bogged down with technical details of the applications. This is a delicate balance, and it is up to the reader to decide how well I managed to steer the exposition between these Scylla and Charybdis.

Apart from correcting misprints and improving the order of exercises I added several small but significant sections that provide links between chapters and make the whole construction of the course more connected. The most notable additions are:

- The chapter on secret sharing (Chap. 6) has now an application to cryptography proper (Chap. 2). By using secret sharing we show how a cryptosystem like RSA can be used by an organisation to share the decryption key between members of that organisation.
- The chapter on polynomials (Chap. 5) was extended by a new section on permutation polynomials which relates this chapter to Chaps. 2 and 3.
- The chapter on compression of information (Chap. 8) was a bit one-sided since it was dealing with encoding of an unknown source but not a known one. The reason was that encoding of an unknown source (universal encoding) has not been adequately reflected in the undergraduate literature while encoding a known source (famous Huffman's codes) was everywhere. However, for the purpose of this book to be self-contained, I wrote a section about Huffman's codes, adding it to Chap. 8.

I also added a number of exercises. Since in the first edition of this book all exercises had solutions (and they still have), I decided to add new exercises (with a few exceptions) without solutions. Those without solutions are marked with a small circle. I also added an index to the book.

Enjoy the book!

Auckland, New Zealand   Arkadii Slinko
February 2020

Preface to the First Edition

The aim of a Lecturer should be, not to gratify his vanity by a shew of originality; but to explain, to arrange, and to digest with clearness, what is already known in the science...

George Pryme (1781–1868)

This book originated from my lecture notes for the one-semester course which I have given many times in The University of Auckland since 1998. The goal of this book is to show the incredible power of algebra and number theory in the real world. It does not advance far in theoretical algebra, theoretical number theory or combinatorics. Instead, we concentrate on concrete objects like groups of points on elliptic curves, polynomial rings and finite fields, study their elementary properties and show their exceptional applicability to various problems in information handling. Among the applications are cryptography, secret sharing, error-correcting, fingerprinting and compression of information.

Some chapters of this book—and especially number-theoretic and cryptographic ones—use GAP for illustrations of the main ideas. GAP is a system for computational discrete algebra, which provides a programming language, a library of thousands of functions implementing algebraic algorithms, written in the GAP language, as well as large data libraries of algebraic objects.

If you are using this book for self-study, then, studying a certain topic, familiarise yourself with the corresponding section of Appendix A, where you will find detailed instructions how to use GAP for this particular topic. As GAP will be useful for most topics, it is not a good idea to skip it completely.

I owe a lot to Robin Christian who in 2006 helped me to introduce GAP to my course and proofread the lecture notes. The introduction of GAP has been the biggest single improvement to this course. The initial version of the GAP notes, which have now been developed into Appendix A, was written by Robin. Stefan Kohl, with the assistance of Eamonn O'Brien, kindly provided us with two programs for GAP that allowed us to calculate in groups of points on elliptic curves. I am grateful to Paul Hafner, Primož Potočnic, Jamie Sneddon and especially to Steven Galbraith who in various years were members of the teaching team for this course and suggested valuable improvements or contributed exercises.

Many thanks go to Shaun White who did a very thorough job proofreading part of the text in 2008 and to Steven Galbraith who improved the section of cryptography in 2009 and commented on the section of compression. However, I bear the sole responsibility for all mistakes and misprints in this book. I would be most obliged if you report any noticed mistakes and misprints to me.

I hope you will enjoy this book as much as I enjoyed writing it.

Auckland   Arkadii Slinko
March 2015

Contents

1 Integers 1
  1.1 Natural Numbers 1
    1.1.1 Basic Principles 1
    1.1.2 Divisibility and Primes 4
    1.1.3 Factoring Integers. The Sieve of Eratosthenes 9
  1.2 Euclidean Algorithm 14
    1.2.1 Divisors and Multiples 14
    1.2.2 Greatest Common Divisor and Least Common
        Multiple 15
    1.2.3 Extended Euclidean Algorithm. Chinese Remainder
        Theorem 18
  1.3 Fermat's Little Theorem and Its Generalisations 23
    1.3.1 Congruences. Fermat's Little Theorem 23
    1.3.2 Euler's ϕ-Function. Euler's Theorem 26
  1.4 The Ring of Integers Modulo n. The Field $\mathbb{Z}_p$ 29
  1.5 Representation of Numbers 34

2 Cryptology 41
  2.1 Classical Secret-Key Cryptology 42
    2.1.1 The One-Time Pad 43
    2.1.2 An Affine Cryptosystem 46
    2.1.3 Hill's Cryptosystem 47
  2.2 Modern Public-Key Cryptology 51
    2.2.1 One-Way Functions and Trapdoor Functions 52
  2.3 Computational Complexity 53
    2.3.1 Orders of Magnitude 54
    2.3.2 The Time Complexity of Several Number-Theoretic
        Algorithms 58
  2.4 The RSA Public-Key Cryptosystem 62
    2.4.1 How Does the RSA System Work? 63
    2.4.2 Why Does the RSA System Work? 66
    2.4.3 Pseudoprimality Tests 68
  2.5 Applications of Cryptology 74

3   Groups 79
   3.1   Permutations 79
        3.1.1   Composition of Mappings. The Group
                of Permutations of Degree n 79
        3.1.2   Block Permutation Cipher 84
        3.1.3   Cycles and Cycle Decomposition 86
        3.1.4   Orders of Permutations 88
        3.1.5   Analysis of Repeated Actions 91
        3.1.6   Transpositions. Even and Odd 93
        3.1.7   Puzzle 15 97
   3.2   General Groups 100
        3.2.1   Definition of a Group. Examples 100
        3.2.2   Powers, Multiples and Orders. Cyclic Groups 103
        3.2.3   Isomorphism 105
        3.2.4   Subgroups 109
   3.3   The Abelian Group of an Elliptic Curve 112
        3.3.1   Elliptic Curves. The Group of Points of an Elliptic
                Curve 113
        3.3.2   Quadratic Residues and Hasse's Theorem 119
        3.3.3   Calculating Large Multiples Efficiently 123
   3.4   Applications to Cryptography 124
        3.4.1   Encoding Plaintext 124
        3.4.2   Additive Diffie–Hellman Key Exchange
                and the ElGamal Cryptosystem 126

4   Fields 129
   4.1   Introduction to Fields 129
        4.1.1   Examples and Elementary Properties of Fields 130
        4.1.2   Vector Spaces 132
        4.1.3   Cardinality of a Finite Field 136
   4.2   The Multiplicative Group of a Finite Field is Cyclic 138
        4.2.1   Lemmas on Orders of Elements 139
        4.2.2   Proof of the Main Theorem 141
        4.2.3   Proof of Euler's Criterion 142
        4.2.4   Discrete Logarithms 143
   4.3   Elgamal Cryptosystem Revisited 144

5   Polynomials 147
   5.1   The Ring of Polynomials 147
        5.1.1   Introduction to Polynomials 147
        5.1.2   Lagrange's Interpolation 152
        5.1.3   Factoring Polynomials 154
        5.1.4   Greatest Common Divisor and Least Common
                Multiple 157

    5.2   Finite Fields 159
          5.2.1 Polynomials Modulo $m(x)$ 159
          5.2.2 Minimal Annihilating Polynomials 164
    5.3   Permutation Polynomials and Applications 167
          5.3.1 Permutation Polynomials 167
          5.3.2 Cryptosystem Based on a Permutation Polynomial 168

6   Secret Sharing 171
    6.1   Introduction to Secret Sharing 172
          6.1.1 Access Structure 172
          6.1.2 Shamir's Threshold Access Scheme 173
    6.2   A General Theory of Secret Sharing Schemes 176
          6.2.1 General Properties of Secret Sharing Schemes 176
          6.2.2 Linear Secret Sharing Schemes 181
          6.2.3 Ideal and Non-ideal Secret Sharing Schemes 186
    6.3   Applications of Secret Sharing 190

7   Error-Correcting Codes 191
    7.1   Binary Error-Correcting Codes 192
          7.1.1 The Hamming Weight and the Hamming Distance 192
          7.1.2 Encoding and Decoding. Simple Examples 195
          7.1.3 Minimum Distance, Minimum Weight.
                 Linear Codes 198
          7.1.4 Matrix Encoding Technique 202
          7.1.5 Parity Check Matrix 207
          7.1.6 The Hamming Codes 212
          7.1.7 Polynomial Codes 215
          7.1.8 Bose–Chaudhuri–Hocquenghem (BCH) Codes 217
    7.2   Non-binary Error-Correcting Codes 221
          7.2.1 The Basics of Non-binary Codes 221
          7.2.2 Reed–Solomon (RS) Codes 224
    7.3   Fingerprinting Codes 227
          7.3.1 The Basics of Fingerprinting 228
          7.3.2 Frameproof Codes 230
          7.3.3 Codes with the Identifiable Parent Property 231

8   Compression 235
    8.1   Encoding a Known Source 236
          8.1.1 Motivating Example 236
          8.1.2 Prefix Codes 237
          8.1.3 Huffman's Optimal Code 240
    8.2   Encoding an Unknown Source 243
          8.2.1 Compressing Binary Sequences (Files) 244
          8.2.2 Information and Information Relative to a Partition 245

	8.2.3	Fitingof's Compression Code. Encoding	248
	8.2.4	Fitingof's Compression Code. Fast Decoding	251
8.3	Information and Uncertainty		254

9   Appendix A: GAP 257
9.1   Computing with GAP 257
      9.1.1   Starting with GAP 257
      9.1.2   The GAP Interface 257
      9.1.3   Programming in GAP: Variables, Lists, Sets
              and Loops 258
9.2   Number Theory 260
      9.2.1   Basic Number-Theoretic Algorithms 260
      9.2.2   Arithmetic Modulo *m* 262
      9.2.3   Digitising Messages 264
9.3   Matrix Algebra 266
9.4   Algebra 267
      9.4.1   Permutations 267
      9.4.2   Elliptic Curves 268
      9.4.3   Finite Fields 271
      9.4.4   Polynomials 272

10 Appendix B: Miscellania 275
10.1 Linear Dependency Relationship Algorithm 275
10.2 The Vandermonde Determinant 276
10.3 Stirling's Formula 277

11 Solutions to Exercises 281
11.1 Solutions to Exercises of Chap. 1 281
11.2 Solutions to Exercises of Chap. 2 295
11.3 Solutions to Exercises of Chap. 3 312
11.4 Solutions to Exercises of Chap. 4 326
11.5 Solutions to Exercises of Chap. 5 331
11.6 Solutions to Exercises of Chap. 6 340
11.7 Solutions to Exercises of Chap. 7 346
11.8 Solutions to Exercises of Chap. 8 358

Literature 363

Index 365

Integers

We must get back to primeval integrity.

Rhinoceros. Eugène Ionesco (1909–1994)

The formula 'Two and two make five' is not without its attractions.

Notes from Underground. Fyodor Dostoevsky (1821–1881).

The theory of numbers is the oldest and the most fundamental mathematical discipline. Despite the old age, it is one of the most active research areas of mathematics due to two main reasons. Firstly, the advent of fast computers has changed number theory profoundly and made it in some ways almost an experimental discipline. Secondly, new important areas of applications, such as cryptography, have emerged. Some of the applications of number theory will be considered in this course.

1.1 Natural Numbers

1.1.1 Basic Principles

The theory of numbers is devoted to studying the set $\mathbb{N} = \{1, 2, 3, 4, 5, 6, \ldots\}$ of positive integers, also called the *natural numbers*. The most important properties of $\mathbb{N}$ are formulated in the following three principles.

The Least Integer Principle Every non-empty set $S \subseteq \mathbb{N}$ of positive integers contains a smallest (least) element.

© Springer Nature Switzerland AG 2020
A. Slinko, *Algebra for Applications*, Springer Undergraduate Mathematics Series,
https://doi.org/10.1007/978-3-030-44074-9_1

The Principle of Mathematical Induction Let $S \subseteq \mathbb{N}$ be a set of positive integers which contains 1 and contains $n + 1$ whenever it contains n. Then $S = \mathbb{N}$.

The Principle of Strong Mathematical Induction Let $S \subseteq \mathbb{N}$ be a set of positive integers which contains 1 and contains $n + 1$ whenever it contains $1, 2, \ldots, n$. Then $S = \mathbb{N}$.

These three principles are equivalent to each other: If you accept one of them, you can prove the remaining two as theorems. Normally one of them, most often the Principle of Mathematical Induction, is taken as an axiom of arithmetic but in proofs we use all of them since one may be much more convenient to use than others.

Example 1.1.1 On planet Tralfamadore there are only 3 cent and 5 cent coins in circulation. Prove that an arbitrary sum of $n \geq 8$ cents can be paid (provided one has a sufficient supply of coins).

Solution Suppose that this statement is not true and there are positive integers $m \geq 8$ for which the sum of m cents cannot be paid by a combination of 3 cent and 5 cent coins. By the Least Integer Principle there is a smallest such positive integer s (the minimal counterexample). It is clear that s is not 8, 9 or 10 as $8 = 3 + 5$, $9 = 3 + 3 + 3$, $10 = 5 + 5$. Thus $s - 3 \geq 8$ and, since s was minimal, the sum of $s - 3$ cents can be paid as required. Adding to $s - 3$ cents one more 3 cent coin we obtain that the sum of s cents can be also paid, which is a contradiction.

Example 1.1.2 Prove that

$$\frac{1}{1^2} + \frac{1}{2^2} + \cdots + \frac{1}{n^2} < 2.$$

Solution Denote the left-hand side of the inequality by $F(n)$. We have a sequence of statements $A_1, A_2, \ldots, A_n, \ldots$ to be proved, where A_n is $F(n) < 2$, and we are going to use the Principle of Mathematical Induction to prove all of them.

The statement A_1 reduces to

$$\frac{1}{1^2} < 2,$$

which is true. Now we have to derive the validity of A_{n+1} from the validity of A_n, that is, to prove that

$$F(n) < 2 \quad \text{implies} \quad F(n) + \frac{1}{(n+1)^2} < 2.$$

Oops! It is not possible because, while we do know that $F(n) < 2$, we do not have the slightest idea how close $F(n)$ is to 2, and we therefore cannot be sure that there will be room for $\frac{1}{(n+1)^2}$. What shall we do?

Surprisingly, the stronger inequality

$$\frac{1}{1^2} + \frac{1}{2^2} + \cdots + \frac{1}{n^2} \leq 2 - \frac{1}{n}$$

can be proved smoothly. Indeed, A_1 is again true as

$$\frac{1}{1^2} = 2 - \frac{1}{1},$$

and

$$F(n) \leq 2 - \frac{1}{n} \quad \text{implies} \quad F(n) + \frac{1}{(n+1)^2} \leq 2 - \frac{1}{n+1} \tag{1.1}$$

is now true. Due to the induction hypothesis, which is $F(n) \leq 2 - \frac{1}{n}$, to show (1.1) it would be sufficient to show that

$$\left(2 - \frac{1}{n}\right) + \frac{1}{(n+1)^2} \leq 2 - \frac{1}{n+1}.$$

This reduces to

$$\frac{1}{(n+1)^2} \leq \frac{1}{n} - \frac{1}{n+1} = \frac{1}{n(n+1)},$$

which is true.

This example shows that we should not expect that someone has already prepared the problem for us so that the Principle of Mathematical Induction can be applied directly.

The reader needs to be familiar with the induction principles. The exercises below concentrate on the use of the Least Integer Principle.

Exercises

1. Verify that each of the following two statements is false:

    (a) Every non-empty set of integers (we do not require the integers in the set to be positive) contains a smallest element.
    (b) Every non-empty set of positive rational numbers contains a smallest element.

2. Prove that, for any integer $n \geq 1$, the integer $4^n + 15n - 1$ is divisible by 9.
3. Prove that $11^{n+2} + 12^{2n+1}$ is divisible by 133 for all $n \geq 0$.
4. Let $F_n = 2^{2^n} + 1$ be the nth *Fermat number*. Show that $F_0 F_1 \ldots F_n = F_{n+1} - 2$.
5. Prove that $2^n + 1$ is divisible by n for all numbers of the form $n = 3^k$.
6. Prove that an arbitrary positive integer N can be represented as a sum of distinct powers chosen from $1, 2, 2^2, \ldots, 2^n, \ldots$.
7. Use the Least Integer Principle to prove that the representation of N as a sum of distinct powers of 2 is unique.

8. Several discs of equal diameter lie on a table so that some of them touch each other but no two of them overlap. Prove that these discs can be painted with four colours so that no two discs of the same colour touch.

9°. Suppose that you begin with a chocolate bar made up of $n \times k$ squares. At each step, you choose a piece of chocolate that has more than two squares and snap it in two along any line, vertical or horizontal. Eventually, it will be reduced to single squares. Show by induction that the number of snaps required to reduce it to single squares is $nk - 1$.

1.1.2   Divisibility and Primes

The set of all *integers*

$$\ldots, -3, -2, -1, 0, 1, 2, 3, \ldots$$

is denoted by $\mathbb{Z}$.

Theorem 1.1.1 (Division with Remainder) *Given any integers a, b, with $a > 0$, there exist $\underline{unique}$ integers q, r such that*

$$b = qa + r, \qquad and \quad 0 \le r < a.$$

In this case we also say that q and r are, respectively, the *quotient* and the *remainder* of b when it is divided by a. It is often said that q and r are the quotient and the remainder of dividing a into b. The notation $r = b \bmod a$ is often used. You can find q and r by using long division, a technique which most students learn at school. If you want to find q and r using a calculator, use it to divide b by a. This will give you a number with decimals. Discard all the digits to the right of the decimal point to obtain q. Then find r as $a - bq$.

Example 1.1.3 (a) $35 = 3 \cdot 11 + 2$, (b) $-51 = (-8) \cdot 7 + 5$; so that $2 = 35 \bmod 11$ and $5 = -51 \bmod 7$.

Definition 1.1.1 An integer b is *divisible* by an integer $a \ne 0$ if there exists an integer c such that $b = ac$, that is, we have $b \bmod a = 0$. We also say that a is a *divisor* of b and write $a|b$.

Let n be a positive integer. Let us denote by $d(n)$ the number of positive divisors of n. It is clear that 1 and n are always divisors of any number n which is greater than 1. Thus we have $d(1) = 1$ and $d(n) \ge 2$ for $n > 1$.

Definition 1.1.2 A positive integer n is called a *prime* if $d(n) = 2$. An integer $n > 1$ which is not prime is called a *composite* number.

Example 1.1.4 (a) $2, 3, 5, 7, 11, 13$ are primes; (b) $1, 4, 6, 8, 9, 10$ are not primes; (c) $4, 6, 8, 9, 10$ are composite numbers.

A composite positive integer n can be always represented as a product of two other positive integers different from 1 and n. Indeed, since $d(n) > 2$, there is a divisor n_1 such that $1 < n_1 < n$. But then $n_2 = n/n_1$ also satisfies $1 < n_2 < n$ and $n = n_1 n_2$. We are ready to prove

Theorem 1.1.2 (The Fundamental Theorem of Arithmetic) *Every positive integer $n > 1$ can be expressed as a product of primes (with perhaps only one factor), that is*

$$n = p_1^{\alpha_1} p_2^{\alpha_2} \cdots p_r^{\alpha_r},$$

where $p_1, p_2, \ldots, p_r$ are distinct primes and $\alpha_1, \alpha_2, \ldots, \alpha_r$ are positive integers. This factorisation is unique apart from the order of the prime factors.

Proof Let us prove first that any number $n > 1$ can be decomposed into a product of primes. We will use the Principle of Strong Mathematical Induction. If $n = 2$, the decomposition is trivial and we have only one factor, which is 2 itself. Let us assume that for all positive integers which are less than n, a decomposition into a product of primes exists. If n is a prime, then $n = n$ is the decomposition required. If n is composite, then $n = n_1 n_2$, where $n > n_1 > 1$ and $n > n_2 > 1$, and by the induction hypothesis there are prime decompositions $n_1 = p_1 \ldots p_r$ and $n_2 = q_1 \ldots q_s$ for n_1 and n_2. Then we may combine them

$$n = n_1 n_2 = p_1 \ldots p_r q_1 \ldots q_s$$

and get a decomposition for n and prove the first statement.

To prove that the decomposition is unique, we shall assume the existence of an integer capable of two essentially different prime decompositions, and from this assumption derive a contradiction. This will show that the hypothesis that there exists an integer with two essentially different prime decompositions cannot be true, and hence the prime decomposition of every integer is unique. We will use the Least Integer Principle.

Suppose that there exists a positive integer with two essentially different prime decompositions, then there will be a smallest such integer

$$n = p_1 p_2 \ldots p_r = q_1 q_2 \ldots q_s, \tag{1.2}$$

where p_i and q_j are primes. By rearranging the order of the factors, if necessary, we may assume that

$$p_1 \leq p_2 \leq \cdots \leq p_r, \qquad q_1 \leq q_2 \leq \cdots \leq q_s.$$

It is impossible that $p_1 = q_1$, for, if it were the case, we would cancel the first factor from each side of Eq. (1.2) and obtain two essentially different prime decompositions for the number n/p_1, which is smaller than n, contradicting the choice of n. Hence either $p_1 < q_1$ or $q_1 < p_1$. Without loss of generality we suppose that $p_1 < q_1$.

We now form the integer

$$n' = n - p_1 q_2 q_3 \ldots q_s. \qquad (1.3)$$

Since $p_1 < q_1$, we have $n' = n - p_1 q_2 q_3 \ldots q_s > n - q_1 q_2 q_3 \ldots q_s = 0$, thus this number is positive. It is obviously smaller than n. The two distinct decompositions of n give the following two decompositions of n':

$$n' = (p_1 p_2 \ldots p_r) - (p_1 q_2 \ldots q_s) = p_1 (p_2 \ldots p_r - q_2 \ldots q_s), \qquad (1.4)$$

$$n' = (q_1 q_2 \ldots q_s) - (p_1 q_2 \ldots q_s) = (q_1 - p_1)(q_2 \ldots q_s). \qquad (1.5)$$

Since n' is a positive integer, which is smaller than n and greater than 1, the prime decomposition for n' must be unique, apart from the order of the factors. This means that if we complete prime factorisations (1.4) and (1.5), the result will be identical. From (1.4) we learn that p_1 is a factor of n' and must appear as a factor in decomposition (1.5). Since $p_1 < q_1 \leq q_i$, we see that $p_1 \neq q_i, i = 2, 3, \ldots, s$. Hence, it is a factor of $q_1 - p_1$, i.e., $q_1 - p_1 = p_1 m$ or $q_1 = p_1(m + 1)$, which is impossible as q_1 is prime and $q_1 \neq p_1$. This contradiction completes the proof of the fundamental theorem of arithmetic.   □

Example 1.1.5 $396 = 2^2 \cdot 3^2 \cdot 11$ and $17 = 17$ are two prime factorisations. The corresponding output of GAP will look as follows:

```
gap> FactorsInt(396);
[ 2, 2, 3, 3, 11 ]
gap> FactorsInt(17);
[ 17 ]
```

GAP conveniently remembers all 168 primes not exceeding 1000. They are stored in the array `Primes` (in Sect. 9.2 all the primes in this array are listed). GAP can also check if a particular number is prime or not.

```
gap> IsPrime(2^(2^4)-1);
false
gap> IsPrime(2^(2^4)+1);
true
```

What GAP cannot answer is whether or not there are infinitely many primes. This is something that can only be proved.

Theorem 1.1.3 (attributed to Euclid[1]) *The number of primes is infinite.*

Proof Suppose there were only finitely many primes $p_1, p_2, \ldots, p_r$. Then form the integer

$$n = 1 + p_1 p_2 \ldots p_r.$$

Since $n > p_i$ for all i, it must be composite. Let q be the smallest prime factor of n. As $p_1, p_2, \ldots, p_r$ represent all existing primes, then q is one of them, say $q = p_1$ and $n = p_1 m$. Now we can write

$$1 = n - p_1 p_2 \ldots p_r = p_1 m - p_1 p_2 \ldots p_r = p_1 (m - p_2 \ldots p_r).$$

We have got that $p_1 > 1$ is a factor of 1, which is an absurdity. So our initial assumption that there were only finitely many primes must be false.   □

In the past many mathematicians looked for a formula that always evaluates to a prime number. Euler[2] noticed that all values of a quadratic polynomial $P(n) = n^2 - n + 41$ are prime for $n = 0, 1, 2, \ldots, 40$. However, $P(41) = 41^2$ is not prime. For the same reason Fermat introduced the numbers $F_m = 2^{2^m} + 1$, $m \geq 0$, which are now called *Fermat numbers*, and prime Fermat numbers are called *Fermat primes*. He checked that $F_0 = 3$, $F_1 = 5$, $F_2 = 17$, $F_3 = 257$ and $F_4 = 65537$ are primes. He believed that all such numbers are primes however he could not prove that $F_5 = 4294967297$ is prime. Euler in 1732 showed that F_5 was composite by presenting its prime factorisation $F_5 = 641 \cdot 6700417$. We can now easily check this with GAP:

```
gap> F5:=2^(2^5)+1;
4294967297
gap> IsPrime(F5);
false
gap> FactorsInt(F5);
[ 641, 6700417 ]
```

---

[1]**Euclid of Alexandria (about 325 BC–265 BC)** is one of the most prominent educators of all times. He is best known for his treatise on mathematics *The Elements* which is divided into 13 books: the first six on geometry, three on number theory, one is devoted to Eudoxus's theory of irrational numbers and the last three to solid geometry. Euclid is not known to make any original discoveries, and the elements are based on the work of the people before him such as Eudoxus, Thales, Hippocrates and Pythagoras. Over a thousand editions of this work have been published since the first printed version appeared in 1482. Very little, however, is known about his life. The enormity of work attributed to him even led some researchers to suggest that Euclid was not a historic character and that the Elements were written by a team of mathematicians at Alexandria who took the name Euclid from the historic character who lived 100 years earlier.

[2]**Leonhard Euler (1707–1783)** was a Swiss mathematician who made enormous contributions in fields as diverse as infinitesimal calculus and graph theory. He introduced much of the modern mathematical terminology and notation [3]. He is also renowned for his work in mechanics, fluid dynamics, optics, astronomy and music theory.

Since then it has been shown that all numbers $F_5, F_6, \ldots, F_{32}$ are composite. The status of F_{33} remains unknown (May 2020). It is not also known if there are infinitely many Fermat primes.

Many early scholars felt that the numbers of the form $2^n - 1$ were prime for all prime values of n, but in 1536 Hudalricus Regius showed that $2^{11} - 1 = 2047 = 23 \cdot 89$ was not prime. French monk Marin Mersenne (1588–1648) gave in the preface to his Cogitata Physica-Mathematica (1644) a list of positive integers $n < 257$ for which the numbers $2^n - 1$ were prime. Several numbers in that list were incorrect. By 1947 Mersenne's range, $n < 257$, had been completely checked, and it was determined that the correct list was

$$n = 2, 3, 5, 7, 13, 17, 19, 31, 61, 89, 107, 127.$$

Mersenne still got his name attached to these numbers.

As of May 2020 there are 48 known Mersenne primes. The last one was discovered in January 2013 by the Great Internet Mersenne Prime Search (GIMPS) project led by Dr. Curtis Cooper.[3] The new prime number is $2^{57,885,161} - 1$; it has 17,425,170 digits. This is the largest known prime to date. We can check with GAP if the number of digits of this prime was reported correctly:

```
gap> n:=57885161;;
gap> 2^n-1;
<integer 581...951 (17425170 digits)>
```

But checking its primality is currently beyond GAP. 4

Exercises

1. Write a GAP program that calculates 2007th prime p_{2007}. Calculate p_{2007}.
2. Write a GAP program that finds the smallest k for which

$$n = p_1 p_2 \ldots p_k + 1$$

   is composite. It should output k, n and its prime factorisation.
3. Find all integers $a \neq 3$ for which $a - 3$ is a divisor of $a^3 - 17$.
4. Prove that the set P of all primes that are greater than 2 is split into two disjoint classes: primes of the form $4k + 1$ and primes of the form $4k + 3$. Similarly, P is split into two other disjoint classes: primes of the form $6k + 1$ and primes of the form $6k + 5$.
5. Prove that any prime of the form $3k + 1$ is also of the form $6k + 1$ (but for a different k, of course).

---

[3] See http://www.mersenne.org/various/57885161.htm.

6. GAP remembers all 168 primes not exceeding 1000. The command `Primes[i];` gives you the ith prime. Using GAP:

   (a) Create two lists of prime numbers, called Primes1 and Primes3, which include in the first list all the primes $p \le 1000$ for which $p = 4k + 1$ and include in the second list all the primes $p \le 1000$ for which $p = 4k + 3$.
   (b) Output the number of primes in each list.
   (c) Output the 32nd prime from the first list and the 53rd prime from the second list.
   (d) Output the positions of 601 and 607 in their respective lists.

7. (a) Use GAP to list all primes up to 1000 representable in the form $6k + 5$.
   (b) Prove that there are infinitely many primes representable in the form $6k + 5$.

8. Give an alternative proof that the number of primes is infinite along the following lines:

   • Assume that there are only k primes $p_1, p_2, \ldots, p_k$.
   • Given n, find an upper bound $f(n)$ for the number of products

   $$p_1^{\alpha_1} p_2^{\alpha_2} \cdots p_k^{\alpha_k}$$

   that do not exceed n by estimating the number of values that $\alpha_1, \alpha_2, \ldots, \alpha_k$ might assume.
   • Show that $f(n)$ grows slower than n for n sufficiently large.

9°. Prove that there exist infinitely many positive integers n such that $4n^2 + 1$ is divisible both by 5 and 13.

10°. Show there is no positive integer n for which $n, n + 10, n + 20$ and $n + 30$ are all prime.

11°. Prove that for every positive integer n there exists a positive integer x such that each of the terms of the infinite sequence $x + 1, x^x + 1, x^{x^x} + 1, \ldots$ is divisible by n. (Note that $a^{b^c} = a^{(b^c)}$.)

12°. Given an integer n, show that a positive integer can be found which contains only digits 0 and 1 (in the decimal representation) and which is divisible by n.

1.1.3   Factoring Integers. The Sieve of Eratosthenes

None of the ideas we have learned up to now will help us to find the prime factorisation of a particular integer n. Finding prime factorisations is not an easy task, and there are no simple ways to do so. The theorem that we will prove in this section is of some help since it says where to look for the smallest prime divisor of n.

Firstly, we have to define the following useful function.

Definition 1.1.3 Let x be a real number. By $\lfloor x \rfloor$ we denote the largest integer n such that $n \le x$. The integer $\lfloor x \rfloor$ is called the *integer part* of x or the *floor* of x.

Example 1.1.6 $\lfloor \pi \rfloor = 3$, $\lfloor \sqrt{19} \rfloor = 4$, $\lfloor -2.1 \rfloor = -3$.

Theorem 1.1.4 *The smallest prime divisor of a composite number n is less than or equal to* $\lfloor \sqrt{n} \rfloor$.

Proof We prove first that n has a divisor which is greater than 1 but not greater than $\sqrt{n}$. As n is composite, we have $n = d_1 d_2$, where $d_1 > 1$ and $d_2 > 1$. If $d_1 > \sqrt{n}$ and $d_2 > \sqrt{n}$, then

$$n = d_1 d_2 > (\sqrt{n})^2 = n,$$

which is impossible. Suppose, $d_1 \leq \sqrt{n}$. Then any of the prime divisors of d_1 will be less than or equal to $\sqrt{n}$. But every divisor of d_1 is also a divisor of n, thus the smallest prime divisor p of n will satisfy the inequality $p \leq \sqrt{n}$. Since p is an integer, $p \leq \lfloor \sqrt{n} \rfloor$.   □

Now we may demonstrate a beautiful and efficient method of listing all primes up to x, called the *Sieve of Eratosthenes*.

Algorithm (*The Sieve of Eratosthenes*) To find all the primes up to x, begin by writing down all the integers from 2 to x in ascending order. The first number on the list is 2. Leave it there and cross out all other multiples of 2. Then use the following iterative procedure. Let d be the smallest number on the list that is not eliminated. Leave d on the list and, if $d \leq \sqrt{x}$, cross out all other multiples of it. If $d > \sqrt{x}$, then stop. The prime numbers up to x are those which have not been crossed out.

For example, if we write all positive integers not exceeding 100 in a 10×10 square table, then at the end of the process our table will look like:

	2	3		5		7			
11		13				17		19	
		23						29	
31						37			
41		43				47			
		53						59	
61						67			
71		73						79	
		83						89	
						97			

The numbers in this table are all primes not exceeding 100. Please note that we had to cross out only multiples of the primes from the first row since $\sqrt{100} = 10$.

The simplest algorithm for factoring integers is Trial Division.

Algorithm (*Trial Division*) Suppose a sufficiently long list of primes is available. Given a positive integer n, divide it with remainder by all primes on the list which do not exceed $\sqrt{n}$, starting from 2. The first prime which divides n (call this prime p_1) will be the smallest prime divisor of n. In this case n is composite. Calculate $n_1 = n/p_1$ and repeat the procedure. If none of the primes, which do not exceed $\sqrt{n}$, divide n, then n is prime, and its prime factorisation is trivial.

Using the list of primes stored by GAP in array `Primes` we can apply Trial Division algorithm to factorise numbers not exceeding one million. Practically, it is virtually impossible to completely factor a large number of about 100 decimal digits only with Trial Division unless it has small prime divisors. Trial Division is very fast for finding small factors (up to about 10^6) of n.

It is important to know how many operation will be needed to factorise n. If we do not know how many operations are needed, it is impossible to estimate the time it would take to use the Trial Division algorithm in the worst possible case—the case in which small factors are absent.

Let $\pi(x)$ denote the number of primes which do not exceed x. Because of the irregular occurrence of the primes, we cannot expect a simple formula for $\pi(x)$. The following simple program calculates this number for $x = 1000$.

```
gap> n:=1000;;
gap> piofx:=0;;
gap> p:=2;;
gap> while p<n do
> p:=NextPrimeInt(p);
> piofx:=piofx+1;
> od;
gap> piofx;
168
```

As we see there are 168 primes not exceeding 1000. GAP stores them in an array `Primes`. For example, the command

```
gap> Primes[100];
541
```

gives you the 100th prime.

One of the most impressive results in advanced number theory gives an asymptotic approximation for $\pi(x)$.

Theorem 1.1.5 (The Prime Number Theorem)

$$\lim_{x \to \infty} \pi(x) \frac{\ln x}{x} = 1, \tag{1.6}$$

where $\ln x$ *is the natural logarithm, to base e.*

The proof is beyond the scope of this book. The first serious attempt towards proving this theorem (which was long conjectured to be true) was made by Chebyshev[4] who proved (1848–1850) that if the limit exists at all, then it is necessarily equal to one. The existence of the limit (1.6) was proved independently by Hadamard[5] and Vallée-Poussin[6] with both papers appearing almost simultaneously in 1896.

Corollary 1.1.1 *For a large positive integer n there exist approximately $n/\ln n$ primes among the numbers $1, 2, \ldots, n$. This can be expressed as*

$$\pi(n) \sim \frac{n}{\ln n}, \tag{1.7}$$

where $\sim$ means approximately equal for large n (In Sect. 2.3.1 we will give it a precise meaning).

Example 1.1.7 Using (1.7) we approximate $\pi(999)$ as $\frac{999}{6.9} \approx 145$. The real value of $\pi(999)$, as we know, is 168. The number 999 is too small for the approximation in (1.7) to be good.

Example 1.1.8 Suppose $n = 999313$. Let us see how much effort will go into finding a prime factorisation for it. We have $\lfloor \sqrt{n} \rfloor = 999$. So, if we try to find a minimal prime divisor of n using Trial Division, then, in the worst case scenario, we might need to perform 168 divisions. However $n = 7 \cdot 142759$, so the fourth division will already reveal the smallest prime divisor of n which is 7. When 7 is factored out, we will need to deal with the prime factorisation of 142759. Since $\lfloor \sqrt{142759} \rfloor = 377$ and $\pi(377) = 74$, we need, in the worst case scenario, to perform 74 additional divisions dividing 142759 by all primes smaller than or equal to 377. Since 142759 is prime all 74 divisions will be indeed needed. So the total number of divisions required will be 78.

The following two facts are also related to the distribution of primes. Both facts are useful to know and easy to remember.

Theorem 1.1.6 (Bertrand's Postulate) *For each positive integer $n > 1$ there is a prime p such that $n < p < 2n - 2$.*

---

[4]**Pafnutii Lvovich Chebyshev (1821–1894)** was a Russian mathematician who is largely remembered for his investigations in number theory. Chebyshev is also famous for the orthogonal polynomials he invented. He had a strong interest in mechanics as well.

[5]**Jacques Salomon Hadamard (1865–1963)** was a French mathematician whose most important result is the prime number theorem which he proved in 1896. He worked on entire functions and zeta functions and became famous for introducing Hadamard matrices and Hadamard transforms.

[6]**Charles Jean Gustave Nicolas, Baron de la Vallée Poussin (1866–1962)** is best known for his proof of the prime number theorem and his major work Cours d'Analyse. He was additionally known for his writings about the zeta function, Lebesgue and Stieltjes integrals, conformal representation, algebraic and trigonometric series.

In 1845 Bertrand[7] conjectured that there is at least one prime between n and $2n - 2$ for every $n > 3$ and checked it for numbers up to at least $2 \cdot 10^6$. This conjecture, similar to one stated by Euler one hundred years earlier, was proved by Chebyshev in 1850.

Theorem 1.1.7 *There are arbitrarily large gaps between consecutive primes.*

Proof This follows from the fact that, for any positive integer n, all numbers

$$n! + 2, \ n! + 3, \ \ldots, \ n! + n$$

are composite. This is true since for any $2 \le k \le n$

$$n! + k = k \left(\frac{n!}{k} + 1 \right).$$

Thus, for any n there are $n - 1$ consecutive composite integers. $\qquad\qquad\square$

Exercises

1. (a) Use the Sieve of Eratosthenes to find the prime numbers up to 210. Hence calculate $\pi(210)$ exactly.
   (b) Calculate the estimate that the prime number theorem gives for $\pi(210)$ and compare your result with the exact value of $\pi(210)$ obtained in (a).
2. Convince yourself that the following program implements the Sieve of Eratosthenes

```
n:=2*10^3;;
set:=Set([2..n]);;
p:=2;;
while p<RootInt(n)+1 do
k:=2;;
while k*p<n+1 do
RemoveSet(set,k*p);
k:=k+1;
od;
p:=NextPrimeInt(p);
od;
```

and stores in an array "set" all primes not exceeding 2000.

---

[7] **Joseph Louis Francois Bertrand (1822–1900)**, born and died in Paris, was a professor at the École Polytechnique and Collège de France. He was a member of the Paris Academy of Sciences and was its permanent secretary for twenty-six years. Bertrand made a major contribution to group theory and published many works on differential geometry and on probability theory.

3. Professor Woodhead has compiled a list of all primes that are less than 10,000 and is very proud of himself. He checks that the number $n = 123123137$ does not have any prime divisors in his list by dividing n by all of the primes that he found.

   (a) Can he claim that n is prime?
   (b) Estimate the number of additional divisions that Professor Woodhead must perform in order to be able to claim that n is prime.

4. A composite number n does not have prime divisors which are less than or equal to $\lfloor \sqrt[3]{n} \rfloor$. Prove that it is a product of two primes.
5. Use Bertrand's postulate to show that any integer greater than 6 is the sum of two relatively prime integers each of which is greater than 1.
6. What would be the output for the following GAP program?

```
n:=10^6;
set:=Set([1..n]);
p:=3;
while p<n+1 do;
k:=1;
while k*p<n+1 do;
RemoveSet(set,k*p);
k:=k+1;
od;
p:=NextPrimeInt(p);
od;
set;
```

   In particular, how many numbers will be displayed?
7°. Prove that the integers $\{1, 2, \ldots, 2k\}$ can be arranged into k disjoint pairs so that the sums of the numbers in each pair is prime. (**Hint::** Use Bertrand's postulate.)

1.2   Euclidean Algorithm

1.2.1   Divisors and Multiples

Let n be a positive integer with the prime factorisation

$$n = p_1^{\alpha_1} p_2^{\alpha_2} \cdots p_r^{\alpha_r}, \tag{1.8}$$

where p_i are distinct primes and α_i are positive integers. How can we find all divisors of n? Let d be a divisor of n. Then $n = dm$, for some m, thus

$$n = dm = p_1^{\alpha_1} p_2^{\alpha_2} \cdots p_r^{\alpha_r},$$

Since the prime factorisation of n is unique, d cannot have in its prime factorisation a prime which is not among the primes $p_1, p_2, \ldots, p_r$. Also, a prime p_i in the prime factorisation of d cannot have an exponent greater than α_i. Therefore

$$d = p_1^{\beta_1} p_2^{\beta_2} \ldots p_r^{\beta_r}, \qquad 0 \le \beta_i \le \alpha_i, \qquad i = 1, 2, \ldots, r. \tag{1.9}$$

Theorem 1.2.1 *The number of positive divisors of n is*

$$d(n) = (\alpha_1 + 1)(\alpha_2 + 1) \ldots (\alpha_r + 1). \tag{1.10}$$

Proof Indeed, we have exactly $\alpha_i + 1$ possibilities to choose β_i in (1.9), namely $0, 1, 2, \ldots, \alpha_i$. Thus the total number of divisors will be exactly the product $(\alpha_1 + 1)(\alpha_2 + 1) \ldots (\alpha_r + 1)$. $\qquad\qquad\square$

It is important to note that Eq. (1.10) does not give us a self-contained algorithm of calculation of $d(n)$ as we need to run the prime factorisation algorithm first. No direct method of calculation is known.

Definition 1.2.1 The numbers in, where $i = 0, \pm 1, \pm 2, \ldots$, are called *multiples* of n.

It is clear that any multiple of n given by (1.8) has the form

$$m = k p_1^{\gamma_1} p_2^{\gamma_2} \ldots p_r^{\gamma_r}, \qquad \gamma_i \ge \alpha_i, \qquad i = 1, 2, \ldots, r,$$

where k has none of the primes $p_1, p_2, \ldots, p_r$ in its prime factorisation. The number of multiples of n is infinite.

1.2.2 Greatest Common Divisor and Least Common Multiple

Let a and b be two positive integers. If d is a divisor of a and also a divisor of b, then we say that d is a common divisor of a and b. As there are only a finite number of common divisors, there is a *greatest common divisor*, denoted by $\gcd(a, b)$. The number m is said to be a common multiple of a and b if m is a multiple of a and also a multiple of b. Among all common multiples there is a minimal one (Least Integer Principle!). It is called the *least common multiple*, and it is denoted by $\mathrm{lcm}(a, b)$.

In the decomposition (1.8), all exponents were positive. However, sometimes it is convenient to allow some exponents to be 0 as in the formulation of the following theorem.

Theorem 1.2.2 *Let*

$$a = p_1^{\alpha_1} p_2^{\alpha_2} \ldots p_r^{\alpha_r}, \qquad b = p_1^{\beta_1} p_2^{\beta_2} \ldots p_r^{\beta_r},$$

where $\alpha_i \geq 0$ and $\beta_i \geq 0$, be two arbitrary positive integers. (We could assume that a and b are expressed using the same primes $p_1, p_2, \ldots, p_r$ because we allowed some exponents to be 0.) Then

$$gcd(a, b) = p_1^{\min(\alpha_1, \beta_1)} p_2^{\min(\alpha_2, \beta_2)} \ldots p_r^{\min(\alpha_r, \beta_r)}, \qquad (1.11)$$

and

$$lcm(a, b) = p_1^{\max(\alpha_1, \beta_1)} p_2^{\max(\alpha_2, \beta_2)} \ldots p_r^{\max(\alpha_r, \beta_r)}. \qquad (1.12)$$

Moreover,

$$gcd(a, b) \cdot lcm(a, b) = a \cdot b. \qquad (1.13)$$

Proof Formulas (1.11) and (1.12) follow from our description of common divisors and common multiples. To prove (1.13), we have to notice that $\min(\alpha_i, \beta_i) + \max(\alpha_i, \beta_i) = \alpha_i + \beta_i$. □

Example 1.2.1 Let

$$a = 136995569568 = 2^5 \cdot 3^{11} \cdot 11 \cdot 13^3 = 2^5 \cdot 3^{11} \cdot 11^1 \cdot 13^3 \cdot 17^0,$$
$$b = 84474819 = 3^5 \cdot 11^2 \cdot 13^2 \cdot 17 = 2^0 \cdot 3^5 \cdot 11^2 \cdot 13^2 \cdot 17^1.$$

Then $gcd(a, b) = 3^5 \cdot 11 \cdot 13^2 = 415737$.

Theorem 1.2.2 gives us an algorithm for calculating the greatest common divisor. However, it depends on the prime factorisation algorithm. However with the existing methods the calculation of prime factorisation is computationally difficult. It is suspected but has not yet been proved that no easy algorithms for prime factorisation exist. So it is desirable in any number-theoretic algorithm to avoid factorisation of the numbers involved. The algorithm given above for finding the greatest common divisor cannot be used unless prime factorisation has already been done. Fortunately the greatest common divisor $gcd(a, b)$ of numbers a and b can be found without knowing the prime factorisations of a and b. Such an algorithm will be presented below. It was known to Euclid; he could even be the first to discover it. The algorithm is based on the following simple observation.

Proposition 1.2.1 *Let a, b, q, r be any integers such that $a = qb + r$. Then $gcd(a, b) = gcd(b, r)$.*

Proof Indeed, if d is a common divisor of a and b, we have $a = a'd$ and $b = b'd$. Then $r = a - qb = a'd - qb'd = (a' - qb')d$ and d is also a common divisor of b and r. Also, if d is a common divisor of b and r, then $b = b'd$, $r = r'd$ and $a = qb + r = qb'd + r'd = (qb' + r')d$, whence d is a common divisor of a and b. □

Now to the algorithm, the idea of it is clear: Start with the pair (a, b) for which the greatest common divisor is sought, and replace it with a "smaller" pair with the same greatest common divisor. Repeat the process (if necessary) until the greatest common divisor is easily seen.

Theorem 1.2.3 (The Euclidean algorithm) *Let a and b be positive integers. We use the division algorithm several times to find:*

$$a = q_1 b + r_1, \ 0 < r_1 < b,$$
$$b = q_2 r_1 + r_2, \ 0 < r_2 < r_1,$$
$$r_1 = q_3 r_2 + r_3, \ 0 < r_3 < r_2,$$

$$\vdots$$

$$r_{s-2} = q_s r_{s-1} + r_s, \ 0 < r_s < r_{s-1},$$
$$r_{s-1} = q_{s+1} r_s.$$

Then $r_s = \gcd(a, b)$.

Proof By Proposition 1.2.1, $\gcd(a, b) = \gcd(b, r_1) = \gcd(r_1, r_2) = \ldots = \gcd(r_{s-1}, r_s) = r_s$. $\qquad\square$

Example 1.2.2 Let $a = 321$, $b = 843$. Find the greatest common divisor $\gcd(a, b)$. The Euclidean algorithm yields

$$321 = 0 \cdot 843 + 321$$
$$843 = 2 \cdot 321 + 201$$
$$321 = 1 \cdot 201 + 120$$
$$201 = 1 \cdot 120 + 81$$
$$120 = 1 \cdot 81 + 39$$
$$81 = 2 \cdot 39 + 3$$
$$39 = 13 \cdot 3 + 0,$$

and therefore $\gcd(321, 843) = 3$ and $\operatorname{lcm}(321, 843) = \dfrac{321 \cdot 843}{3} = 107 \cdot 843 = 90201$.

Definition 1.2.2 If $\gcd(a, b) = 1$, the numbers a and b are said to be *relatively prime* (or *coprime*).

For example, the numbers $200 = 2^3 \cdot 5^2$ and $567 = 3^4 \cdot 7$ are coprime.

Exercises

1. How many divisors does the number $2^2 \cdot 3^3 \cdot 4^4 \cdot 5^5$ have? (No GAP, please.)
2. How many divisors does the number 123456789 have?
3. Find all common divisors of 10650 and 6750.
4. (a) Find the greatest common divisor and the least common multiple of $m = 2^4 \cdot 3^2 \cdot 5^7 \cdot 11^2$ and $n = 2^2 \cdot 5^4 \cdot 7^2 \cdot 11^3$.
   (b) Use GAP to check up the identity $\text{lcm}(m, n) \cdot \gcd(m, n) = m \cdot n$.
5. Find all positive integers $n \leq 10000$ with exactly 33 distinct positive divisors.
6. Calculate $d(d(246^{246}))$, where $d(n)$ is the number of divisors of n.
7. Show that $\gcd(a, b) = \gcd(a, a - b)$.
8. Show that the fraction $\dfrac{8n + 13}{13n + 21}$ is in lowest possible terms for every $n \geq 1$.
9. Suppose two positive integers a and b are relatively prime.

   (a) Prove that $\gcd(a^2, a + b) = 1$.
   (b) Suppose $a + b$ and $a^2 + b^2$ are not relatively prime. Find the greatest common divisor of this pair and give an example of two such integers.

10. Prove that $\gcd(2^a - 1, 2^b - 1) = 2^{\gcd(a,b)} - 1$.
11. Show that any two distinct Fermat numbers are coprime. (Use Exercise 4 of Sect. 1.1.1).
12. Use Fermat numbers to give an alternative proof that the number of primes is infinite.
13°. Let a, b, c be positive integers. Determine whether the following assertions are true or false. If true, prove the result, and if false, give a counterexample.

   (a) If $a^3 \mid c^2$, then $a \mid c$;
   (b) If $a^2 \mid c^3$, then $a \mid c$;
   (c) $\text{lcm}(a^2, ab, b^2) = \text{lcm}(a^2, b^2)$.

14°. Let p be an odd prime and m, n be positive integers. Show that

$$\gcd(p^m - 1, p^n - 1) = p^{\gcd(m,n)} - 1.$$

1.2.3   Extended Euclidean Algorithm. Chinese Remainder Theorem

Given two integers a and b we can consider all their possible linear combinations $k_1 a + k_2 b$, where $k_1, k_2 \in \mathbb{Z}$. Let us denote this set as $<a, b>$. We note that a and b belong to this set since $a = 1 \cdot a + 0 \cdot b$ and $b = 0 \cdot a + 1 \cdot b$. We also note that when we add two numbers from $<a, b>$, even with some coefficients, we always remain in $<a, b>$. Indeed, suppose we have linear combinations $k_1 a + k_2 b$ and $k_1' a + k_2' b$. Then

$$u(k_1 a + k_2 b) + v(k_1' a + k_2' b) = (uk_1 + vk_1')a + (uk_2 + vk_2'),$$

which is an element of $<a, b>$.

Analysing the chain of divisions with remainder in the formulation of Theorem 1.2.3, we come to the conclusion that all remainders r_i, $i = 1, 2, \ldots, s$ belong to $<a, b>$. In particular, $\gcd(a, b)$ belongs to $<a, b>$. This is an important fact, and we formulate it as theorem for further references.

Theorem 1.2.4 *Let a and b be positive integers. Then there exist integers m and n such that*

$$\gcd(a, b) = ma + nb. \tag{1.14}$$

The numbers m and n in (1.14) are not unique, moreover there exist infinitely many such pairs However, sometimes, knowing even one pair of such numbers is more important than knowing the greatest common divisor itself. One pair of numbers m and n satisfying (1.14) can be easily obtained from the Euclidean algorithm by back substitution. The following theorem provides us with a convenient way of calculating them. It also gives an alternative proof of the existence of m and n based on linear algebra.

Theorem 1.2.5 (The Extended Euclidean algorithm) *Let us write the following matrix with two rows R_1, R_2, and three columns C_1, C_2, C_3:*

$$[C_1 \, C_2 \, C_3] = \begin{bmatrix} R_1 \\ R_2 \end{bmatrix} = \begin{bmatrix} a & 1 & 0 \\ b & 0 & 1 \end{bmatrix}.$$

In accordance with the Euclidean algorithm above, we perform elementary row operations $R_3 := R_1 - q_1 R_2$, $R_4 := R_2 - q_2 R_3$, $\ldots$, each time creating a new row, so as to obtain:

$$[C_1' \, C_2' \, C_3'] = \begin{bmatrix} a & 1 & 0 \\ b & 0 & 1 \\ r_1 & 1 & -q_1 \\ r_2 & -q_2 & 1 + q_1 q_2 \\ \vdots & & \\ r_s & m & n \end{bmatrix}.$$

Then $\gcd(a, b) = r_s = ma + nb$.

Proof Note that $C_1 = aC_2 + bC_3$. In linear algebra you have learned that elementary row operations do not change linear relationships between columns. Since new rows were obtained by means of elementary row operations on the existing rows, the relationships between the columns of C_1, C_2, C_3 must be exactly the same as those between the columns of C_1', C_2', C_3' (see Sect. 10.1 of Appendix B for the justification of this claim). Thus we conclude that $C_1' = aC_2' + bC_3'$. In particular, $r_s = ma + nb$. $\qquad\square$

Example 1.2.3 Let $a = 321$, $b = 843$. Find a linear presentation of the greatest common divisor in the form $\gcd(a, b) = ma + nb$.

The Euclidean algorithm on these numbers was performed in Example 1.2.2, and we know that $\gcd(321, 843) = 3$ and all the quotients obtained at each division. The Extended Euclidean algorithm yields

$$
\begin{array}{rrr|r}
321 & 1 & 0 \\
843 & 0 & 1 & 0 \\
321 & 1 & 0 & 2 \\
201 & -2 & 1 & 1 \\
120 & 3 & -1 & 1 \\
81 & -5 & 2 & 1 \\
39 & 8 & -3 & 2 \\
3 & -21 & 8 & 13
\end{array}
$$

where for convenience of performing row operations the quotients are placed on the right of the bar. Thus we obtain the linear presentation $\gcd(321, 843) = 3 = (-21) \cdot 321 + 8 \cdot 843$. So $m = -21$ and $n = 8$.

The following properties of relatively prime numbers are often used. They are gathered in the following

Lemma 1.2.1 *Let a and b be relatively prime positive integers. Then*

(a) *a and b do not have common primes in their prime factorisations;*
(b) *If c is a multiple of a and c is also a multiple of b, then c is a multiple of ab;*
(c) *If ac is a multiple of b, then c is a multiple of b;*
(d) *There exist integers m, n such that ma + nb = 1.*

Proof Suppose as in the proof of Theorem 1.2.2 that

$$
a = p_1^{\alpha_1} p_2^{\alpha_2} \dots p_r^{\alpha_r}, \qquad b = p_1^{\beta_1} p_2^{\beta_2} \dots p_r^{\beta_r},
$$

where $\alpha_i \geq 0$ and $\beta_i \geq 0$ are non-negative integers. Then by (1.11)

$$
\gcd(a, b) = p_1^{\min(\alpha_1, \beta_1)} p_2^{\min(\alpha_2, \beta_2)} \dots p_r^{\min(\alpha_r, \beta_r)} = 1,
$$

which implies $\min(\alpha_i, \beta_i) = 0$ for all $i = 1, 2, \dots, r$. This means that either prime p_i does not enter the prime factorisation of a or it does not enter the prime factorisation of b. Thus a and b do not have primes in common. This proves (a).

Let us prove (b). As we know from (a) numbers a and b do not have primes in common in their prime factorisations. Hence

$$
a = p_1^{\alpha_1} p_2^{\alpha_2} \dots p_r^{\alpha_r}, \qquad b = q_1^{\beta_1} q_2^{\beta_2} \dots q_s^{\beta_s},
$$

where $p_i \neq q_j$ for all i and j. We have

$$
c = p_1^{\alpha_1} p_2^{\alpha_2} \dots p_r^{\alpha_r} k = q_1^{\beta_1} q_2^{\beta_2} \dots q_s^{\beta_s} m,
$$

Since prime factorisation is unique k must be divisible by $q_1^{\beta_1} q_2^{\beta_2} \ldots q_s^{\beta_s}$, which is b, and m must be divisible by $p_1^{\alpha_1} p_2^{\alpha_2} \ldots p_r^{\alpha_r}$, which is a. As a result, c is divisible by ab,

Let us prove (c). We have $ac = bd$ for some positive integer d so

$$ac = p_1^{\alpha_1} p_2^{\alpha_2} \ldots p_r^{\alpha_r} c = bd = q_1^{\beta_1} q_2^{\beta_2} \ldots q_s^{\beta_s} d,$$

Due to the uniqueness of prime factorisation of ac the number c must be divisible by $q_1^{\beta_1} q_2^{\beta_2} \ldots q_s^{\beta_s}$ which is b.

Now (d) follows from Theorem 1.2.5. □

The following result is extremely important. Its author is not known exactly but it could be Sun Tzu (or Sun Zi)[8] in which book it was first mentioned.

Theorem 1.2.6 (The Chinese remainder theorem) *Let a and b be two relatively prime numbers, $0 \le r < a$ and $0 \le s < b$. Then there exists a unique number N such that $0 \le N < ab$ and*

$$r = N \bmod a \quad and \quad s = N \bmod b, \tag{1.15}$$

that is, N has remainder r on dividing by a and remainder s on dividing by b.

Proof Let us prove, first, that there exists at most one integer N with the conditions required. Assume, on the contrary, that for two integers N_1 and N_2 we have $0 \le N_1 < ab, 0 \le N_2 < ab$ and

$$r = N_1 \bmod a \quad s = N_1 \bmod b, \qquad r = N_2 \bmod a \quad s = N_2 \bmod b.$$

Without loss of generality let us assume that $N_1 > N_2$. Then the number $M = N_1 - N_2$ satisfies $0 \le M < ab$ and

$$0 = M \bmod a \quad 0 = M \bmod b. \tag{1.16}$$

By Lemma 1.2.1(b), condition (1.16) implies that M is divisible by ab, whence $M = 0$ and $N_1 = N_2$.

Now we will find an integer N such that $r = N \bmod a$ and $s = N \bmod b$, ignoring the condition $0 \le N < ab$. By Theorem 1.2.4 there are integers m, n such that $\gcd(a, b) = 1 = ma + nb$. Multiplying this equation by $r - s$ we get the equation

$$r - s = (r - s)ma + (r - s)nb = m'a + n'b.$$

---

[8]**Sun Tzu (3rd–5th century AD)** (or Sun Zi) was a Chinese mathematician and astronomer. He investigated Diophantine equations. He authored "Sun Tzu's Calculation Classic," which contained, among other things, the Chinese remainder theorem.

Now we define the number

$$N = r - m'a = s + n'b$$

It clearly satisfies condition (1.15). If N does not satisfy $0 \le N < ab$, we divide N by ab with remainder. Let $N = q \cdot ab + N_1$, where N_1 is the remainder. Then $0 \le N_1 < ab$ and N_1 satisfies (1.15) since N_1 has the same remainder as N on division by a and also by b. The theorem is proved.   □

Example 1.2.4 Find the smallest positive integer N such that

$$5 = N \bmod 991, \qquad 8 = N \bmod 441.$$

Using the Extended Euclidean algorithm we find

$$
\begin{array}{rrr|r}
991 & 1 & 0 & \\
441 & 0 & 1 & 2 \\
109 & 1 & -2 & 4 \\
5 & -4 & 9 & 21 \\
4 & 85 & -191 & 1 \\
1 & -89 & 200 & 4 \\
0 & & &
\end{array}
$$

thus yielding $1 = (-89) \cdot 991 + 200 \cdot 441$. We may write $8 - 5 = 3 = (-267) \cdot 991 + 600 \cdot 441$ and obtain the number $N = -264592 = 8 - 600 \cdot 441 = 5 + (-267) \cdot 991$, which satisfies all the requirements apart from being between 0 and $437031 = 991 \cdot 441$. We divide N by 437031 with remainder. We have $-264592 = (-1) \cdot 437031 + 172439$, and the remainder $N_1 = 172439$ will be the number required.

Exercises

1. Use the Extended Euclidean algorithm to find the greatest common divisor d of 3773 and 3596 and find any integers x and y such that $d = 3773x + 3596y$.
2. Using the Extended Euclidean algorithm, find at least one pair of integers (x, y) satisfying $1840x + 1995y = 5$, and at least three pairs of integers (z, w) satisfying $1840z + 1995w = -10$.
3. Let a, b, c and d be non-negative integers with $c > 1$ and $d > 1$. Suppose that there exists an integer N such that

$$a = N \bmod c \quad \text{and} \quad b = N \bmod d.$$

Prove that $a - b$ is a multiple of $\gcd(c, d)$.

4. (a) Find any integer y such that

$$y = 9 \bmod 26 \quad \text{and} \quad y = 35 \bmod 68.$$

(note that 26 and 68 are not relatively prime.)
(b) Find the unique integer x such that $0 \le x < 3550$ and

$$x = 4 \bmod 50 \quad \text{and} \quad x = 19 \bmod 71.$$

5. (a) Prove that there are no integers x, y satisfying $1840x + 1995y = 3$.
(b) Let a and b be nonzero integers. Describe the set of integers c for which there exist integers x and y satisfying the equation $ax + by = c$.
6°. Find three consecutive positive integers which are not square-free. (Recall that n is said to be square-free if it is not divisible by m^2 for any $m > 1$.)

1.3 Fermat's Little Theorem and Its Generalisations

1.3.1 Congruences. Fermat's Little Theorem

Definition 1.3.1 Let a and b be integers and m be a positive integer. We say that a is *congruent* to b modulo m and write $a \equiv b \bmod m$ if a and b have the same remainder on dividing by m, that is $a \bmod m = b \bmod m$.

For example, $41 \equiv 80 \bmod 13$ since the numbers 41 and 80 both have remainder 2 when divided by 13. Also, $41 \equiv -37 \bmod 13$. When a and b are not congruent, we write $a \not\equiv b \bmod m$. For example, $41 \not\equiv 7 \bmod 13$ because 41 has remainder 2, when divided by 13, and 7 has remainder 7.

Lemma 1.3.1 (Criterion) *Let a and b be two integers and m be a positive integer. Then $a \equiv b \bmod m$, if and only if $a - b$ is divisible by m.*

Proof By the division algorithm

$$a = q_1 m + r_1, \quad 0 \le r_1 < m, \quad \text{and} \quad b = q_2 m + r_2, \quad 0 \le r_2 < m.$$

Thus $a - b = (q_1 - q_2)m + (r_1 - r_2)$, where $-m < r_1 - r_2 < m$. We see that $a - b$ is divisible by m if and only if $r_1 - r_2$ is divisible by m but this can happen if and only if $r_1 - r_2 = 0$, or $r_1 = r_2$, which is the same as $a \equiv b \bmod m$. $\square$

Lemma 1.3.2 *Let a and b be two integers and m be a positive integer. Then*

(a) *If $a \equiv b \bmod m$ and $c \equiv d \bmod m$, then $a + c \equiv b + d \bmod m$;*
(b) *If $a \equiv b \bmod m$ and $c \equiv d \bmod m$, then $ac \equiv bd \bmod m$;*

(c) If a ≡ b mod m and n is a positive integer, then $a^n \equiv b^n$ mod m;
(d) If ac ≡ bc mod m and c is relatively prime to m, then a ≡ b mod m.

Proof (a) is an exercise.
(b) If $a \equiv b$ mod m and $c \equiv d$ mod m, then $m|(a-b)$ and $m|(c-d)$, i.e., $a-b = im$ and $c-d = jm$ for some integers i, j. Then

$$ac - bd = (ac - bc) + (bc - bd) = (a-b)c + b(c-d) = icm + jbm = (ic + jb)m,$$

whence $ac \equiv bd$ mod m;
(c) follows immediately from (b).
(d) suppose that $ac \equiv bc$ mod m and $\gcd(c, m) = 1$. Then, by the criterion, $(a - b)c = ac - bc$ is a multiple of m. As $\gcd(c, m) = 1$, by Lemma 1.2.1(c) $a - b$ is a multiple of m, and by the criterion $a \equiv b$ mod m.

□

Theorem 1.3.1 (Fermat's little theorem) *Let p be a prime. If an integer a is not divisible by p, then $a^{p-1} \equiv 1$ mod p. Also $a^p \equiv a$ mod p for all a.*

Proof Let a be relatively prime to p. Consider the numbers $a, 2a, \ldots, (p-1)a$. They all have different remainders on dividing by p. Indeed, suppose that for some $1 \leq i < j \leq p-1$ we have $ia \equiv ja$ mod p. Then by Lemma 1.3.2(d) a can be cancelled and $i \equiv j$ mod p, which is impossible. Therefore these remainders are $1, 2, \ldots, p-1$ and by repeated application of Lemma 1.3.2(b) we have

$$a \cdot 2a \cdot \ldots \cdot (p-1)a \equiv (p-1)! \text{ mod } p,$$

which is

$$(p-1)! \cdot a^{p-1} \equiv (p-1)! \text{ mod } p.$$

Since $(p-1)!$ is relatively prime to p it can be cancelled by Lemma 1.3.2(d) and we get $a^{p-1} \equiv 1$ mod p. When a is relatively prime to p, the last statement follows from the first one. If a is a multiple of p, the last statement is also clear.   □

Example 1.3.1 Find 328^{2013} mod 7.
   Firstly we note that $6 = 328$ mod 7 and using Lemma 1.3.2(c) we find that 328^{2013} mod $7 = 6^{2013}$ mod 7. Now we have to reduce 2013. We can do this using Fermat's little theorem. Since for all a relatively prime to 7 we have $a^6 \equiv 1$ mod 7 we can replace 2013 with its remainder on division by 6. Since $3 = 2013$ mod 6 we obtain

$$328^{2013} \text{ mod } 7 = 6^{2013} \text{ mod } 7 = 6^3 \text{ mod } 7 = 6.$$

The latter follows from the calculation $6^3 = 36 \cdot 6 \equiv 1 \cdot 6$ mod $7 = 6$.

Fermat's little theorem is a powerful (but not perfect) tool for checking primality. Let

$p :=2074722246773485207821695222107608587480996474721117292752992589912196 6847$
$50549658310084416732550077$

be a random 100-digit prime. Then the calculation

```
gap> PowerMod(3,p-1,p);
1
gap> q:=p^2;;
gap> PowerMod(3,q-1,q)=1;
false
```

shows that $3^{p-1} \equiv 1 \bmod p$ but for $q = p^2$ we have $3^{q-1} \not\equiv 1 \bmod q$ thus revealing the compositeness of q. We will discuss thoroughly primality checking in Sect. 2.4.3.

Despite its usefulness, Fermat's little theorem has limited applicability since the modulus p must be a prime. In the next section we will prove Euler's theorem that generalises Fermat's little theorem to an arbitrary positive integer n. It will be very important in cryptographic applications.

Exercises

1. Find the remainder of $2^{(2^{2013})}$ on division by 5.
2. Using Fermat's little theorem find the remainder on dividing by 7 of the number

$$333^{555} + 555^{333}.$$

3. Let $n = 1234567890987654321$ and $a = 111111111$. Calculate $a^{n-1} \bmod n$. Is the result consistent with the hypothesis that n is prime?
4. Let $p > 2$ be a prime. Prove that all prime divisors of $2^p - 1$ have the form $2kp + 1$.
5°. Prove that $2^{n+4} + 3^{3n+2} \equiv 0 \bmod 25$ for any positive integer n.
6°. Let $A_n = 111\ldots 11$ (n ones). Prove that for any prime $p > 5$
   (a) A_p is not divisible by p;
   (b) A_{p-1} is divisible by p.
7°. Suppose that p and q are distinct primes, $a^p \equiv a \bmod q$, and $a^q \equiv a \bmod p$. Prove that $a^{pq} \equiv a \bmod pq$.

1.3.2   Euler's ϕ-Function. Euler's Theorem

Definition 1.3.2 Let n be a positive integer. The number of positive integers not exceeding n and relatively prime to n is denoted by $\phi(n)$. This function is called *Euler's ϕ-function* or *Euler's totient function*.

Let us denote $\mathbb{Z}_n = \{0, 1, 2, \ldots, n-1\}$ and by $\mathbb{Z}_n^*$ the set of those positive numbers from $\mathbb{Z}_n$ that are relatively prime to n. Then $\phi(n)$ is the number of elements of $\mathbb{Z}_n^*$, i.e., $\phi(n) = |\mathbb{Z}_n^*|$.

Example 1.3.2 Let $n = 20$. Then $\mathbb{Z}_{20}^* = \{1, 3, 7, 9, 11, 13, 17, 19\}$ and $\phi(20) = 8$.

Lemma 1.3.3 *If $n = p^k$, where p is prime, then $\phi(n) = p^k - p^{k-1} = p^k \left(1 - \dfrac{1}{p}\right)$.*

Proof It is easy to list all positive integers that are less than or equal to p^k and not relatively prime to p^k. They are $1 \cdot p, \ 2 \cdot p, \ 3 \cdot p, \ldots, (p^{k-1} - 1) \cdot p$. They are all multiples of p, and we have exactly $p^{k-1} - 1$ of them. To obtain $\mathbb{Z}_n^*$, we have to remove from $\mathbb{Z}_n$ all these $p^{k-1} - 1$ numbers and also 0. Therefore $\mathbb{Z}_n^*$ will contain $p^k - (p^{k-1} - 1) - 1 = p^k - p^{k-1}$ numbers. $\square$

An important consequence of the Chinese remainder theorem is that the function $\phi(n)$ is multiplicative in the following sense:

Theorem 1.3.2 *Let m and n be any two relatively prime positive integers. Then*

$$\phi(mn) = \phi(m)\phi(n).$$

Proof Let $\mathbb{Z}_m^* = \{r_1, r_2, \ldots, r_{\phi(m)}\}$ and $\mathbb{Z}_n^* = \{s_1, s_2, \ldots, s_{\phi(n)}\}$. Let us consider an arbitrary pair (r_i, s_j) of numbers, one from each of these sets. By the Chinese remainder theorem there exists a unique positive integer N_{ij} such that $0 \leq N_{ij} < mn$ and

$$r_i = N_{ij} \bmod m, \qquad s_j = N_{ij} \bmod n,$$

that is, N_{ij} has remainder r_i on dividing by m, and remainder s_j on dividing by n, so

$$N_{ij} = am + r_i, \qquad N_{ij} = bn + s_j. \tag{1.17}$$

We have $\gcd(N_{ij}, m) = \gcd(m, r_i) = 1$ and $\gcd(N_{ij}, n) = \gcd(n, s_j) = 1$, that is N_{ij} is relatively prime to m and also relatively prime to n. Since m and n are relatively prime, N_{ij} is relatively prime to mn, i.e, $N_{ij} \in \mathbb{Z}_{mn}^*$. Clearly, different pairs $(i, j) \neq (k, l)$ yield different numbers, that is $N_{ij} \neq N_{kl}$ for $(i, j) \neq (k, l)$. We note that there are $\phi(m)\phi(n)$ of the numbers N_{ij}, exactly as many as there are pairs of the form (r_i, s_j). This shows $\phi(m)\phi(n) \leq \phi(mn)$.

Suppose now that a number $N \in \mathbb{Z}_{mn}$ is different from N_{ij} for all i and j. Then

$$r = N \bmod m, \qquad s = N \bmod n,$$

where either r does not belong to $\mathbb{Z}_m^*$ or s does not belong to $\mathbb{Z}_n^*$. Assuming the former, we get $\gcd(r, m) > 1$. But then $\gcd(N, m) = \gcd(m, r) > 1$, in which case $\gcd(N, mn) > 1$ too. Thus N does not belong to $\mathbb{Z}_{mn}^*$. This shows that the numbers N_{ij}—and only these numbers—form $\mathbb{Z}_{mn}^*$. Therefore $\phi(mn) = \phi(m)\phi(n)$.   □

Theorem 1.3.3 *Let n be a positive integer with the prime factorisation*

$$n = p_1^{\alpha_1} p_2^{\alpha_2} \cdots p_r^{\alpha_r},$$

where p_i are primes and α_i are positive integers. Then

$$\phi(n) = n \left(1 - \frac{1}{p_1}\right) \left(1 - \frac{1}{p_2}\right) \cdots \left(1 - \frac{1}{p_r}\right).$$

Proof We use Lemma 1.3.3 and Theorem 1.3.2 to compute $\phi(n)$. Repeatedly applying Theorem 1.3.2 we get

$$\phi(n) = \phi\left(p_1^{\alpha_1}\right) \phi\left(p_2^{\alpha_2}\right) \cdots \phi\left(p_r^{\alpha_r}\right).$$

By Lemma 1.3.3 this can be rewritten as

$$\phi(n) = p_1^{\alpha_1} \left(1 - \frac{1}{p_1}\right) p_2^{\alpha_2} \left(1 - \frac{1}{p_2}\right) \cdots p_r^{\alpha_r} \left(1 - \frac{1}{p_r}\right)$$

$$= p_1^{\alpha_1} p_2^{\alpha_2} \cdots p_r^{\alpha_r} \left(1 - \frac{1}{p_1}\right) \left(1 - \frac{1}{p_2}\right) \cdots \left(1 - \frac{1}{p_r}\right) = n \left(1 - \frac{1}{p_1}\right) \left(1 - \frac{1}{p_2}\right) \cdots \left(1 - \frac{1}{p_r}\right),$$

as required.   □

Example 1.3.3 $\phi(264) = \phi(2^3 \cdot 3 \cdot 11) = 264 \left(\frac{1}{2}\right) \left(\frac{2}{3}\right) \left(\frac{10}{11}\right) = 80$. We also have $\phi(269) = 268$ as 269 is prime.

The following corollary will be important in the cryptography section.

Corollary 1.3.1 *If $n = pq$, where p and q are primes, then $\phi(n) = (p - 1)(q - 1) = pq - p - q + 1$.*

There are no known methods for computing $\phi(n)$ in situations where the prime factorisation of n is not known. If n is so big that modern computers cannot factorise it, you can publish n and keep $\phi(n)$ secret.

Euler's theorem is a generalisation of Fermat's little theorem.

Theorem 1.3.4 (Euler's theorem) *Let n be a positive integer. Then*

$$a^{\phi(n)} \equiv 1 \ mod \ n$$

for all a relatively prime to n.

Proof Let $\mathbb{Z}_n^* = \{z_1, z_2, \ldots, z_{\phi(n)}\}$. Consider the numbers $z_1 a, z_2 a, \ldots, z_{\phi(n)} a$. Both z_i and a are relatively prime to n, therefore $z_i a$ is also relatively prime to n. Suppose that $r_i = z_i a \ mod \ n$, i.e., r_i is the remainder on dividing $z_i a$ by n. Since $\gcd(z_i a, n) = \gcd(r_i, n)$, one has $r_i \in \mathbb{Z}_n^*$. These remainders are all different. Indeed, suppose that $r_i = r_j$ for some $1 \le i < j \le n$. Then $z_i a \equiv z_j a \ mod \ n$. By Lemma 1.3.2(d) a can be cancelled and we get $z_i \equiv z_j \ mod \ n$, which is impossible. Therefore the remainders $r_1, r_2, \ldots, r_{\phi(n)}$ coincide with $z_1, z_2, \ldots, z_{\phi(n)}$, apart from the order in which they are listed. Thus

$$z_1 a \cdot z_2 a \cdot \ldots \cdot z_{\phi(n)} a \equiv r_1 \cdot r_2 \cdot \ldots \cdot r_{\phi(n)} \equiv z_1 \cdot z_2 \cdot \ldots \cdot z_{\phi(n)} \ mod \ n,$$

which is

$$Z \cdot a^{\phi(n)} \equiv Z \ mod \ n,$$

where $Z = z_1 \cdot z_2 \cdot \ldots \cdot z_{\phi(n)}$. Since Z is relatively prime to n it can be cancelled by Lemma 1.3.2(d), and we get $a^{\phi(n)} \equiv 1 \ mod \ n$. $\qquad\qquad\Box$

Example 1.3.4 Using Euler's theorem compute the last decimal digit (unit's digit) of the number 3^{2007}.

Since the last decimal digit of 3^{2007} is equal to $3^{2007} \ mod \ 10$ we have to calculate this remainder. As $\gcd(3, 10) = 1$ and $\phi(10) = 4$ we have $3^4 \equiv 1 \ mod \ 10$. As $3 = 2007 \ mod \ 4$ we obtain

$$3^{2007} \equiv 3^3 \equiv 7 \ mod \ 10.$$

Hence the last digit of 3^{2007} is 7.

Exercises

1. Show that:
   (a) Both sides of the congruence and its modulus can be simultaneously divided by a common positive divisor.
   (b) If a congruence holds modulo m, then it also holds modulo d, where d is an arbitrary divisor of m.
   (c) If a congruence holds for moduli m_1 and m_2, then it also holds modulo $\text{lcm}(m_1, m_2)$.

2. Without using mathematical induction show that $72^{2n+2} - 47^{2n} + 28^{2n-1}$ is divisible by 25 for any $n \ge 1$.
3. Compute $\phi(125)$, $\phi(180)$ and $\phi(1001)$.
4. Factor $n = 4386607$, which is a product of two primes, given $\phi(n) = 4382136$.

5. Find $m = p^2q^2$, given that p and q are primes and $\phi(m) = 11424$.
6. Find all positive integer solutions x, y to the equation $\phi(3^x 5^y) = 600$, where ϕ is the Euler totient function.
7. List all positive integers a such that $0 \le a \le 242$ for which the congruence $x^{162} \equiv a \mod 243$ has a solution.
8. Without resorting to GAP, factorise n if it is known that it is a product of two primes and that $\phi(n) = 3308580$.
9°. Use GAP to find all solutions to the equation $\phi(x) = 24$, where ϕ is the Euler's totient function. Justify the correctness of your algorithm.
10°. Show that $\phi(n) = 14$ is impossible.

1.4   The Ring of Integers Modulo n. The Field $\mathbb{Z}_p$

> God of infinity finds refuge in a ring,
> Birds of eternity sing there.
> And you too find a ring in your heart.
>
> *Velemir Khlebnikov (1885–1922).*

We will consider the set $\mathbb{Z}_n = \{0, 1, 2, \ldots, n-1\}$ as an algebraic object, called *integers modulo m*, by introducing two algebraic operations on it. First, given $a, b \in \mathbb{Z}_n$, we define a new addition $a \oplus b$ by

$$a \oplus b := a + b \mod n. \tag{1.18}$$

According to the definition, $a \oplus b$ is the remainder on dividing $a + b$ by n and therefore it is always in $\mathbb{Z}_n$.

Example 1.4.1 In $\mathbb{Z}_{11}$ the following identities hold: $3 \oplus 4 = 7$, $5 \oplus 9 = 3$, $4 \oplus 7 = 0$.

Theorem 1.4.1 *The new addition satisfies the following properties:*

1. *It is commutative, $a \oplus b = b \oplus a$, for all $a, b \in \mathbb{Z}_n$.*
2. *It is associative, $a \oplus (b \oplus c) = (a \oplus b) \oplus c$, for all $a, b, c \in \mathbb{Z}_n$.*
3. *Element 0 (zero) is the unique element such that $a \oplus 0 = 0 \oplus a = a$, for every $a \in \mathbb{Z}_n$.*
4. *For each $a \in \mathbb{Z}_n$ there exists a unique element $(-a) \overset{df}{=} n - a \in \mathbb{Z}_n$ such that $a \oplus (-a) = (-a) \oplus a = 0$.*

Proof Only the second property is not completely obvious. We prove it by noting that $a \oplus b \equiv a + b \mod n$. Then by Lemma 1.3.2(a)

$$(a \oplus b) \oplus c \equiv (a \oplus b) + c \equiv (a + b) + c \mod n$$

and

$$a \oplus (b \oplus c) \equiv a \oplus (b + c) \equiv a + (b + c) \bmod n,$$

whence $(a \oplus b) \oplus c \equiv a \oplus (b \oplus c) \bmod n$. But these numbers are in $\mathbb{Z}_n$, and the difference between them is less than n. Therefore $(a \oplus b) \oplus c = a \oplus (b \oplus c)$.   □

Properties 1–4 in algebra axiomatically define a *commutative group*.

Definition 1.4.1 An algebraic system $< G, + >$ which consists of a set G together with an algebraic operation $+$ defined on it is said to be a commutative group if the following axioms are satisfied:

CG1   The operation is commutative, $a + b = b + a$, for all $a, b \in G$.
CG2   The operation is associative, $a + (b + c) = (a + b) + c$, for all $a, b, c \in G$.
CG3   There exists a unique element 0 such that $a + 0 = 0 + a = a$, for all $a \in G$.
CG4   For every element $a \in G$ there exists a unique element $-a$ such that $a + (-a) = (-a) + a = 0$, for all $a \in G$.

Thus we can reformulate Theorem 1.4.1 by saying that $< \mathbb{Z}_n, \oplus >$ is a commutative group.

Corollary 1.4.1 *An equation $a \oplus x = b$ has a unique solution in $\mathbb{Z}_n$, namely $x = (-a) \oplus b$.*

Proof Suppose that $a \oplus x = b$, where x is a solution. Add $(-a)$ to both sides of the equation. We get

$$(-a) \oplus (a \oplus x) = (-a) \oplus b,$$

from where, by using properties 1–4, we can find that $(-a) \oplus (a \oplus x) = ((-a) \oplus a) \oplus x = 0 \oplus x = x$, hence $x = (-a) \oplus b$. Similar computations show that $x = (-a) \oplus b$ is indeed a solution.   □

Example 1.4.2 Equation $18 \oplus x = 13$ in $\mathbb{Z}_{26}$ has a solution $x = (-18) \oplus 13 = 8 \oplus 13 = 21$.

Now, given $a, b \in \mathbb{Z}_n$, we define a new multiplication $a \odot b$ by

$$a \odot b := ab \bmod n. \tag{1.19}$$

According to the definition, $a \odot b$ is the remainder on dividing ab by n and therefore it is always in $\mathbb{Z}_n$.

Example 1.4.3 In $\mathbb{Z}_{12}$ the following identities hold: $5 \odot 5 = 1$, $2 \odot 4 = 8$, $4 \odot 6 = 0$.

Theorem 1.4.2 *The new multiplication modulo n satisfies the following properties:*

5. *It is commutative, $a \odot b = b \odot a$, for all $a, b \in \mathbb{Z}_n$.*
6. *It is associative, $a \odot (b \odot c) = (a \odot b) \odot c$, for all $a, b, c \in \mathbb{Z}_n$.*
7. *It is distributive relative to the addition, $a \odot (b \oplus c) = (a \odot b) \oplus (a \odot c)$ and $(a \oplus b) \odot c = (a \odot c) \oplus (b \odot c)$, for all $a, b, c \in \mathbb{Z}_n$.*
8. *There is a unique element 1 in $\mathbb{Z}_n$ such that $a \odot 1 = 1 \odot a = a$, for every $a \in \mathbb{Z}_n$.*

Proof Statements 5 and 8 are clear. The two other can be proved as in Theorem 1.4.1.
□

Properties 1–8 in algebraic terms mean that $\mathbb{Z}_n$ together with operations $\oplus$ and $\odot$ is a commutative ring with a unity element 1 which is defined by the following set of axioms.

Definition 1.4.2 An algebraic system $< R, +, \cdot >$ which consists of a set R together with two algebraic operations $+$ and $\cdot$ defined on it is said to be a *commutative ring* if the following axioms are satisfied:

CR1 $< R, + >$ is a commutative group.
CR2   The operation $\cdot$ is commutative, $a \cdot b = b \cdot a$, for all $a, b \in R$.
CR3   The operation $\cdot$ is associative, $a \cdot (b \cdot c) = (a \cdot b) \cdot c$, for all $a, b, c \in R$.
CR4   There exists a unique element $1 \in R$ such that $a \cdot 1 = 1 \cdot a = a$, for all nonzero $a \in R$.
CR5   The distributive law holds, that is, $a \cdot (b + c) = a \cdot b + a \cdot c$, for all $a, b, c \in R$.

Example 1.4.4 Other commutative rings include the ring of polynomials $\mathbb{Z}[x]$ with integer coefficients or else with rational or real coefficients. The set of all $n \times n$ matrices over the integers $\mathbb{Z}_{n \times n}$ is also a ring but not commutative since axiom CR2 is not satisfied.

Definition 1.4.3 An element a of a ring R is called *invertible* if there exists an element b in R such that $a \cdot b = b \cdot a = 1$. An element b in this case is called a *multiplicative inverse* of a.

Lemma 1.4.1 *If $a \in \mathbb{Z}_n$ possesses a multiplicative inverse, then this inverse is unique.*

Proof Suppose that we have two inverses of a, say b and c, so that $a \odot b = b \odot a = 1$ and $a \odot c = c \odot a = 1$. Then

$$b \odot (a \odot c) = b \odot 1 = b,$$

and

$$(b \odot a) \odot c = 1 \odot c = c,$$

hence $b = c$ due to the associative law of multiplication.   □

Definition 1.4.4 If a multiplicative inverse of a exists, it is denoted a^{-1}. In this case the element a is called *invertible*.

Theorem 1.4.3 *All elements from $\mathbb{Z}_n^*$ are invertible in $\mathbb{Z}_n$.*

Proof Let $a \in \mathbb{Z}_n^*$. Then $\gcd(a, n) = 1$, and we can write a linear presentation of this greatest common divisor $1 = ua + vn$. Let us divide u by n with remainder w. We have $u = qn + w$, where $0 \le w < n$, and we substitute $qn + w$ instead of u:

$$1 = (qn + w)a + vn = wa + (qa + v)n.$$

It is clear now that $w \in \mathbb{Z}_n$ and that $w \odot a = 1$, which means $w = a^{-1}$.   □

Example 1.4.5 Find 11^{-1} in $\mathbb{Z}_{26}$ and solve $11 \odot x \oplus 5 = 3$.

Solution We use the Extended Euclidean algorithm

$$
\begin{array}{rrr|r}
26 & 1 & 0 & \\
11 & 0 & 1 & 2 \\
4 & 1 & -2 & 2 \\
3 & -2 & 5 & 1 \\
1 & 3 & -7 & 3
\end{array}
$$

to find a linear presentation $1 = 3 \cdot 26 + (-7) \cdot 11$. As $-7 = (-1) \cdot 26 + 19$ we have

$$1 = 3 \cdot 26 + (-7) \cdot 11 = 3 \cdot 26 + ((-1) \cdot 26 + 19) \cdot 11 = 2 \cdot 26 + 19 \cdot 11.$$

Thus $11^{-1} = 19$ and $11 \odot x \oplus 5 = 3$ can be solved as follows: $11 \odot x = 3 \oplus (-5) = 3 \oplus 21 = 24$. Finally, $x = 11^{-1} \odot 24 = 19 \odot 24 = 14$.   □

Definition 1.4.5 A nonzero element $a \in \mathbb{Z}_n$ is called a *zero divisor* if there exists another nonzero element $b \in \mathbb{Z}_n$ such that $a \odot b = 0$.

Example 1.4.6 $4 \odot 5 = 0$ in $\mathbb{Z}_{10}$.

Lemma 1.4.2 *A divisor of zero in $\mathbb{Z}_n$ is never invertible.*

Proof Suppose that $a \odot b = 0$, $a \neq 0$, $b \neq 0$ and a is invertible, that is a^{-1} exists. Then we have $a^{-1} \odot (a \odot b) = a^{-1} \odot 0 = 0$. The left-hand side is equal to $a^{-1} \odot (a \odot b) = (a^{-1} \odot a) \odot b = 1 \odot b = b$. hence $b = 0$, a contradiction. $\square$

Theorem 1.4.4 *If $gcd(a, n) = d > 1$ for some $0 \neq a \in \mathbb{Z}_n$, then a is a zero divisor in $\mathbb{Z}_n$. All elements from $\mathbb{Z}_n^*$ are invertible in $\mathbb{Z}_n$ and all other elements are not invertible.*

Proof Let $a = bd$ and $n = md$. Then $am = bdm = bn$, which is a multiple of n. Thus $a \odot m = 0$ and a is a zero divisor. Thus, in $\mathbb{Z}_n$, aside from $\mathbb{Z}_n^*$, we have the zero element and the zero divisors. On the other hand, by Theorem 1.4.3 all elements of $\mathbb{Z}_n^*$ are invertible. $\square$

Hence, depending on n, the following property may be or may not be true for $\mathbb{Z}_n$:

9. *For every nonzero $a \in \mathbb{Z}_n$ there is a unique element $a^{-1} \in \mathbb{Z}_n$ such that $a \odot a^{-1} = a^{-1} \odot a = 1$.*

Definition 1.4.6 A commutative ring $< R, +, \cdot >$ is called a *field* if the following axiom is satisfied.

F1   For every nonzero $a \in R$ there is a unique element $a^{-1} \in R$ such that $a \cdot a^{-1} = a^{-1} \cdot a = 1$.

Theorem 1.4.4 gives us a complete answer on when $\mathbb{Z}_n$ is a field.

Theorem 1.4.5 *$\mathbb{Z}_n$ is a field if and only if n is prime.*

Proof If p is prime, then $\mathbb{Z}_p^*$ consists of all nonzero elements of $\mathbb{Z}_p$. And by Theorem 1.4.3 all elements of $\mathbb{Z}_p^*$ are invertible. Hence $\mathbb{Z}_p$ is a field. Suppose $\mathbb{Z}_n$ is a field. Since all nonzero elements of any field are invertible, by Theorem 1.4.4 $\mathbb{Z}_n^* = \mathbb{Z}_n \setminus \{0\}$, that is all integers smaller than n are relatively prime to n. This is possible only when n is prime. $\square$

Exercises
1. Prove that in any commutative ring R a divisor of zero is not invertible. (Hint: prove first that for any $a \in R$ we have $a \cdot 0 = 0$. Then follow the proof of Lemma 1.4.2.)
2. (a) List all invertible elements of $\mathbb{Z}_{16}$ and for each invertible element a give its inverse a^{-1}.
   (b) List all zero divisors of $\mathbb{Z}_{15}$ and for each zero divisor a give all nonzero elements b such that $a \odot b = 0$.
3. (a) Which one of the two elements 74 and 77 is invertible in $\mathbb{Z}_{111}$ and which one is a zero divisor? For the invertible element a, give the inverse a^{-1} and for the zero divisor b give the element $c \in \mathbb{Z}_{111}$ such that $b \odot c = c \odot b = 0$.

(b) Solve the equations $77 \odot x \oplus 21 = 10$ and $74 \odot x \oplus 11 = 0$ in $\mathbb{Z}_{111}$.

4. Let a and b be two elements of the ring $\mathbb{Z}_{21}$ and let $f : \mathbb{Z}_{21} \rightarrow \mathbb{Z}_{21}$ be a linear function defined by $f(x) = a \odot x \oplus b$ (where the operations are computed in $\mathbb{Z}_{21}$).

   (a) Describe the set of all pairs (a, b), for which the function f is one-to-one.
   (b) Find the range of the function f for the case $a = 7, b = 3$.
   (c) Suppose $a = 4$ and $b = 15$. Find the inverse function $f^{-1}(x) = c \odot x \oplus d$ which satisfies $f^{-1}(f(x)) = x$ for each $x \in \mathbb{Z}_{21}$.

5. How many solutions in $\mathbb{Z}_{11}$ does the equation $x^{102} = 4$ have? List them all.

6. Given an odd number $m > 1$, find the remainder when $2^{\phi(m)-1}$ is divided by m. This remainder should be expressed in terms of m.

7. (Wilson's Theorem) Let p be an integer greater than one. Prove that p is prime if and only if $(p - 1)! = -1$ in $\mathbb{Z}_p$. (Hint: 1 and $-1 = p - 1$ are the only self-inverse elements of $\mathbb{Z}_p^*$.)

8°. Find 299^{-1} in 328.

9°. Find all solutions to the equation $205 \odot x = 287$ has in $\mathbb{Z}_{328}$.

10°. Prove that in a ring with identity all invertible elements form a group relative to the multiplication.

11°. Show that a finite commutative ring without zero divisors is a field.

1.5   Representation of Numbers

There is an important distinction between numbers and their representations. In the decimal system the zero and the first nine positive integers are denoted by symbols $0, 1, 2, \ldots, 9$, respectively. These symbols are called *digits*. The same symbols are used to represent all the integers. The tenth integer is denoted as 10, and an arbitrary integer N can now be represented in the form

$$N = a_n \cdot 10^n + a_{n-1} \cdot 10^{n-1} + \ldots + a_1 \cdot 10 + a_0, \tag{1.20}$$

where $a_1, a_2, \ldots, a_n$ are integers that can be represented by a single digit $0, 1, 2, \ldots, 9$. For example, the year this course was first taught can be written as

$$1 \cdot 10^3 + 9 \cdot 10^2 + 9 \cdot 10 + 8.$$

We shorten this expression to $(1998)_{(10)}$ or simply 1998, having the decimal system in mind. In this notation the meaning of a digit depends on its position. Thus two-digit symbols "9" are situated in the tens and the hundreds places and their meaning is different. In general, for the number N given by (1.20) we write

$$N = (a_n a_{n-1} \ldots a_1 a_0)_{(10)}$$

to emphasise the exceptional role of 10. This notation is called *positional*, and its invention has been attributed to the Sumerians or the Babylonians. It was further development by Hindus and proved to be of enormous significance to civilisation. In Roman symbolism, for example, one wrote

MCMXCVIII = (thousand) + (nine hundreds) + (ninety) + (five) + (one) + (one),

It is clear that more and more new symbols such as I, V, X, C, M are needed as numbers get larger while with the Hindu positional system, now in use, we need only ten "Arabic numerals" $0, 1, 2, \ldots, 9$, no matter how large is the number. The positional system was introduced into medieval Europe by merchants, who learned it from the Arabs. It is exactly this system which is to blame for the fact that the ancient art of computation, once confined to a few adepts, has become a routine algorithmic skill that can be done automatically by a machine and is now taught in primary school.

Mathematically, there is nothing special about the decimal system. The use of ten as the base goes back to the dawn of civilisation and is attributed to the fact that we have ten fingers on which to count. Other numbers could be used as the base, and undoubtedly some of them were used. The number words in many languages show remnants of other bases, mainly twelve, fifteen and twenty. For example, in English the words for 11 and 12 and in Spanish the words for 11, 12, 13, 14 and 15 are not constructed on the decimal principle. In French the word for 20—vingt—suggests that number had a special role at some time in the past. The Babylonian astronomers had a system of notation with base 60. This is believed to be the reason for the customary division of the hour and the angular degree into 60 min. In the theorem that follows we show that an arbitrary positive integer $b > 1$ can be used as a base.

Theorem 1.5.1 *Let $b > 1$ be a positive integer. Then every positive integer N can be uniquely represented in the form*

$$N = d_0 + d_1 b + d_2 b^2 + \ldots + d_n b^n, \tag{1.21}$$

where "the digits" $d_0, d_1, \ldots, d_n$ lie in the range $0 \le d_i \le b-1$, for all i.

Proof The proof is by induction on N, the number being represented. Clearly, the representation $1 = 1$ for 1 is unique. Suppose, inductively, that every integer $1, 2, \ldots, N-1$ is uniquely representable. Now consider the integer N. Let $d_0 = N \bmod b$. Then $N - d_0$ is divisible by b and let $N_1 = (N - d_0)/b$. Since $N_1 < N$, by the induction hypothesis N_1 is uniquely representable in the form

$$N_1 = \frac{N - d_0}{b} = d_1 + d_2 b + d_3 b^2 + \ldots + d_n b^{n-1},$$

Then clearly

$$N = d_0 + N_1 b = d_0 + d_1 b + d_2 b^2 + \ldots + d_n b^n$$

is the representation required.

Finally, suppose that N has some other representation in this form, i.e.,

$$N = d_0 + d_1 b + d_2 b^2 + \ldots + d_n b^n = e_0 + e_1 b + e_2 b^2 + \ldots + e_n b^n.$$

Then $d_0 = e_0 = r$ as they are equal to the remainder of N on dividing by b. Now the number

$$N_1 = \frac{N - r}{b} = d_1 + d_2 b + d_3 b^2 + \ldots + d_n b^{n-1} = e_1 + e_2 b + e_3 b^2 + \ldots + e_n b^{n-1}$$

has two different representations which contradicts the inductive assumption, since we have assumed the truth of the result for all $N_1 < N$.   □

Corollary 1.5.1 *We use the notation*

$$N = (d_n d_{n-1} \ldots d_1 d_0)_{(b)} \tag{1.22}$$

to express (1.21). The digits d_i can be found by the repeated application of the division algorithm as follows:

$$
\begin{aligned}
N &= q_1 b + d_0, & (0 \le d_0 < b) \\
q_1 &= q_2 b + d_1, & (0 \le d_1 < b) \\
&\;\;\vdots \\
q_n &= 0 \cdot b + d_n & (0 \le d_n < b)
\end{aligned}
$$

For example, the positional system with base 5 employs the digits $0, 1, 2, 3, 4$, and we can write

$$1998_{(10)} = 3 \cdot 5^4 + 0 \cdot 5^3 + 4 \cdot 5^2 + 4 \cdot 5 + 3 = 30443_{(5)}.$$

But in the era of computers it is the binary (or dyadic) system (base 2) that has emerged as the most important. This system has only two digits, 0 and 1 and a very simple multiplication table for them. But under the binary system, representations of numbers get longer quickly. For example,

$$150_{(10)} = 1 \cdot 2^7 + 0 \cdot 2^6 + 0 \cdot 2^5 + 1 \cdot 2^4 + 0 \cdot 2^3 + 1 \cdot 2^2 + 1 \cdot 2 + 0 = 10010110_{(2)}. \tag{1.23}$$

Leibniz[9] was one of the ardent proponents of the binary system. According to Laplace: "Leibniz saw in his binary arithmetic the image of creation. He imagined that Unity represented God, and zero the void; that the Supreme Being drew

---

[9] **Gottfried Wilhelm von Leibniz (1646–1716)** was a German mathematician and philosopher who developed infinitesimal calculus independently of Isaac Newton, and Leibniz's mathematical notation has been widely used ever since it was published. He invented an early mechanical calculating machine.

all beings from the void, just as unity and zero express all numbers in his system of numeration."

Let us look at the binary representation of a number from the information point of view. Information is measured in bits. One *bit* is a unit of information expressed as a choice between two possibilities 0 and 1. The number of binary digits in the binary representation of a number N is therefore the number of bits we need to transmit N through an information channel (or input into a computer). For example, the Eq. (1.23) shows that we need 8 bits to transmit or convey the number 150.

Theorem 1.5.2 *To input a number N by converting it into its binary representation we need $\lfloor \log_2 N \rfloor + 1$ bits of information.*

Proof Suppose that N has n binary digits in its binary representation. That is

$$N = 2^{n-1} + a_{n-2}2^{n-2} + \ldots + a_1 2^1 + a_0 2^0, \qquad a_i \in \{0, 1\}.$$

Then $2^n > N \geq 2^{n-1}$ or $n > \log_2 N \geq n - 1$, which is equivalent to $\lfloor \log_2 N \rfloor = n - 1$. Hence $n = \lfloor \log_2 N \rfloor + 1$. $\qquad\qquad\square$

Example 1.5.1 To input 150, we need $\lfloor \log_2 150 \rfloor + 1 = \lfloor 7.2 \rfloor + 1 = 8$ bits.

Example 1.5.2 The input is the number 15011. Convert it to binary. What is the length of this input?

Solution Let $15011 = (a_n a_{n-1} \ldots a_1 a_0)_{(2)}$ be the binary representation of 15111. We can find the binary digits of 15011 recursively by a series of divisions with remainder:

$$
\begin{aligned}
15011 &= 2 \cdot 7505 + 1 &\longrightarrow\quad & a_0 = 1, \\
7505 &= 2 \cdot 3752 + 1 &\longrightarrow\quad & a_1 = 1, \\
3752 &= 2 \cdot 1876 + 0 &\longrightarrow\quad & a_2 = 0, \\
1876 &= 2 \cdot 938 + 0 &\longrightarrow\quad & a_3 = 0, \\
938 &= 2 \cdot 469 + 0 &\longrightarrow\quad & a_4 = 0, \\
469 &= 2 \cdot 234 + 1 &\longrightarrow\quad & a_5 = 1, \\
234 &= 2 \cdot 117 + 0 &\longrightarrow\quad & a_6 = 0, \\
117 &= 2 \cdot 58 + 1 &\longrightarrow\quad & a_7 = 1, \\
58 &= 2 \cdot 29 + 0 &\longrightarrow\quad & a_8 = 0, \\
29 &= 2 \cdot 14 + 1 &\longrightarrow\quad & a_9 = 1, \\
14 &= 2 \cdot 7 + 0 &\longrightarrow\quad & a_{10} = 0, \\
7 &= 2 \cdot 3 + 1 &\longrightarrow\quad & a_{11} = 1, \\
3 &= 2 \cdot 1 + 1 &\longrightarrow\quad & a_{12} = 1, \\
1 &= 2 \cdot 0 + 1 &\longrightarrow\quad & a_{13} = 1,
\end{aligned}
$$

we see that $15011 = 11101010100011_{(2)}$, reading the binary digits from the column of remainders from bottom to the top. Hence the length of the input is 14 bits. $\quad\square$

Example 1.5.3 To estimate from above and from below the number of bits required to input an integer N which has 100 digits in its decimal representation, we may use GAP command `LogInt(N,2)` to calculate $\lfloor \log_2 N \rfloor$. A 100-digit integer is between 10^{99} and 10^{100}, so we have

```
gap> LogInt(10^100,2)+1;
333
gap> LogInt(10^99,2)+1;
329
```

So the number in this range will need between 329 and 333 bits.

The negative powers of 10 are used to express those real numbers which are not integers. This also works in other bases. For example,

$$\frac{1}{8} = 0.125_{(10)} = \frac{1}{10} + \frac{2}{10^2} + \frac{5}{10^3} = \frac{0}{2} + \frac{0}{2^2} + \frac{1}{2^3} = 0.001_{(2)}$$

$$\frac{1}{7} = 0.142857142857\ldots_{(10)} = 0.(142857)_{(10)} = 0.001001\ldots_{(2)} = 0.(001)_{(2)}$$

The binary expansions of irrational numbers, such as

$$\sqrt{5} = 10.001111000110111\ldots_{(2)},$$

are used sometimes in cryptography for simulating a random sequence of bits. But this method is considered to be insecure. The number, $\sqrt{5}$ in the example above, can be guessed after knowing the initial segment, which will reveal the whole sequence.

Exercises

1. Find the binary representation of the number $2002_{(10)}$ and the decimal representation of the number $1100101_{(2)}$.
2. (a) Find the binary representation of the number whose decimal representation is 2011.
   (b) Find the decimal representation of the number whose binary representation is 101001000.
3. Use Euler's theorem to find the last three digits in the binary representation of 75^{1015}.
4. How many nonzero digits are there in the binary representation of the integer

$$\underbrace{100\ldots001}_{n}{}_{(2)} \cdot \underbrace{100\ldots001}_{m}{}_{(2)}?$$

5. The integer n has base 7 representation $n = abcd_{(7)}$, where a, b, c, d are base 7 digits of n. Prove that n is divisible by 6 if and only if the sum $a + b + c + d$ of its digits is divisible by 6.

6. The symbols A, B, C, D, E and F are used to denote the digits 10, 11, 12, 13, 14 and 15, respectively, in the hexadecimal representation (i.e., to base 16).

   (a) Find the decimal representation of $2A4F_{(16)}$.
   (b) Find the hexadecimal representation of $1000_{(10)}$.

7°. Prove that the sequence of Fibonacci numbers mod m is periodic with period of length at most $m^2 - 1$.

Cryptology

2

> *Enigmatic words—they are all full of meaning.*
>
> *Nikolai Roerich (1874–1947)*

Cryptology (or cryptography) is about communication in the presence of adversaries or potential adversaries. In medieval times diplomats had to communicate with their superiors using a messenger. Messengers could be killed and letters could be captured and read by adversaries. During times of war, orders from military headquarters needed to be sent to the line officers without being intercepted and understood by the enemy. The case of a war is an extreme example where the adversary is clearly defined. But there are also situations where the existence of an "adversary" is less obvious. For example, corporate deals and all negotiations must remain secret until completed. Sometimes two parties want to communicate privately even if they do not have any adversaries. For example, they wish to exchange love letters, and confidentiality of messages for them remains a very high priority. Thus, a classic goal of cryptography is privacy. Authentication is another goal of cryptography which is any process by which you verify that someone is indeed who they claim they are. We use passwords to ensure that only certain people have access to certain resources (e.g., if you do your banking on the Internet you do not want other people to know your financial situation or to tamper with your accounts). Digital signatures are a special technique for achieving authentication. Apart from signing your encrypted emails, digital signatures are used for other applications, for example, to ensure that automatic software updates originate from the company they are supposed to, rather than being viruses. Digital signatures are to electronic communication what handwritten signatures are to paper-based communication. Nowadays cryptography has matured and it is addressing an ever increasing number of other goals.

© Springer Nature Switzerland AG 2020

A. Slinko, *Algebra for Applications*, Springer Undergraduate Mathematics Series,

https://doi.org/10.1007/978-3-030-44074-9_2

In his Chap. 13 of the Handbook of Theoretical Computer Science, Ronald Rivest writes: "The invention of radio gave a tremendous impetus to cryptography, since an adversary can eavesdrop easily over great distance. The course of World War II was significantly affected by use, misuse, and breaking of cryptographic systems used for radio traffic. It is intriguing the computational engines designed and built by the British to crack the German *Enigma* cipher are deemed by some to be the first real "computers"; one could argue that cryptography is the mother (or at least the midwife) of computer science". (This chapter can be downloaded from Ron Rivest's web page.)

Rivest mentioned the famous computers "Colossus" here. Until recently all information about them was classified. The Colossus computers were built by a dedicated team of British mathematicians and engineers led by Alan Turing and Tommy Flowers. It was extensively used in the cryptanalysis of high-level German communications. It is believed that this heroic effort shortened the Second World War by many months. Recently Colossus was recreated and outperformed a modern computer (in deciphering messages which had been encrypted using the Lorenz SZ 40/42 cipher machine).[1] Due to secrecy that surrounded everything related to Colossus, there arose a myth that the ENIAC was the first large-scale electronic digital calculator in the world. It was not.

2.1   Classical Secret-Key Cryptology

One of the oldest ciphers known is *Atbash*. It even appears in the Hebrew Scriptures of the Bible. Any occurrence of the first letter of the alphabet is replaced by the last letter, occurrences of the second letter are replaced by the second to last, etc. Atbash is a specific example of a general technique called inversion.

Caesar is also a very old cipher used by Gaius Julius Caesar (130 BC–87 BC). Letters are simply replaced by letters three steps further down the alphabet. This way "a" becomes "d", "b" becomes "e", etc. In fact, any cipher using a displacement of any size is now known as a Caesar. Caesar is a specific example of a general technique called displacement.

These two ciphers are examples of the so-called *substitution methods* which use a mapping of an alphabet onto itself that replace a character with the one it mapped onto. If the mapping does not change within the message, the scheme is known as a *mono-alphabet scheme*. Such cryptosystems were not very secure but were sufficient enough when literacy was not widespread.

For both of these cryptosystems it is essential to keep the method of encryption secret, because, even publicising the idea on which it is based on, might give away an essential part of the security of the system, especially if the adversary managed to intercept sufficiently many encrypted messages.

---

[1] Read about this exciting project and many other historical information about Colossus at http://www.codesandcyphers.org.uk/lorenz/rebuild.htm.

By the end of nineteenth century it became clear that security must be introduced differently. In 1883 Auguste Kerckhoff's[2] wrote two journal articles titled La Cryptographie Militaire, in which he stated six design principles for military ciphers. His main idea was—which is called now Kerckhoffs' principle—that the security must be a result of not keeping the encryption mechanism secret but as a result of keeping a changeable part of the encryption mechanism—called secret key—secret. Depending on the secret key the encryption mechanism should encrypt messages differently. So, even if the adversary knows the encryption method but does not know the key, they will not know how to decrypt messages.

Thus, until recently, a standard cryptographic solution to the privacy problem was a *secret-key cryptosystem*, which consisted of the following:

- A message space $\mathcal{M}$: a set of strings (plaintext messages) over some alphabet (e.g., binary alphabet, English, Cyrillic or Arabic alphabets);
- A ciphertext space $\mathcal{C}$: a set of strings (ciphertext messages) over some alphabet (e.g., the alphabet of the dancing men in one of the Arthur Conan Doyle's stories of Sherlock Holmes);
- A key space $\mathcal{K}$: a set of strings (keys) over some alphabet;
- An encryption algorithm $E: \mathcal{M} \times \mathcal{K} \to \mathcal{C}$, which to every pair $m \in \mathcal{M}$ and $k \in \mathcal{K}$ puts in correspondence a ciphertext $E(m, k)$;
- A decryption algorithm $D: \mathcal{C} \times \mathcal{K} \to \mathcal{M}$ with the property that $D(E(m, k), k) = m$ for all $m \in \mathcal{M}$ and $k \in \mathcal{K}$.

The meaning of the last condition is that if a message is encrypted with a key k, then the same key, when used in the decryption algorithm, will decrypt this message from the ciphertext.

To use a secret-key cryptosystem the parties wishing to communicate privately agree on a key $k \in \mathcal{K}$, which they must keep secret. They communicate a message $m \in \mathcal{M}$ by sending the ciphertext $c = E(m, k)$. The recipient can decrypt the ciphertext to obtain the message m by means of the key k and the decryption algorithm D since $m = D(c, k)$. The cryptosystem is considered to be secure if it is infeasible in practice for an eavesdropper, who has discovered $E(m, k)$ but does not know k, to deduce m.

Below we present three examples.

2.1.1   The One-Time Pad

The one-time pad is a nearly perfect solution to the privacy problem. It was invented in 1917 by Gilbert Vernam (D. Kahn, The Codebreakers, Macmillan, New-York,

---

[2]**Auguste Kerckhoffs** (1835–1903) was a Dutch linguist and cryptographer who was professor of languages at the Ecole des Hautes Etudes Commerciales in Paris.

1967) for use in telegraphy. In this secret-key cryptosystem the key is as long as the message being encrypted. The key, once used, is discarded and never reused.

Suppose that parties managed to generate a very long string k of randomly chosen 0's and 1's. Suppose that they also managed to secretly deliver k to all parties involved with the intention to use it as a key. If a party A wishes to send a telegraphic message m to other parties, then it writes the message as a string of zeros and ones $m = m_1 m_2 \ldots m_n$, takes the first n numbers from k, that is $k' = k_1 k_2 \ldots k_n$ and adds these two strings component-wise mod 2 to get the encrypted message

$$c = m \oplus k' = c_1 c_2 \ldots c_n, \quad \text{where} \quad c_i = m_i \oplus k_i.$$

Then A destroys the first n numbers of the key. On the receiving end all other parties decrypt the message c by computing $m = c \oplus k'$ and also destroy the first n numbers of the key. When another message is to be sent, another part of the key will be used—hence, the name "one-time pad". This system is unconditionally secure in the following sense. If $c = c_1 c_2 \ldots c_n$ is the ciphertext, then an arbitrary message $m = m_1 m_2 \ldots m_n$ could be sent. Indeed, if the key were $m \oplus c$, then $m \oplus (m \oplus c) = c$ and the ciphertext is c.

For written communication this system can be modified as follows. Each letter of the alphabet is given a number in $\mathbb{Z}_{26}$:

$$A \ B \ C \ D \ E \ F \ G \ H \ I \ J \ K \ L \ M$$
$$0 \ \ 1 \ \ 2 \ \ 3 \ \ 4 \ \ 5 \ \ 6 \ \ 7 \ \ 8 \ \ 9 \ 10 \ 11 \ 12$$

$$N \ O \ P \ Q \ R \ S \ T \ U \ V \ W \ X \ Y \ Z$$
$$13 \ 14 \ 15 \ 16 \ 17 \ 18 \ 19 \ 20 \ 21 \ 22 \ 23 \ 24 \ 25$$

You then agree to use a book, little-known to the general public (considered as a very long string of letters), as the secret key. For example, "The Complete Poems of Emily Dickinson" would be a good choice.[3] Then you do the same as we did with telegraphic messages except that we add messages mod 26. Suppose we need to send a message

BUY TELECOM SHARES

and the first unused poem from the book is

> Best Witchcraft is Geometry
> To the magician's mind -
> His ordinary acts are feats
> To thinking of mankind.

---

[3]This choice reflects author's fascination with Dickinson's poetry.

Then the message will be represented as the following string of 16 numbers:

$$B \ U \ Y \ T \ E \ L \ E \ C \ O \ M \ S \ H \ A \ R \ E \ S$$
$$1 \ 20 \ 24 \ 19 \ 4 \ 11 \ 4 \ 2 \ 14 \ 12 \ 18 \ 7 \ 0 \ 17 \ 4 \ 18$$

and the first 16 letters from the poem will be

$$B \ E \ S \ T \ W \ I \ T \ C \ H \ C \ R \ A \ F \ T \ I \ S$$
$$1 \ 4 \ 18 \ 19 \ 22 \ 8 \ 19 \ 2 \ 7 \ 2 \ 17 \ 0 \ 5 \ 19 \ 8 \ 18$$

Adding these two messages mod 26 we get

$$2 \ 24 \ 16 \ 12 \ 0 \ 19 \ 23 \ 4 \ 21 \ 14 \ 9 \ 7 \ 5 \ 10 \ 12 \ 10$$
$$C \ Y \ Q \ M \ A \ T \ X \ E \ V \ O \ J \ H \ F \ K \ M \ K$$

so the cryptotext will be

CYQMATXEVOJHFKMK.

This version of the one-time pad is much less secure as it is vulnerable to frequency analysis.

Exercises

1. Use Khlebnikov's poem

   > Today I will go once again
   > Into life, into haggling, into the flea market,
   > And lead the army of my songs
   > To duel against the market tide.[4]

   as the key to encrypt **BUY MORE PROPERTY** and to decrypt **RCXRN-WOAPDYWCAUERKYWHZRGSXQJW**.

2. Use GAP command **Random([0..25]);** to generate a sequence of random letters of the alphabet of length 20.

3. Using the sequence, obtained in the previous exercise, as a key for one-time pad cryptosystem, encrypt and then decrypt back the sentence by Emily Dickinson **"I HAVE NO TIME TO HATE"**. The GAP programmes LettertoNumber and NumbertoLetter found in Sect. 9.2.3 can help you to convert messages into the digital format and back.

---

[4]**Velemir Khlebnikov** (1885–1922) was one of the key poets in the Russian Futurist movement but his work and influence stretch far beyond it. He was educated as a mathematician and his poetry is very abstract and mathematical. He experimented with the Russian language, drawing deeply upon its roots.

4°. Explain why the one-time pad becomes insecure if the same key is used more than once.

5°. A Latin square of order n is an $n \times n$ matrix $L = (\ell_{ij})$, where $\ell_{ij} \in \{1, 2, \ldots, n\}$, such that each element of the set $\{1, 2, \ldots, n\}$ appears exactly once in each row and in each column of L. A Latin square defines a cipher with the message space $\mathcal{M} = \{1, 2, \ldots, n\}$ and the key space $\mathcal{K} = \{1, 2, \ldots, n\}$ defined as $E(m, k) = \ell_{mk}$.

(a) Construct a Latin square L of order 4.
(b) Encode the message 1111 with key $k = 3$.

2.1.2 An Affine Cryptosystem

This is a substitution cipher which is also based on modular arithmetic. The key to this cryptosystem is a pair $k = (a, b)$ of numbers $a \in \mathbb{Z}_{26}^*$, $b \in \mathbb{Z}_{26}$. Under this system a number in $\mathbb{Z}_{26}$ is assigned to every letter of the alphabet as in the previous section. Each letter is encoded into the corresponding number x it is assigned to and then into the letter to which the number $a \odot x \oplus b$ is assigned. For instance, if $a = 3$ and $b = 5$, then the letter "H" will have a numerical encoding "7". Then $3 \odot 7 \oplus 5 = 0$ is computed, and we note that "0" is the numerical encoding for "A", which shows that "H" is encrypted into "A". Using the key $k = (3, 5)$, the message

BUY TELECOM SHARES

will be encrypted into

INZKRMRLVPHAFERH

The requirement that $a \in \mathbb{Z}_{26}^*$, i.e., $\gcd(a, 26) = 1$ is needed to ensure that the encryption function is one-to-one. Indeed, this is equivalent to having the function

$$E(x) = a \odot x \oplus b$$

be one-to-one. If a is a divisor of zero and $a \odot d = 0$ for some $d \neq 0$, then $E(x) = E(x \oplus d)$ and E is not one-to-one. In particular, $E(0) = E(d)$ and unambiguous decryption is impossible. On the other hand, if a is invertible, then $a \odot x \oplus b = y$ implies $x = a^{-1} \odot (y \oplus (-b))$ and the decryption function exists:

$$D(y) = a^{-1} \odot (y \oplus (-b)).$$

Since the key is very short, this system is not secure: one can simply use all keys one by one and see which key gives a meaningful result. However it can be meaningfully used in combination with other cryptosystems. For example, if we use this encryption first and then use the one-time pad (or the other way around), the frequency analysis will be very much hampered.

Exercises

1. Is it possible to use $k = (13, 11)$ as a key in an affine cryptosystem?
2. Using the affine cryptosystem with the key $k = (11, 13)$ encrypt the message **CRYPTO** and decrypt the message **DRDOFP**.
3. In the affine cryptosystem with an unknown key Eve guessed that the letter **F** was encrypted as **N** and the letter **K** was encrypted as **O**. Help Eve to calculate the key.
4. A plaintext (in English) has been encrypted using the affine cryptosystem. The obtained ciphertext is:

   ljpcc puxya nip ljc cbhcx quxya wxrcp ljc aqo achcx nip ljc rskpn bipra ux ljcup jkbba in alixc xuxc nip miplkb mcx riimcr li ruc ixc nip ljc rkpq bipr ix jua rkpq ljpixc ux ljc bkxr in miprip sjcpc ljc ajkrisa buc ixc puxy li pwbc ljcm kbb ixc puxy li nuxr ljcm ixc puxy li vpuxy ljcm kbb kxr ux ljc rkpqxcaa vuxr ljcm ux ljc bkxr in miprip sjcpc ljc ajkrisa buc

   Find the original plaintext. The following estimation on relative frequencies of the 26 letter in English texts may be of some help. You are encouraged to use any help of computers you find useful.

letter	rel. freq.	letter	rel. freq.	letter	rel. freq.
a	.082	j	.002	s	.063
b	.015	k	.008	t	.091
c	.028	l	.040	u	.028
d	.043	m	.024	v	.010
e	.127	n	.067	w	.023
f	.022	o	.075	x	.001
g	.020	p	.019	y	.020
h	.061	q	.001	z	.001
i	.070	r	.060		

2.1.3 Hill's Cryptosystem

We now consider a slightly more sophisticated encryption procedure, sometimes called The Hill Cipher which was invented in 1929 by Lester S. Hill. Instead of substituting letters it substitutes blocks of letters of fixed length m. The whole message is divided into such m-tuples and each m-tuple is encrypted separately as follows. The key for this cryptosystem is an invertible $m \times m$ matrix over $\mathbb{Z}_{26}$. Both matrix operations, addition and multiplication, are defined by means of addition and multiplication modulo 26. Since we do not use any other operations, it is no longer appropriate to write symbols $\oplus$ and $\odot$ for modular operations. To simplify things we will use ordinary notation. We will consider the case $m = 2$ and therefore pairs of letters and 2×2 matrices. Let

$$K = \begin{bmatrix} a & b \\ c & d \end{bmatrix}$$

be the key matrix. The encryption of a pair of letters (P_1, P_2) is carried out by

$$(P_1, P_2) \rightarrow \begin{bmatrix} x_1 \\ x_2 \end{bmatrix} \rightarrow K \begin{bmatrix} x_1 \\ x_2 \end{bmatrix} = \begin{bmatrix} y_1 \\ y_2 \end{bmatrix} \rightarrow (C_1, C_2),$$

where x_1, x_2 are the numerical codes for P_1, P_2 and y_1, y_2 are the numerical codes for C_1, C_2. The invertibility of K is needed for unambiguous recovery of x_1, x_2 from y_1, y_2.

Example 2.1.1 Let

$$K = \begin{bmatrix} 3 & 3 \\ 2 & 5 \end{bmatrix}$$

and suppose the plaintext message is **HELP**. Then this plaintext is represented by two pairs

$$\mathbf{HELP} \rightarrow \begin{bmatrix} H \\ E \end{bmatrix}, \begin{bmatrix} L \\ P \end{bmatrix} \rightarrow \begin{bmatrix} 7 \\ 4 \end{bmatrix}, \begin{bmatrix} 11 \\ 15 \end{bmatrix}.$$

Then we compute

$$\begin{bmatrix} 3 & 3 \\ 2 & 5 \end{bmatrix} \begin{bmatrix} 7 \\ 4 \end{bmatrix} = \begin{bmatrix} 7 \\ 8 \end{bmatrix}, \quad \begin{bmatrix} 3 & 3 \\ 2 & 5 \end{bmatrix} \begin{bmatrix} 11 \\ 15 \end{bmatrix} = \begin{bmatrix} 0 \\ 19 \end{bmatrix}$$

and continue encryption as follows:

$$\begin{bmatrix} 7 \\ 8 \end{bmatrix}, \begin{bmatrix} 0 \\ 19 \end{bmatrix} \rightarrow \begin{bmatrix} H \\ I \end{bmatrix}, \begin{bmatrix} A \\ T \end{bmatrix} \rightarrow \mathbf{HIAT}$$

so the cryptotext is **HIAT**.

The matrix K is invertible, hence, an inverse K^{-1} exists such that $KK^{-1} = K^{-1}K = I_2$, where I_2 is the identity matrix of order 2. It follows that

$$K^{-1}K \begin{bmatrix} x_1 \\ x_2 \end{bmatrix} = I_2 \begin{bmatrix} x_1 \\ x_2 \end{bmatrix} = \begin{bmatrix} x_1 \\ x_2 \end{bmatrix},$$

and decryption is possible. To implement decoding, we compute

$$K^{-1} = 9^{-1} \begin{bmatrix} 5 & 23 \\ 24 & 3 \end{bmatrix} = 3 \begin{bmatrix} 5 & 23 \\ 24 & 3 \end{bmatrix} = \begin{bmatrix} 15 & 17 \\ 20 & 9 \end{bmatrix}.$$

Let us see how we can decrypt the ciphertext **HIAT**:

$$\mathbf{HIAT} \rightarrow \begin{bmatrix} H \\ I \end{bmatrix}, \begin{bmatrix} A \\ T \end{bmatrix} \rightarrow \begin{bmatrix} 7 \\ 8 \end{bmatrix}, \begin{bmatrix} 0 \\ 19 \end{bmatrix}$$

Then we compute

$$\begin{bmatrix} 15 & 17 \\ 20 & 9 \end{bmatrix}\begin{bmatrix} 7 \\ 8 \end{bmatrix} = \begin{bmatrix} 7 \\ 4 \end{bmatrix}, \qquad \begin{bmatrix} 15 & 17 \\ 20 & 9 \end{bmatrix}\begin{bmatrix} 0 \\ 19 \end{bmatrix} = \begin{bmatrix} 11 \\ 15 \end{bmatrix}$$

and continue decryption as follows:

$$\begin{bmatrix} 7 \\ 4 \end{bmatrix}, \begin{bmatrix} 11 \\ 15 \end{bmatrix} \rightarrow \begin{bmatrix} H \\ E \end{bmatrix}, \begin{bmatrix} L \\ P \end{bmatrix} \rightarrow \textbf{HELP}.$$

We need a criterion of invertibility of a matrix over $\mathbb{Z}_{26}$. The standard criterion of invertibility for matrices over $\mathbb{R}$ is a nonzero determinant. Since $\mathbb{Z}_{26}$ has divisors of zero, we have to slightly modify the standard criterion.

Theorem 2.1.1 *An $n \times n$ matrix K over $\mathbb{Z}_{26}$ is invertible if and only if $\det(K)$ is an invertible element in $\mathbb{Z}_{26}$.*

Proof We will prove this theorem only for $n = 2$. Let us consider a 2×2 matrix $K = \begin{bmatrix} a & b \\ c & d \end{bmatrix}$ whose determinant is $\Delta = \det(K) = ad - bc$. Let us compute

$$\begin{bmatrix} a & b \\ c & d \end{bmatrix}\begin{bmatrix} d & -b \\ -c & a \end{bmatrix} = \begin{bmatrix} \Delta & 0 \\ 0 & \Delta \end{bmatrix} \qquad (2.1)$$

If Δ is a divisor of zero, say $\Delta\Gamma = 0$, then

$$\begin{bmatrix} a & b \\ c & d \end{bmatrix}\begin{bmatrix} d & -b \\ -c & a \end{bmatrix}\begin{bmatrix} \Gamma & 0 \\ 0 & \Gamma \end{bmatrix} = \begin{bmatrix} 0 & 0 \\ 0 & 0 \end{bmatrix}.$$

Let us denote the product of the two rightmost matrices as

$$L = \begin{bmatrix} d & -b \\ -c & a \end{bmatrix}\begin{bmatrix} \Gamma & 0 \\ 0 & \Gamma \end{bmatrix}.$$

If $L \neq 0$, then $KL = 0$ and K is a left divisor of zero. As in Lemma 1.4.2 it can be shown that K cannot be invertible. If however

$$L = \begin{bmatrix} d & -b \\ -c & a \end{bmatrix}\begin{bmatrix} \Gamma & 0 \\ 0 & \Gamma \end{bmatrix} = \begin{bmatrix} \Gamma d & -\Gamma b \\ -\Gamma c & \Gamma a \end{bmatrix} = \begin{bmatrix} 0 & 0 \\ 0 & 0 \end{bmatrix},$$

then $\Gamma a = \Gamma b = \Gamma c = \Gamma d = 0$ and then

$$\begin{bmatrix} a & b \\ c & d \end{bmatrix}\begin{bmatrix} \Gamma & 0 \\ 0 & \Gamma \end{bmatrix} = \begin{bmatrix} 0 & 0 \\ 0 & 0 \end{bmatrix},$$

whence K is again a left divisor of zero.

On the other hand, Eq. 2.1 shows that if Δ is invertible, then

$$\begin{bmatrix} a & b \\ c & d \end{bmatrix}^{-1} = \Delta^{-1} \begin{bmatrix} d & -b \\ -c & a \end{bmatrix}$$

is the inverse.   □

Hill's cryptosystem is not considered secure. In particular, it is vulnerable to the so-called *known plaintext attack*. Indeed, if an $k \times k$ matrix K is a key, then it is normally enough to know that message fragments $m_1, m_2, \ldots, m_k$ are encrypted as $c_1, c_2, \ldots, c_k$. Indeed, if the ith column of a matrix X represents the numerical encodings of the plain text fragment m_i and the ith column of a matrix Y represents the numerical encodings of the cipher text fragment c_i, then $Y = KX$ from which the key can be found as $K = YX^{-1}$. In rare cases the matrix X may appear degenerate, in which case we will not be able to find K exactly but still will have much information about it.

In cryptanalysis which is an art of breaking ciphers the so-called method of cribs is widely used. This term was introduced by cryptographers in Bletchley Park and it means a suspected plaintext. For example, an English language text contains the word "that" with high probability and a letter often starts with the word "dear".

Exercises

1. (a) Which one of the two matrices being considered as matrices over $\mathbb{Z}_{26}$

$$\begin{bmatrix} 1 & 12 \\ 12 & 1 \end{bmatrix}, \qquad \begin{bmatrix} 1 & 6 \\ 6 & 1 \end{bmatrix}$$

is invertible and which is not? Find the inverse for the invertible matrix.
   (b) Let M be the invertible matrix and found in part (a). Use it as a key in the Hill's cryptosystem to encrypt **YEAR** and to decrypt **ROLK**.
2. In the Hill's cryptosystem with the key

$$K = \begin{bmatrix} 11 & 12 \\ 12 & 11 \end{bmatrix}$$

find all pairs of letters **XY** which do not change after being encoded twice, i.e., if we encode **XY** we get a pair **ZT** which is being encoded as **XY**.
3. You have captured the ciphertext

NWOLBOTEPEHKICNSHR.

You know it has been encrypted using the Hill cipher with a 2×2 matrix and you know the first 4 letters of the message are the word "DEAR". Find the secret key and obtain the message.

4. The key for Hill's cryptosystem is the following matrix over $\mathbb{Z}_{26}$

$$K = \begin{bmatrix} 1 & 2 & 3 & 4 & 5 \\ 9 & 11 & 18 & 12 & 4 \\ 1 & 2 & 8 & 23 & 3 \\ 7 & 14 & 21 & 5 & 1 \\ 5 & 20 & 6 & 5 & 0 \end{bmatrix}.$$

Use GAP to decrypt the message

WGVUUTGEPVRIMFTXMXMHCYTNGYMJJE

EZKEWHLQQISDJYJCTYEUBYKFBWPBBE

5. (advanced linear algebra required) Prove that a square $n \times n$ matrix A over $\mathbb{Z}_{26}$ is invertible if and only if its determinant $\det(A)$ is invertible element of $\mathbb{Z}_{26}$.
6°. Generate randomly 4×4 invertible matrix K over $\mathbb{Z}_{26}$. Use it as a key to encode and then decode a message of your choice.
7°. Eve eavesdrops on the correspondence of Ark, Pip and Tom who use Hill's cryptosystem to communicate. She discovered that they split messages into segments which are three letters long and that the names **ARK**, **PIP** and **TOM** are encoded as **GCB**, **APM** and **BWZ**, respectively. Find the matrix K which is used as the key.

2.2 Modern Public-Key Cryptology

Traditional secret-key cryptology assumes that both the sender and the receiver must be in possession of the same secret key which they use both for encryption and decryption. This secret key must be delivered all around the world, to all the correspondents. This is a major weakness of this system.

The modern public-key cryptology breaks the symmetry between the sender and the receiver. It requires two separate keys for each user, one of which is secret (or private) and one of which is public. The public key is used to encrypt plaintext, whereas the private key is used to decrypt ciphertext. Since the public key is no longer secret it can be widely distributed without compromising security. The term "asymmetric" stems from the use of different keys to perform the encryption and the decryption, each operation being the inverse of the other—while the conventional ("symmetric") cryptography uses the same key to perform both.

The computational complexity is the main reason why the system works. The adversary will know how to decrypt messages but will still be unable to do it due to the extremely high complexity of the task.

2.2.1   One-Way Functions and Trapdoor Functions

The idea of public-key cryptology is closely related to the concept of one-way function.

Definition 2.2.1 A function $n \mapsto f(n)$ is said to be a *one-way function* if the computation of $f(n)$, given n, is computationally easy while the computation of n, given $f(n)$, is intractable.

Example 2.2.1 Given the availability of ordinary telephone books the function

$$\text{TELEPHONE NUMBER} = f(\text{NAME})$$

can be easily performed in seconds as it is easy to find the name in the book since they are listed in the alphabetical order but the function

$$\text{NAME} = f^{-1}(\text{TELEPHONE NUMBER})$$

can hardly be performed at all, since in the worst case you need to read the whole book in order to find the name corresponding to a given number. You might need a month to do that.

We note that Definition 2.2.1 should not be treated as a rigorous mathematical definition. It contains references to "easy" and "intractable" tasks which may be dependent on the computing resources available.

A publicly available one-way function f has a number of useful applications. In time-shared computer systems, instead of storing a table of login passwords, one can store, for each password w, the value $f(w)$. Passwords can be easily checked for correctness at login, but even the system administrator cannot deduce any user's password by examining the stored table.

Definition 2.2.2 Let $n \mapsto f(n)$ be a one-way function and let t be an additional parameter (it can be a number, a graph, a function, anything) such that it is computationally easy to compute n, given $f(n)$ and t. Then t is called a *trapdoor* and f is called a *trapdoor function*.

Example 2.2.2 Imagine that you have taken the time to enter the telephone directory into your computer, sorted all phone numbers in increasing order, and printed them. Suppose that it took one month of your time. Then you possess a trapdoor to the one-way function f described in Example 2.2.1. For you it is easy to compute f or f^{-1} and you are the only person (at least for the next month) who can compute f^{-1}.

We describe the idea of a public-key cryptosystem on the following example. Imagine that Alice possesses a trapdoor function

$$f(\text{TEXT}) = \text{CIPHERTEXT}$$

with a secret trapdoor t. Then she puts this function f in the public domain, where it is accessible to everyone, and asks everybody to send her f(TEXT) each time when the necessity arises to send a message TEXT confidentially. Knowing the trapdoor t, it is an easy job for her to compute the TEXT from f(TEXT) while it is infeasible to compute it for anybody else. The function f (or a certain parameter which determines f uniquely) is called Alice's *public key* and the trapdoor t is called her *private key*.

Example 2.2.3 Let us see how we can use the trapdoor function of Example 2.2.1 to construct a public-key cryptosystem. Take the University of Auckland telephone directory and announce the method of encryption as follows. Your correspondent must take a letter of her message, find a name in the directory starting with this letter, and assign to this letter the phone number of the person with the chosen name. She must do it with all letters of her message. Then all these phone numbers combined will form a ciphertext. For example, the message SELL BRIERLY, sent to you, will be encrypted as follows:

S	SCOTT	8751
E	EVANS	8057
L	LEE	8749
L	LEE	5999
B	BANDYOPADHYAY	7439
R	ROSENBERG	5114
I	ITO	7518
E	ESCOBAR	6121
R	RAWIRI	7938
L	LEE	6346
Y	YU	5125

The message in encrypted form will look like this:

$$8751805787495999743951147518612179386 3465125$$

For decryption you must use your private key, which is the inverse telephone directory.

2.3   Computational Complexity

In this section we will develop several rigorous concepts necessary for implementing the idea of the previous section. To measure the running time of an algorithm we need first to choose a unit of work, say one multiplication, or one division with remainder, etc.; we will often call them *steps*.

It is often the case that not all instances of a problem under consideration are equally hard even if the two inputs are of the same length. For example, if we feed an algorithm two different—but equally long—inputs (and we feed them in one at a time, not both at once), then the algorithm might require an astronomical number of operations to deal with the first input, but only a handful of operations to deal with the second input. The *(worst case) time complexity* of an algorithm is a function that for each length of the input shows the maximal number of units of work that may be required. We say that an algorithm is of time complexity $f(n)$ if for all n and for all inputs of n bits, the execution of the algorithm requires at most $f(n)$ steps. The worst-case complexity takes into consideration only the hardest instances of the problem. It is relevant when people are pessimistic, and think that it is very likely that a really hard instance of the problem will crop up.

Average case complexity, on the other hand, estimates how difficult the relevant problem is "on average". An optimist, thinking that hard instances of the problem are rare, will be more interested in the average-case than the worst-case complexity. At present, much less is known about the average-case complexity than about the worst-case one, so we concentrate on the former.

We need a language to compare the time complexity functions of different algorithms.

2.3.1   Orders of Magnitude

Firstly, we will say what it means to be asymptotically equal.

Definition 2.3.1 Let $f(x)$ and $g(x)$ be two real-valued functions. We say that $f(n) \sim g(n)$ (read "f is *asymptotically equal* to g") if

$$\lim_{n \to \infty} \frac{f(n)}{g(n)} = 1.$$

Example 2.3.1 Let $f(x) = \sum_{k=0}^{d} a_k x^{d-k}$ be a polynomial of degree d. Then $f(n) \sim a_0 n^d$. Indeed, when $n \to \infty$

$$\frac{f(n)}{a_0 n^d} = \frac{a_0 n^d + a_1 n^{d-1} + \ldots + a_d}{a_0 n^d} = 1 + \frac{a_1}{a_0} \cdot \frac{1}{n} + \ldots + \frac{a_d}{a_0} \cdot \frac{1}{n^d} \to 1.$$

Example 2.3.2 The famous *Stirling's formula*

$$n! \sim \sqrt{2\pi n} \cdot n^n e^{-n} \tag{2.2}$$

gives us a tool to compare the factorial growth with the others.

For comparing growth of functions we use the "little-oh," "big-Oh" and "big-Theta" notation.

Definition 2.3.2 We say that $f(n) = o(g(n))$ (read "f is little-oh of g") if

$$\lim_{n \to \infty} \frac{f(n)}{g(n)} = 0.$$

Informally, this means that f grows slower than g when n gets large.

Example 2.3.3 $1000n^{2.9} = o(n^3)$.
It is almost obvious since

$$\frac{1000n^{2.9}}{n^3} = \frac{1000}{n^{0.1}} \to 0.$$

However not all comparisons can be done so easily. To compare the rate of growth of two functions one often needs *L'Hospital's rule*. We formulate it in the form that suits our applications.

Theorem 2.3.1 (L'Hospital's rule) *Let $f(x)$ and $g(x)$ be two differentiable functions such that $\lim_{x \to \infty} f(x) = \infty$, and $\lim_{x \to \infty} g(x) = \infty$. Suppose that*

$$\lim_{x \to \infty} \frac{f'(x)}{g'(x)}$$

exists. Then

$$\lim_{x \to \infty} \frac{f(x)}{g(x)} = \lim_{x \to \infty} \frac{f'(x)}{g'(x)}.$$

Example 2.3.4 $\ln n = o(\sqrt{n})$.
Let us justify this using L'Hospital's rule. Indeed,

$$\lim_{x \to \infty} \frac{\ln x}{\sqrt{x}} = \lim_{x \to \infty} \frac{(\ln x)'}{(\sqrt{x})'} = \lim_{x \to \infty} \frac{1/x}{1/2\sqrt{x}} = \lim_{x \to \infty} \frac{2}{\sqrt{x}} = 0.$$

Example 2.3.5 (a) $n^{1999} = o(e^n)$, (b) $c^n = o(n!)$.
(a) again follows from L'Hospital's rule and we leave it as an exercise. (b) follows from Stirling's formula. Indeed,

$$\lim_{n \to \infty} \frac{c^n}{n!} = \lim_{n \to \infty} \frac{c^n}{\sqrt{2\pi n} \cdot n^n e^{-n}} = \lim_{n \to \infty} \frac{1}{\sqrt{2\pi n}} \cdot \frac{(ec)^n}{n^n} = \lim_{n \to \infty} \frac{1}{\sqrt{2\pi n}} \cdot \left(\frac{ec}{n}\right)^n = 0.$$

Definition 2.3.3 Let $g(n)$ be a function taking only non-negative values. We say that $f(n) = O(g(n))$ (read "f is big-Oh of g") if there exists a number $C > 0$ and an integer n_0 such that for $n > n_0$

$$|f(n)| < Cg(n).$$

Informally, it means that f does not grow at a faster rate than g when n gets large.

Example 2.3.6 (a) $\sin n = O(1)$, (b) $1000n^3 + \sqrt{n} = O(n^3)$.

In the first case $\sin n \leq 1 \cdot 1$ so we can take $C = 1$. In the second we note that $\sqrt{n} \leq n^3$, hence

$$1000n^3 + \sqrt{n} \leq 1001n^3$$

and we can take $C = 1001$.

Proposition 2.3.1 *Let* $f(x) = \sum_{k=0}^{d} a_k x^k$ *be a polynomial of degree* d. *Then* $f(n) = O(n^d)$.

Proof Let $C = |a_0| + |a_1| + \cdots + |a_d|$. Then $x^i < x^d$ for sufficiently large x, and

$$|f(x)| = |a_0 + a_1 x + a_2 x^2 + \cdots + a_d x^d| \leq (|a_0| + |a_1| + |a_2| + \cdots + |a_d|)x^d = Cx^d,$$

which proves the statement. □

Definition 2.3.4 Let $g(n)$ be a function taking only non-negative values. We say that $f(n) = \Theta(g(n))$ (read "f is big-Theta of g") if there exist two numbers $c, C > 0$ and an integer n_0 such that for $n > n_0$

$$cg(n) < |f(n)| < Cg(n).$$

Informally, it means that f grows as fast as g does when n gets large.

Example 2.3.7 $\pi n + \sin(n) = \Theta(n)$ since $2n < |\pi n + \sin(n)| < 5n$ so we can choose $c = 2$ and $C = 5$.

The functions $1, \log n, n, n^d, c^n$ ($c > 1$), $n!$ are considered to be standard and we measure the growth of other functions by comparing their growth against the standard ones:

$O(1)$	at most constant		$\Theta(1)$	constant
$O(\log n)$	at most logarithmic		$\Theta(\log n)$	logarithmic
$O(n)$	at most linear		$\Theta(n)$	linear
$O(n^2)$	at most quadratic		$\Theta(n^2)$	quadratic
$O(n^3)$	at most cubic		$\Theta(n^3)$	cubic
$O(n^d)$	at most polynomial		$\Theta(n^d)$	polynomial
$O(c^n)$	at most exponential		$\Theta(c^n)$	exponential
$O(n!)$	at most factorial		$\Theta(n!)$	factorial

These functions are listed in increasing order of the rapidity of growth. Of course there are some intermediate cases like $O(\log \log n)$ and $O(n \log n)$. The table below provides estimates of the running times of algorithms for certain orders of complexity. Here we have problems with input strings of 2, 16 and 64 bits.

Problem size	Order	of	complexity			
n	$\log n$	n	$n \log_2 n$	n^2	2^n	$n!$
2	1	2	2	4	4	2
16	4	16	64	256	6.5×10^4	2.1×10^{13}
64	6	64	384	4096	1.8×10^{19}	$> 10^{89}$

If we assume that one operation (unit of labour) requires $1\,\mu s$ $(= 10^{-6}\,\text{s})$, then it is worth noting that a problem with exponential complexity will require on input of 64 bits:

$$1.84 \times 10^{19}\ \mu s = 5845 \text{ centuries.}$$

Problems, which can only be solved by algorithms whose time complexity is exponential, quickly become intractable when the size of the input grows. That is why mathematicians and computer scientists consider polynomial growth as the upper bound of what can be practically computed. Everything which is beyond polynomial growth is considered to be intractable (though there are some interesting intermediate cases, such as the subexponential time complexity algorithms for factorisation of integers).

Exercises

1. Prove that $(\log n)^2 = o(\sqrt{n})$.
2. Use L'Hospital's rule to compare the growth of the two functions:

$$f(n) = n^{2007}, \qquad g(n) = 2^{\sqrt{n}}.$$

3. Let $f(x) = \sum_{k=0}^{d} a_k x^k$ be a polynomial of degree d. Prove that $f(n) = \Theta(n^d)$.
4. It has been experimentally established that the function $\psi(x) = \int_{2}^{x} \frac{dt}{\ln t}$ approximates the function $\pi(x)$ introduced in Sect. 1.1.3 even better than $x/\ln x$. Using L'Hospital's rule, prove that

$$\psi(x) \sim \frac{x}{\ln x}.$$

5. List the following functions in the increasing order of magnitude, when $n \to \infty$:
   (a) $f(n) = (\ln n)^{1000}$, $\quad g(n) = n^{10}$, $\quad h(n) = \sqrt[3]{e^n}$.
   (b) $f(n) = e^{\sin n}$, $\quad g(n) = n^2$, $\quad h(n) = \ln n!$
6. We say that ℓ is a perfect power if there are positive integers $m > 1$ and $k > 1$ such that $\ell = m^k$. Suppose that the unit of work is execution of one GAP command $\text{RootInt}(x,y)$ and that multiplication is costless. Write a GAP programme that has a polynomial complexity and determines if the given integer n is a perfect power or not. Find out if the following number n is a perfect power

3216591599079596080620115649769213179918945365883182177751170074891356872908523398835627858363307507667451980912979425575549941566762328495958107942767427463876601038320227540205184142004885083069045762860916300473260617321314772376006202261722385053673443941918742352729861843482679785060898180075920878659088367693192622340064634811419535028889335540064440165586139725678645254602330925876521569202612057875582421892741493318951011726830528228072784935869965845514150622272147684764562970500861499137153642010326348634959615993459063845793313984237722143683892937148998975391746809877568851727623360135437006245741745750242447915272819370.

If it is a perfect power, output m and k such that $n = m^k$.

7. Use Stirling's formula to establish the character of growth of the following binomial coefficients:
   (a) $\binom{n}{k}$, where k is fixed,
   (b) $\binom{n}{k}$, where $k \sim \alpha n$, and α is a fixed real number with $0 < \alpha < 1$.

8°. Let us call a number N *almost prime* if does not have prime divisors smaller than $(\log_2 N)^2$.
   (a) Prove that there exist almost prime numbers that are composite.
   (b) Using GAP find the smallest almost prime number that is composite.

2.3.2   The Time Complexity of Several Number-Theoretic Algorithms

In a Number-Theoretic algorithm the input is often a number (or several numbers). So what is the length of the input in bits if it is an integer N? In other words, we are asking how many zeros and ones one needs to express N. This question was solved in Sect. 1.5, where we learned how to represent numbers in binary. By Theorem 1.5.2 to express N in binary we need $n = \lfloor \log_2 N \rfloor + 1$ bit of information. For most calculations it would be sufficient to use the following approximations: $N \approx 2^n$ and $n \approx \log_2 N$.

Now we will consider two algorithms for calculating $c^N \bmod m$, where c and m are fixed numbers. Here N is the input and $c^N \bmod m$ is an output. The running time of the algorithm will be measured by the number of modular multiplications required. In ordinary arithmetic this measure might not be satisfactory since the numbers grow in size and some multiplications are much more labour intensive than the others. However in modular arithmetic all numbers are of approximately equal size and our assumption is realistic.

Algorithm 1 is given by the formula $c^N = (\ldots(((c \cdot c) \cdot c) \cdot c) \ldots) \cdot c$. That is we calculate powers of c recursively by setting $c^1 = c$ and $c^{i+1} = c^i \cdot c$. To calculate c^N by this method we require $N - 1$ multiplications. Hence the complexity function $f(n)$ for this algorithm is $f(n) = N - 1 \approx 2^n - 1$, where $n = \lfloor \log_2 N \rfloor + 1$ is the length of the input. Since $\frac{1}{2} 2^n < f(n) < 2^n$ we have $f(n) = \Theta(2^n)$. This algorithm has exponential complexity.

We have been too straightforward in calculating $c^N \bmod m$ and the result was appalling. We can be much more clever and do much better.

Algorithm 2 (*Square and Multiply*) Let us represent N in binary

$$N = 2^k + a_{k-1}2^{k-1} + \ldots + a_1 2^1 + a_0 2^0, \qquad a_i \in \{0, 1\},$$

where $k = \lfloor \log_2 N \rfloor = n - 1$. We can rewrite this as

$$N = 2^{i_0} + 2^{i_1} + \ldots + 2^{i_s}, \qquad k = i_0 > i_1 > \ldots > i_s, \qquad s \leq k + 1 = n,$$

By successive squaring we compute

$$c^{2^1} = c^2 \bmod m, \ c^{2^2} = (c^2)^2 \bmod m, \ c^{2^3} = (c^{2^2})^2 \bmod m, \ldots, \ c^{2^k} = (c^{2^{k-1}})^2 \bmod m$$

using $k = n - 1$ modular multiplications. At most, another $s \leq n$ additional modular multiplications may be required to calculate

$$c^N = c^{2^{i_0} + 2^{i_1} + \ldots + 2^{i_s}} = c^{2^{i_0}} \cdot c^{2^{i_1}} \cdot \ldots \cdot c^{2^{i_s}} \bmod m.$$

So $n - 1 \leq f(n) \leq 2n - 1$. This means that $f(n) = \Theta(n)$ and the algorithm has linear complexity. We have now proven the following theorem.

Theorem 2.3.2 *Let c and m be positive integers. Then for every positive integer N we can calculate $c^N \bmod m$ using at most $2 \log_2 N$ multiplications modulo m. Algorithm 2 (Square and Multiply) has linear complexity.*

Example 2.3.8 How many multiplications are needed to calculate c^{29} using Algorithms 1 and 2?

The binary representation for 29 is as follows:

$$29 = 16 + 8 + 4 + 1 = 11101_{(2)}.$$

and $n = \lfloor \log_2 29 \rfloor + 1 = 5$. Hence we need 4 multiplications to calculate c^2, c^4, c^8, c^{16} by successive squaring, and then we will need 3 more to calculate $c^{29} = c^{16} \cdot c^8 \cdot c^4 \cdot c$. Thus Algorithm 2 would use 7 multiplications in total. Algorithm 1 would use 28 multiplications.

The complexity of the Euclidean algorithm will also be important for us. So we prove:

Theorem 2.3.3 *For any two positive integers a and b the Euclidean algorithm will find their greatest common divisor after at most* $2 \log_2 N + 1$ *integer divisions with remainder, where* $N = \max(a, b)$.

Proof Let us make one observation first. Suppose $a = qb + r$ is a division with remainder, $a' = a/\gcd(a, b)$, $b' = b/\gcd(a, b)$, and $r' = r/\gcd(a, b)$. Then $a' = qb' + r'$ is also a division with remainder. Hence the number of steps that the Euclidean algorithm (Theorem 1.2.3) requires is the same for the pair (a, b) as for the pair (a', b'). This allows us to assume that $\gcd(a, b) = 1$. Let us also assume that a is not smaller than b.

We will first prove that if $a \geq b$ (as we just assumed) then on dividing a by b with remainder

$$a = qb + r, \qquad (0 \leq r < b),$$

we get $r < a/2$. Indeed, if $q \geq 2$, then $r < b < a/q \leq a/2$, and when $q = 1$, then $b > a/2$, hence $r = a - b < a/2$.

Let us perform the Euclidean algorithm on a and b

$$
\begin{aligned}
a &= q_1 b + r_1, & 0 < r_1 < b, \\
b &= q_2 r_1 + r_2, & 0 < r_2 < r_1, \\
r_1 &= q_3 r_2 + r_3, & 0 < r_3 < r_2, \\
&\;\;\vdots \\
r_{s-2} &= q_s r_{s-1} + r_s, & 0 < r_s < r_{s-1}, \\
r_{s-1} &= q_{s+1} r_s.
\end{aligned}
$$

Then $r_s = \gcd(a, b)$. Due to the observation in the beginning of the proof we can conclude that

$$r_3 < r_1/2 < a/4, \quad r_5 < r_3/2 < a/8,$$

and by induction $r_{2k+1} < \dfrac{a}{2^{k+1}}$ and $r_{2k} < \dfrac{b}{2^k}$. Suppose the algorithm stops at step s, i.e., after calculating that $r_s = 1$. Then if $s = 2k + 1$, we have $2^{k+1} < a$ and $k < \log_2 a$, whence $s = 2k + 1 < 2 \log_2 a + 1$. Hence $s \leq 2 \log_2 a = 2 \log_2 N$. If $s = 2k$, then $2^k < b$, whence $k < \log_2 b \leq \log_2 N$, and $s = 2k < 2 \log_2 N$.

If a is smaller than b, then we will need an additional step, and the number of steps will be not greater than $2 \log_2 N + 1$.   □

Now we can make conclusions about the time complexity of the Euclidean algorithm. For one unit of work we will adopt the execution of a single $a \bmod b$ operation that is division of a by b with remainder.

Corollary 2.3.1 *The Euclidean algorithm has linear worst-case complexity.*

Proof The upper bound in the Theorem 2.3.3 can be interpreted as follows. The number $\log_2 N$, where $N = \max(a, b)$, is almost exactly the number of bits, say k, in the binary representation of N. So the length of the input, n, (numbers a and b) is at least k and at most $2k$ while the number of units of work is at most $2k$. So for the time complexity function $f(n)$ we have $f(n) \leq 2n$. Thus $f(n) = O(n)$.   □

In Sect. 1.1.3 we saw that the Trial Division algorithm for factoring an integer n (which, we recall, could involve performing as many divisions as there are primes between 2 and $\lfloor \sqrt{n} \rfloor$), was computationally difficult. Now we can state this precisely. It has exponential time complexity!

Theorem 2.3.4 (A worst-case time complexity for factoring) *Trial Division algorithm for factoring integers has exponential complexity.*

Proof Let the unit of work be one division. Let us assume that we have an infinite memory and that all primes are stored there: $p_1, p_2, \ldots, p_m, \ldots$. Given a positive integer N we have to try to divide it by all primes which do not exceed $M = \lfloor \sqrt{N} \rfloor$. According to the prime number theorem there are approximately

$$\frac{M}{\ln M} \approx 2\frac{\sqrt{N}}{\ln N}$$

such primes. This means that in the worst-case scenario we have to try all of them and thus perform $2\sqrt{N}/\ln N$ divisions. Since $N \approx 2^n$, where n is the number of input bits, the worst-case complexity function takes the form

$$f(n) \approx \frac{2}{n \ln 2}\sqrt{2^n} \approx \frac{2}{\ln 2} \cdot \frac{1}{n} \cdot \left(\sqrt{2}\right)^n.$$

Let $\sqrt{2} = \alpha\beta$, where $\alpha > 1$ and $\beta > 1$. Then

$$\frac{1}{n} \cdot \left(\sqrt{2}\right)^n = \frac{\alpha^n}{n} \cdot \beta^n > \beta^n.$$

Thus $f(n)$ is growing faster than β^n.   □

In the case of calculating Nth powers we know one efficient and one inefficient algorithm. For factoring integers we know only one and it is inefficient. All attempts of researchers in number theory and computer science to come up with a more efficient algorithm have resulted in only very modest improvements. Several algorithms are known that are subexponential with the time complexity function, for example, $f(n) = e^{cn^{1/3}(\ln n)^{2/3}}$ (see [14]). This growth is still very fast. At the moment of writing it is not feasible to factor a 200 digit integer unless it has many small divisors.

Exercises

1. (a) Estimate the number of bits required to input an integer N which has 100 digits in its decimal representation.

   (b) Represent $n = 1234567$ in binary and decide how many multiplications mod m the Square-and-Multiply algorithm would require to calculate $c^n \bmod m$.

2. The Bubble Sort algorithm takes a finite list of numbers and arranges them in increasing order. Given a list of numbers, it compares each item in the list with the item next to it, and swaps them if the former is larger than the latter. The algorithm repeats this process until it makes a pass all the way through the list without swapping any items (in other words, all items are in the correct order). This causes larger values to "bubble" to the end of the list while smaller values "sink" towards the beginning of the list.

   Assume that one needs 100 bits to input any number on the list (so the length of the input is $100n$). Take one swap as one unit of work. Determine the worst case complexity of the Bubble Sort algorithm. Use the appropriate notation (big-oh, little-oh, etc.) to express the character of the growth.

3. The input of an algorithm is a positive integer N. The algorithm tries to divide N by the first $(\log_2 N)^3$ primes and, if one of them divides N, it declares N composite. If none of those primes divide N, the algorithm declares N interesting. What is the worst-case complexity of this algorithm?

4. Let (f_n) be the sequence of Fibonacci numbers given by $f_0 = f_1 = 1$ and $f_{n+2} = f_{n+1} + f_n$.

   (a) Prove that

   $$f_n < 2f_{n-1} \quad \text{and} \quad f_{n+5} > 10 f_n.$$

   (b) Using part (a), prove Lamé's theorem that the number of divisions with remainder required by the Euclidean algorithm for finding $\gcd(a, b)$ is at most five times the number of decimal digits in the smaller of a or b.

5°. Algorithms A and B spend exactly $f_A(n) = c_A n \log_2 n$ and $f_B(n) = c_B n^2$ microseconds, respectively, for a problem of size n. Find the best algorithm for processing $n = 2^{20}$ data items if the algorithm A spends $10\,\mu s$ to process 1024 items and the algorithm B spends only $1\,\mu s$ to process 1024 items.

2.4   The RSA Public-Key Cryptosystem

Alice wishes to receive confidential messages from her correspondents. For this purpose she may use the public key RSA cryptosystem, named after Rivest, Shamir and Adelman [18], who invented it in 1977. It is widely used now. It is based on the fact that the mapping

$$f : x \mapsto x^e \bmod n$$

for a specially selected very large number n and exponent e is a one-way function.

2.4.1 How Does the RSA System Work?

Alice creates her public and private keys as follows:

1. She generates two large primes $p \neq q$ of roughly the same size;
2. Calculates $n = pq$ and $\phi = (p - 1)(q - 1)$, where ϕ is the value of the Euler's ϕ-function, $\phi(n)$;
3. Using trial and error method, selects a random integer e with $1 < e < \phi$ and $\gcd(e, \phi) = 1$;
4. Computes d such that $ed \equiv 1 \bmod \phi$ and $1 < d < \phi$.

We will later discuss how Alice can generate two large primes. She can then do steps 2–4 because the complexity of the extended Euclidean algorithm is so low that it easily works for very large numbers. Note that finding d is also done by extended Euclidean algorithm.

Alice uses a certain public domain which is accessible for all her correspondents, for example, her home page, to publish her *public key* (n, e), keeping everything else secret; in particular, d which is Alice's *private key* (which will be used for decryption). It must be clear for everybody that (n, e) is indeed Alice's public key and nobody but Alice could publish it.

She then instructs how to use her public key to convert text into ciphertext. In the first instance all messages must be transformed into numbers by some convention specified by Alice, e.g., we may use "01" instead of "a", "02" instead of "b", etc; for simplicity, let us not distinguish between upper and lower case, and denote a space by "27". Thus a message for us is a non-negative integer. The public key (n, e) stipulates that Alice may receive messages, which are non-negative integers m which are smaller than n. (If the message is longer it should be split into several shorter messages.) The message m must be encrypted applying the following function to the message:

$$f(m) = m^e \bmod n.$$

This function is uniquely determined by Alice's public key. It is a one-way function to everybody but Alice who has a trapdoor d (we will see later how it used for decryption). For example, when Bob wishes to send a private message to Alice, he obtains Alice's public key (n, e) from the public domain and uses it as follows:

- Turns the message text into a non-negative integer $m < n$ (or several of them if breaking the text into blocks of smaller size is necessary);

- Computes the ciphertext $c = m^e \bmod n$;
- Sends the ciphertext c to Alice.

Alice then recovers the plaintext m using her private key d (which is the trapdoor for f) by calculating

$$m = c^d \bmod n.$$

This may seem to be a miracle at this stage but it can (and, below, will) be explained.

This system can work only because of the clever choice of the primes p and q. Indeed, p and q should be chosen so that their product $n = pq$ is infeasible to factorise. This secures that p and q are known only to Alice, while at the same time n and her public exponent e are known to everybody. This implies that Alice's private exponent d is also known only to her. Indeed, to calculate d from publicly known parameters, one needs to calculate $\phi(n)$ first. But the only known method of calculating $\phi(n)$ requires calculation of the prime factorisation of n. Since it is infeasible, we can publish n but keep $\phi(n)$, and hence d, secret.

Example 2.4.1 This is of course very small example (too small for practical purposes), just to illustrate the algorithms involved. Suppose the Alice's arrangements were as follows:

1. $p = 101, q = 113$;
2. $n = pq = 11413, \phi = (p - 1)(q - 1) = 11200$;
3. $e = 4203$ (picked at random from the interval $(1, \phi)$, making sure that $\gcd(e, \phi) = 1$);
4. $d = 3267$ (the inverse of e in $\mathbb{Z}_\phi$);
5. The public key is therefore $(11413, 4203)$, the private key is 3267.

If Bob wants to send the message "Hello Alice" he transforms it into a number as described. The message is then represented by the integer

$$0805121215270112090305.$$

This is too large (≥ 11413), so we break the message text into chunks of 2 letters at a time.

A. The first message fragment is $m = 0805$;
B. Bob computes $c = m^e = 805^{4203} \equiv 6134 \bmod 11413$;
C. Alice decrypts this message fragment by calculating $c^d = 6134^{3267} \equiv 805 \bmod 11413$.

If Bob wants to receive an encrypted answer from Alice he has to set up a similar scheme. In practice people do not set up cryptosystems individually but use a trusted provider of such services. Such a company would create a public domain and place there all public keys attributed to participating individuals. Such a company creates an infrastructure that makes encrypted communication possible. The infrastructure that is needed for such cryptosystem to work is called a *public-key infrastructure* (PKI) and the company that certifies that a particular public key belongs to a certain person or organisation is called a *certification authority* (CA). The most known such companies are Symantec (which bought VeriSign's business), Comodo, GlobalSign, Go Daddy, etc. Furthermore, we will show in Sect. 2.5 that the PKI also allows Alice and Bob to sign their letters with digital signatures.

Exercises

1. With the primes given in Example 2.4.1 decide which one of the two numbers $e_1 = 2145$ and $e_2 = 3861$ can be used as a public key and calculate the matching private key for it.
2. Alice and Bob agreed to use RSA cryptosystem to communicate in secret. Each message consist of a single letter which is encoded as

$$A = 11, \ B = 12, \ \ldots, \ Z = 36.$$

   Bob's public key is $(n, e) = (143, 113)$ and Alice sent him the message 97. Which letter did Alice sent to Bob in this message?
3. Alice's public exponent in RSA is $e = 41$ and the modulus is $n = 13337$. How many multiplications mod n Bob needs to perform to encrypt his message $m = 2619$? (Do not do the actual encryption, just count.)
4. Set up your own RSA cryptosystem. Demonstrate how a message addressed to you can be encrypted and how you can decrypt it using your private key.
5. Alice and Bob have the public RSA keys $(20687, 17179)$ and $(20687, 4913)$, respectively. Bob sent an encrypted message to Alice, Eve found out that the encrypted message was 353. Help Eve to decrypt the message, suspecting that the modulus 20687 might be a product of two three-digit primes. Try to do it with an ordinary calculator first, then check your answer with GAP.
6. Alice and Bob encrypt their messages using the RSA method. Bob's public key is $(n, e) = (24613, 1003)$.
   (a) Alice would like to send Bob the plaintext $m = 183$. What ciphertext should she send?
   (b) Bob knows that $\phi(n) = 24300$ but has forgotten his private key d. Help Bob to calculate d.
   (c) Bob has received the ciphertext 16935 from Casey addressed to him. Show how he finds the original plaintext.
7°. Suppose that Alice and Bob use RSA public keys with the same modulus n but different encryption exponents e_1 and e_2. Prove that Eve can decrypt a message sent simultaneously to Alice and Bob provided that $\gcd(e_1, e_2) = 1$.

2.4.2 Why Does the RSA System Work?

There are five issues here:

1. Why is $m = (m^e)^d \bmod n$?
2. Can $m^e \bmod n$ and $c^d \bmod n$ be calculated efficiently?
3. To what extent can the RSA cryptosystem be considered secure?
4. How can the encryption exponent e and the decryption exponent d be found?
5. How can large primes p and q be found?

Let us address these issues one by one.

1. First we consider the question why the text recovered by Alice via her private decryption key is actually the original plaintext. This means we must consider $(m^e)^d \bmod n$. We note that since $ed \equiv 1 \bmod \phi$ and $\phi = \phi(n) = (p-1)(q-1)$ we have $ed = 1 + \phi(n)k$ for some integer k. Suppose first that m and n are coprime. Then by Euler's theorem $m^{\phi(n)} \equiv 1 \bmod n$ and

$$\left(m^e\right)^d = m^{ed} = m^{1+\phi(n)k} = m \cdot \left(m^{\phi(n)}\right)^k \equiv m \bmod n. \tag{2.3}$$

There is a very small probability that m will be divisible by p or q but even in this unlikely case we still have $m = (m^e)^d \bmod n$. To prove this we have to consider $(m^e)^d \bmod p$ and $(m^e)^d \bmod q$ separately. Indeed,

$$\left(m^e\right)^d = m^{ed} = m^{1+(p-1)(q-1)x} = m \cdot m^{(p-1)(q-1)x}$$
$$\equiv \begin{cases} m \ \bmod p & \text{if } \gcd(m, p) = 1, \\ 0 \ \bmod p & \text{if } p | m. \end{cases}$$

since in the first case by Fermat's Little Theorem $m^{(p-1)} \equiv 1 \bmod p$. In both cases we see that $m \equiv (m^e)^d \bmod p$.

Similarly we find $(m^e)^d \equiv m \bmod q$. Then the statement follows from the Chinese remainder theorem (Theorem 1.2.6). According to this theorem, there is a unique integer N in the interval $[0, pq)$ such that $N \equiv m \bmod p$ and $N \equiv m \bmod q$. We have two numbers with such property, namely m and $(m^e)^d \bmod n$. Hence they coincide and $m = (m^e)^d \bmod n$.

We have established that the decrypted message is identical to the message that was encrypted. This resolves the first issue.

2. To resolve the second issue we considered the computational problem of raising a number to a power. The complexity of this operation is very low, in fact it is linear (see Theorem 2.3.2). Hence $m^e \bmod n$ and $c^d \bmod n$ can be calculated efficiently.

3. It is evident that if the prime factorisation of the number n in the public key is known then anybody can compute ϕ and thus d. In this case encrypted messages are not secure. But for *large* values of n the task of factorisation is too difficult and time consuming to be feasible. So the encryption function (*raise to power e* mod n) is a one-way function, with d as a trapdoor.

To illustrate how secure the system is Rivest, Shamir and Adelman encrypted a sentence in English. This sentence was converted into a number as we did before (the only difference was that they denoted a space as "00". Then they encrypted it further using $e = 9007$ and

$$n = 114381625757888867669325779976146612010218296721242362562561843293$$
$$5706935245733897830597123563958705058989075147599290026879543541.$$

These two numbers were published, and it was made known that $n = pq$, where p and q are primes which contain 64 and 65 digits in their decimal representations, respectively. Also published was the message

$$f(m) = 96869613754622061477140922254355882905759991124574319874695120930$$
$$816298225145708356931476622883989628013391990551829945157815154.$$

An award of \$100 was offered for decrypting it. This award was only paid 17 years later, in 1994, when Atkins, Graff, Lenstra and Leyland [3] reported that they decrypted the sentence. This sentence—"The magic words are squeamish ossifrage",—was placed in the title of their paper. For decrypting, they factored n and found p, and q which were

$$p = 3490529510847650949147849619903898133417764638493387843990820577$$

and

$$q = 32769132993266709549961988190834461413177642967992942539798288533.$$

In this work 600 volunteers participated. They worked 220 hours on 1600 computers to achieve this result! Recently another effort concluded in 2009 by several researchers who factored a 232-digit number (RSA-768) utilising hundreds of machines over a span of two years.[5] Of course, doable does not mean practical but for very sensitive information one would now want to choose primes as large as containing 150 digits and even more.

It can be shown that finding d is just as hard as factoring n, and it is believed that finding any trapdoor is as hard as factoring n, although this has not been proven. 30 years have passed since RSA was invented and so far all attacks on RSA have been unsuccessful.

4. To find e and d we need only the Euclidean and the extended Euclidean algorithms. Indeed, first we try different numbers between 1 and $\phi(n)$ at random until we find one which is relatively prime to $\phi(n)$ (the fact that it can be done quickly we leave here without a proof). This will be taken as e. Since d is the inverse of e modulo $\phi(n)$, we find d using the extended Euclidean algorithm. This can be done because the Euclidean algorithm is very fast (Corollary 2.3.1).

---

[5]See, http://eprint.iacr.org/2010/006.pdf for details.

5. One may ask: if we cannot factorise positive integers efficiently, then surely we will not be able to say if a number is prime or not. If so, our wonderful system is in danger because two big primes cannot be efficiently found. However this is not the case and it is easier to establish if a number is prime or not than to factorise it. We devote the next section to checking primality.

In the case of RSA it is preferable to use the following encodings for letter:

$$A\ B\ C\ D\ E\ F\ G\ H\ I\ J\ K\ L\ M$$
$$11\ 12\ 13\ 14\ 15\ 16\ 17\ 18\ 19\ 20\ 21\ 22\ 23$$

$$N\ O\ P\ Q\ R\ S\ T\ U\ V\ W\ X\ Y\ Z$$
$$24\ 25\ 26\ 27\ 28\ 29\ 30\ 31\ 32\ 33\ 34\ 35\ 36$$

The advantage of it is that a letter has always a two-digit encoding which resolves some ambiguities. We will use it from now on and, in particular, in exercises.

Exercises

1. In RSA Bob has been using a product of two large primes n and a single public exponent e. In order to increase security, he now chooses two public exponents e_1 and e_2 which are both relatively prime to $\phi(n)$. He asks Alice to encrypt her messages twice: once using the first exponent and then using another one. That is, Alice is supposed to calculate $c_1 = m^{e_1} \pmod{n}$, then $c_2 = c_1^{e_2} \pmod{n}$, and send c_2 to Bob. He has prepared also two decryption exponents d_1 and d_2 for decrypting her messages. Does this double encryption increase security over single encryption?

2. Eve intercepted the following message from Bob to Alice:

    5272281348, 21089283929, 3117723025, 26844144908, 22890519533,

    26945939925, 27395704341, 2253724391, 1481682985, 2163791130,

    13583590307, 5838404872, 12165330281, 28372578777, 7536755222.

    In the public domain Eve learns that this message was sent using the encryption modulus $n = pq = 30796045883$. She also observes that Alice's public key is $e = 48611$. Decode the message which was encoded using the encodings $A = 11, B = 12, \ldots, Z = 36$.

2.4.3   Pseudoprimality Tests

In this section we will discuss four probabilistic tests that might be used for testing the compositeness of integers. Their sophistication and quality will gradually increase. And only the last one will be practical.

Definition 2.4.1 By a *pseudoprimality test* we mean a test that is applied to a pair of integers (b, n), where $2 \leq b \leq n - 1$, and that has the following characteristics:

(a) The possible outcomes of the test are: "n is composite" or "inconclusive".
(b) If the test reports "n is composite" then n is composite.
(c) The test runs in a time that is polynomial in $\log n$ (i.e., in the number of bits necessary to input n).

If n is prime, then the outcome of the test will be "inconclusive" for every b. If the test result is "inconclusive" for one particular b, then we say that n is *a pseudoprime to the base b* (which means that n is so far acting like a prime number).

The outcome of the test for the primality of n depends on the base b that is chosen. In a good pseudoprimality test there will be many bases b that will reveal that n is composite in case it is composite. More precisely, a good pseudoprimality test will, with high probability (i.e., for a large number of choices of the base b) declare that a composite number n is composite. More formally, we define

Definition 2.4.2 We say that a pseudoprimality test applied to a pair of integers (b, n) is *good* if there is a fixed positive real number t such that $0 < t \leq 1$, and every composite integer n is declared to be composite for at least $t(n - 2)$ choices of the base b, in the interval $[2, n - 1]$.

A good pseudoprimality test will find the compositeness of n with probability at least t and, most importantly, this number t does not depend on n. This is, in fact, sufficient for practical purposes since we can increase this probability by running this test several times for several different bases. Indeed, if the probability of missing the compositeness of n is p, then the probability of missing the compositeness running it for two different bases will be p^2 and for k different bases p^k. For $k \to \infty$ this value quickly tends to 0, hence, we can make our test as reliable as we want it to be.

Of course, given an integer n, it is silly to say that "there is a high probability that n is prime". Either n is prime or it is not, and we should not blame our ignorance on n itself. Nonetheless, the abuse of language is sufficiently appealing and it is often said that a given integer n is *very probably prime* if it was subjected it to a good pseudoprimality test, with a large number of different bases b, and have found that it is pseudoprime to all of those bases.

Here are four examples of pseudoprimality tests, only one of which is good.

Test 1 *Given b, n. Output "n is composite" if b divides n, else "inconclusive".*

If n is composite, the probability that it will be so declared is the probability that we happen to have found an integer b that divides n. The probability of this event, if b is chosen at random uniformly from $[2, n - 1]$, is

$$p(n) = \frac{d(n) - 2}{n - 2},$$

where $d(n)$ is the number of divisors of n. Certainly $p(n)$ is not bounded from below by a positive constant t, if n is composite. Indeed, if $n_i = p_i^2$, where p_i is the ith prime, then $d(n_i) = 3$, and

$$p(n_i) = \frac{1}{n_i - 2} \to 0$$

as $i \to \infty$.

Example 2.4.2 Suppose $n = 44 = 2^2 \cdot 11$. Then $d(n) = 3 \cdot 2 = 6$, and

$$p(n) = \frac{4}{42} = \frac{2}{21}.$$

This test is not good.

Test 2 *Given b, n, where $2 \le b \le n - 1$. Output "n is composite" if $\gcd(b, n) \neq 1$, else output "inconclusive".*

This test runs in linear time and it is a little better than Test 1, but not yet good. If n is composite, the number of bases b for which Test 2 will produce the result "composite" is $n - \phi(n) - 1$, where ϕ is Euler's totient function. Indeed, we have $\phi(n)$ numbers b that are relatively prime to n; for those numbers b and only for those we have $\gcd(b, n) = 1$. We also have to exclude $b = n$ which is outside of the range. Hence the probability of declaring a composite n composite will be

$$p(n) = \frac{n - \phi(n) - 1}{n - 2}.$$

For this test the number of useful bases will be large if n has some small prime factors, but in that case it is easy to find out that n is composite by other methods. If n has only a few large prime factors, say if $n = p^2$, then the proportion of useful bases is very small, and we have the same kind of inefficiency as in Test 1. Indeed, if $n_i = p_i^2$, then $\phi(n_i) = p_i(p_i - 1)$ and

$$p(n_i) = \frac{n_i - \phi(n_i) - 1}{n_i - 2} = \frac{p_i^2 - p_i(p_i - 1) - 1}{p_i^2 - 2} = \frac{p_i - 1}{p_i^2 - 2} \sim \frac{1}{p_i} \to 0$$

if $p_i \to \infty$.

Example 2.4.3 Suppose $n = 44 = 2^2 \cdot 11$. Then $\phi(n) = 44 \left(1 - \frac{1}{2}\right)\left(1 - \frac{1}{11}\right) = 20$, and

$$p(n) = \frac{44 - 20 - 1}{42} = \frac{23}{42}.$$

Test 3 *Given* b, n. *If* b *and* n *are not relatively prime or if* $b^{n-1} \not\equiv 1 \bmod n$ *then output* "n *is composite*," *else output* "*inconclusive*".

This test rests on Fermat's Little Theorem. Indeed, if $\gcd(b, n) > 1$ or $\gcd(b, n) = 1$ and $b^{n-1} \not\equiv 1 \bmod n$, then n cannot be prime since, if n was prime, by Fermat's Little Theorem in the latter case we must have $b^{n-1} \equiv 1 \bmod n$. It runs also in linear time if we use the Square-and-Multiply algorithm to calculate b^{n-1}. And it works much better than the previous two.

Example 2.4.4 To see how this test works let us calculate $2^{32} \bmod 33$. We obtain:

$$2^{32} = 2^5 \cdot 2^5 \cdot 2^5 \cdot 2^5 \cdot 2^5 \cdot 2^5 \cdot 2^2 \equiv (-1)^6 \cdot 2^2 \equiv 4 \bmod 33.$$

Hence 33 is not prime.

Unfortunately, this test is still not good. It works well for most but not for all numbers. The weak point of it is that there exist composite numbers n, called *Carmichael numbers*, with the property that the pair (b, n) produces the output "inconclusive" for *every* integer b in $[2, n-1]$ that is relatively prime to n. An example of such a Carmichael number is $n = 561$, which is composite ($561 = 3 \cdot 11 \cdot 17$), but for which Test 3 gives the result "inconclusive" on every integer $b < 561$ that is relatively prime to 561 (i.e., that is not divisible by 3 or 11 or 17). On Carmichael numbers Test 3 behaves exactly like Test 2 which we know is unsatisfactory. Moreover, it was proved recently that there are infinitely many Carmichael numbers [2] which means that the drawback is serious. The first ten Carmichael numbers[6] are:

$$561, 1105, 1729, 2465, 2821, 6601, 8911, 10585, 15841, 29341 \ldots$$

Despite such occasional misbehavior, the test usually seems to perform quite well. When $n = 169$ (a difficult integer for Tests 1 and 2) it turns out that there are 158 different b's in $[2, 168]$ that produce the "composite" outcome from Test 3, namely every such b except for 19, 22, 23, 70, 80, 89, 99, 146, 147, 150, 168.

Finally, we will describe a good pseudoprimality test. The idea was suggested in 1976 by Miller (see the details in [2]).

Test 4 (Rabin–Miller) *Given* b, n, *where* $2 \leq b \leq n-1$ *we firstly calculate* $\gcd(b, n)$. *If* $\gcd(b, n) > 1$ *then we output* "*composite*". *If* $\gcd(b, n) = 1$, *let us represent* $n - 1$ *as* $n - 1 = 2^s t$, *where* t *is an odd integer. If*

---

[6]Sequence A002997 from The On-Line Encyclopedia of Integer Sequences http://oeis.org/.

(a) $b^t \not\equiv 1$ mod n, and
(b) for every integer i in $[0, s-1]$

$$b^{2^i t} \not\equiv -1 \; mod \; n,$$

then return "composite", else return "inconclusive".

Let us convince ourselves that Test 4 works. For this we need the identity

$$(a-1)(a+1)(a^2+1) \cdot \ldots \cdot (a^{2^{s-1}}+1) = a^{2^s} - 1, \qquad (2.4)$$

which can be easily proved by induction.

Suppose that conditions (a) and (b) are satisfied but n is prime. Then $\gcd(b,n) = 1$. Substituting $a = b^t$ into the identity (2.4) and, using Fermat's Little Theorem, we will obtain

$$(b^t - 1)(b^t + 1)(b^{2t} + 1) \cdot \ldots \cdot (b^{2^{s-1}t} + 1) = b^{2^s t} - 1 = b^{n-1} - 1 \equiv 0 \bmod n.$$

However by (a) and (b) every bracket is nonzero modulo n. Hence there are zero divisors in $\mathbb{Z}_n$ which contradicts to primality of n. This means that if the test outputs "composite", the number n is composite.

What is the computational complexity of this test? Part (a) of the test by Theorem 2.3.3 can be done in $O(\log n)$ divisions with remainder. and the complexity of this is at most linear. Similarly, in part (b) of the test there are $O(\log n)$ of possible values of i to check, and for each of them we do a single multiplication of two integers calculating $b^{2^i t} = b^{2^{i-1} t} \cdot b^{2^{i-1} t}$, each of which has $O(\log n)$ bits. Hence the overall complexity is still linear.

Theorem 2.4.1 (Rabin) *If n is composite then for at least $\frac{3}{4}(n-2)$ of the integers b, such that $2 \le b \le n-1$, Test 4 gives the result "n is composite".*

This means that Test 4 is a good pseudoprimality test and, if we choose b at random to prove the compositeness of n, then we will find the required b with probability greater than $3/4$. Hence we can set $t = 3/4$. The proof of this result cannot be considered in this book.

Example 2.4.5 If $n = 169$, then it turns out that for 157 of the possible 167 bases b in $[2, 168]$ Test 4 will output "169 is composite". The only bases b that 169 can fool are 19, 22, 23, 70, 80, 89, 99, 146, 147, 150, 168. In this case the performance of Tests 4 and 3 are identical. However, there are no analogues of the Carmichael numbers for Test 4.

How can this pseudoprimality test be used to find large primes? Suppose that you want to generate an n-digit prime. You generate an arbitrary n-digit number r and subject it to a good pseudoprimality test (for example, Rabin–Miller Test) repeating the test several times. Suppose that we have done k runs of the Test 4 with different random bs and each time got the answer "inconclusive". If r is composite, then, the probability that we get the answer "inconclusive" once, is less than $1/4$. If we run this test k times, the probability that we get the answer "inconclusive" k times is less than $1/4^k$. For $k = 5$ this probability is less than 10^{-3}. For $k = 10$ it is less than 10^{-6}, which is a very small number already. Since Test 4 is very fast to perform we may run this test 100 times. If we got answer inconclusive all 100 times, the probability that n is composite is negligible.

In 2002 Manindra Agrawal, Neeraj Kayal and Nitin Saxena [1] came up with a polynomial deterministic algorithm (AKS algorithm) for primality testing. It is based on the following variation of Fermat's Little Theorem for polynomials:

Theorem 2.4.2 *Let* $gcd(a, n) = 1$ *and* $n > 1$*. Then* n *is prime if and only if*

$$(x - a)^n \equiv (x^n - a) \bmod n.$$

The authors received the 2006 Gödel Prize and the 2006 Fulkerson Prize for this work. Originally the AKS algorithm had complexity $O((\log n)^{12})$, where n is the number to be tested, but in 2005 C. Pomerance and H. W. Lenstra, Jr. demonstrated a variant of AKS algorithm that runs in $O((\log n)^6)$ operations Ð a marked improvement over the bound in the original algorithm. Despite all the efforts it is still not yet practical but a number of researchers are actively working on improving this algorithm. See [22] for more information on the algorithm and a proof of Theorem 2.4.2.

Exercises

1. We implement the first and the second pseudoprimality tests by choosing at random b in the interval $1 < b < n$ and applying it to the pair (b, n).
   (a) What is the probability that the first pseudoprimality tests finds that 91 is composite?
   (b) What is the probability that the second pseudoprimality tests finds that 91 is composite?
2. Show that the third pseudoprimality test finds that 91 is composite for the pair $(5, 91)$.
3. Prove that any number $F_n = 2^{2^n} + 1$ is either a prime or a pseudoprime to the base 2. (Use Exercise 4 Sect. 1.1.1.)
4. Write a GAP programme that checks if a number n is a Carmichael number. Use it to find out if the number 15841 is a Carmichael number.
5. Prove without using GAP that 561 is a Carmichael number, i.e., $a^{560} \equiv 1 \bmod 561$ for all a relatively prime to 561.

6. Show that 561 is a pseudoprime to the base 7 (i.e., $n = 561$ passes the third pseudoprimality test with $b = 7$) but not a pseudoprime to the base 7 relative to the Rabin–Miller test.
7. Show that the Rabin–Miller test with $b = 2$ proves that $n = 294409$ is composite (despite 294409 being a Carmichael number).
8. Show that a power of a prime is never a Carmichael number.
9°. Calculate the probability that Test 2 (the second pseudoprimality test) finds the number 111111111 composite.
10°. Choose 100 bases b at random to estimate the probability that Test 3 (the third pseudoprimality test) finds the number 1000001 composite.
11°. Let k be a positive integer such that the three numbers $6k + 1, 12k + 1, 18k + 1$ are all prime numbers. Prove that their product $n = (6k + 1)(12k + 1)(18k + 1)$ is a Carmichael number. Use GAP to find first five Carmichael numbers of this kind.

2.5   Applications of Cryptology

1. Diffie–Hellman exponential secret-key exchange. This idea was suggested in 1976 by Diffie and Hellman [9], and it triggered the development of public-key cryptography. Two parties A and B openly agree on two parameters: positive integer n and $g \in \mathbb{Z}_n^*$. They secretly choose two exponents a and b, respectively. Then A sends g^a to B and B sends g^b to A. After that, B takes the received g^a to the exponent b to get g^{ab} and A takes g^b to the exponent a and also gets g^{ab}. Then they use g^{ab} as their secret key. An eavesdropper has to compute g^{ab} from g, g^a and g^b which for n sufficiently large is intractable. This is called the *Diffie–Hellman problem*. ElGamal cryptosystem, which we will study later, develops this idea further.

2. Digital signatures. The notion of a digital signature may prove to be one of the most fundamental and useful inventions of modern cryptography. A signature scheme provides a way for each user to sign messages so that the signatures can be verified by anyone. More specifically, each user can create a matched pair of private and public keys so that only they can create a signature for a message (using their private key) but anyone can verify the signature for the message (using the signer's public key). The verifier can convince himself that the message content have not been altered since the message was signed. Also, the signer cannot later repudiate having signed the message, since no one but the signer possesses the signer's private key.

For example, when your computer receives a software update, say from Adobe, it checks the digital signature to make sure that this is a genuine update from Adobe and not a virus or Trojan.

At this stage the only public-key cryptosystem that we know is the RSA but as we will see the idea can also be used for other cryptosystems. If in RSA $n = pq$ is the product of two large primes p and q, then the message space $\mathcal{M}$ is the set $\{0, 1, 2, \ldots, n - 1\}$. We have functions E_U and D_U (encryption and decryption in

RSA) as

$$E_U : m \mapsto m^{e_U} \bmod N_U, \qquad D_U : m \mapsto m^{d_U} \bmod N_U,$$

where e_U and d_U are the public exponent and the private exponent of user U, respectively. One can turn this around to obtain a digital signature. If m is a document which is to be signed by the user U then she computes her signature as $s = D_U(m)$. The user sends m together with the signature s. Anyone can now verify the signature by testing whether $E_U(s) \equiv m \bmod N_U$ or not.

This idea was first proposed by Diffie and Hellman [9]. The point is that if the message m was changed then the old signature would be no longer valid, and the only person who can create a new signature, matching the new message, should be someone who knows the private key D_U and we assume that only user U possess D_U.

By analogy with the paper world, where Alice might sign a letter and seal it in an envelope addressed to Bob, Alice can sign her electronic letter m to Bob by appending her digital signature $D_A(m)$ to m, and then seal it in an "electronic envelope" with Bob's address by encrypting her signed message with Bob's public key, sending the resulting message $E_B(m|D_A(m))$ to Bob. Only Bob can open this "electronic envelope" by applying his private key to it to obtain $D_B(E_B(m|D_A(m))) = m|D_A(m)$. After that he will apply Alice's public key to the signature obtaining $E_A(D_A(m))$. On seeing that $E_A(D_A(m)) = m$, Bob can be really sure that the message m came from Alice and its content was not altered by a third party.

These applications of public-key technology to electronic mail are likely to become widespread in the near future. For simplicity, we assumed here that the message m was short enough to be transmitted in one piece. If the message is long there are methods to keep the signature short. We will not dwell on this here.

3. Pay-per-view movies. It is common these days that cable TV operators with all-digital systems encrypt their services. This lets cable operators activate and deactivate cable service without sending a technician to your home. The set-up involves each subscriber having a set-top box which is a device, which is connected to a television set at the subscribers' premises and which allows a subscriber to view encrypted channels of his choice on payment. The set-top box contains a set of private keys of the user. A 'header' broadcast in advance of the movie contains keys sufficient to download the actual movie. This header is in turn encrypted with the relevant user public keys.

4. Friend-or-foe identification. Suppose A and B share a secret key K. Later, A is communicating with someone and he wishes to verify that he is communicating with B. A simple challenge-response protocol to achieve this identification is as follows:

- A generates a random value r and transmits r to the other party.
- The other party (assuming that it is B) encrypts r using their shared secret key K and transmits the result back to A.
- A compares the received ciphertext with the result he obtains by encrypting r himself using the secret key K. If the result agrees with the response from B, A

knows that the other party is B; otherwise, he assumes that the other party is an impostor.

This protocol is generally more useful than the transmission of an unencrypted shared password from B to A, since the eavesdropper could learn the password and then pretend to be B later. With the challenge-response protocol an eavesdropper presumably learns nothing about K by hearing many values of r encrypted with K as key.

An interesting exercise is to consider whether the following variant of the above idea is secure: A sends the encryption of a random r, B decrypts it and sends the value r to A, and A verifies that the response is correct.

Exercises

1. Alice and Bob agreed to use Diffie–Hellman secret-key exchange to come up with a secret key for their secret-key cryptosystem. They openly agreed on the prime

$$p = 1001408894420628141404347115571$$

and an element $g = 13 \in \mathbb{Z}_p$. Alice has decided on her private key by choosing $a=123456789$. She also got a message $g^b = 926392043987322765326424490482$ from Bob. Calculate their shared secret key.

2. Alice and Bob have the following RSA parameters:

$$n_A = 1710247041836161097008180669251978415166711277, \qquad e_A = 1571,$$

$$n_B = 8390735427343693592608713559390626227476633109, \qquad e_B = 87697.$$

Bob knows his two primes which are

$$p_B = 8495789457893457345793, \qquad q_B = 9876345769783456893461 3.$$

Alice signs a message m by calculating her signature $s = m^{d_A} \pmod{n_A}$. She then encrypts the pair (m, s) using Bob's public key by calculating (m_1, s_1), where $m_1 = m^{e_B} \pmod{n_B}$ and $s_1 = s^{e_B} \pmod{n_B}$. She obtains

$$m_1 = 1195704414418897497050318965573868438834 75475,$$

$$s_1 = 4436824304931024869780797195075967956577 29083$$

and sends the pair (m_1, s_1) to Bob. Show how Bob can find the message m and verify that it came from Alice. (Do not try to convert digits of m into letters, the message is meaningless.)

3°. Eve has intercepted the following message from Bob to Alice

[42784996824075900722849497863977508809,
  498308250136673589542748543030806629941,
  925288105342943743271024837479707225255,
  95024328800414254907217356783906225740]

She knows Alice uses the RSA cryptosystem with the modulus

$$n = 956331992007843552652604425031376690367.$$

and that Alice's public exponent is $e = 12398737$. She also knows that, to convert their messages into numbers, Bob and Alice usually use the encodings: space $= 00$, $A = 11$, $B = 12, \ldots, Z = 36$. Help Eve to break the code and decrypt the message.

$4°$. Alice and Bob use RSA cryptosystem with the following parameters:

$n_A = 116843187579509698439177769751386474940457877351734068668377$, $e_A = 1234567$,

$n_B = 41989230468376560622264958706569326838563915099140031193663$, $e_B = 7654321$.

Bob creates a message to Alice splitting it into two parts m_1 and m_2, then signs the last bit applying his decryption key to m_2 obtaining m_3. Then he encrypts the whole message with Alice's public key obtaining the cryptotext $c = [c[1], c[2], c[3]]$ where

$c[1] = 113438632352422763265675742513046812673537179044234633006538$,

$c[2] = 45013089611237457780987479205118742551558572798511252012986$,

$c[3] = 111491725228799790475033209306493492319899494161842598114763$

Demonstrate how Eve will decrypt this message, when intercepted, and check that it has come indeed from Bob.

Groups

3

The concept of a *group* helps to unify a great variety of different mathematical structures which at first sight might appear unrelated. In this chapter, we will start by looking at groups of permutations from which groups take their origin. We will then give a general definition of a group, and move on to studying the multiplicative group of $\mathbb{Z}_n$ and the group of points of an elliptic curve. The latter two groups have recently gained cryptographic significance. Group theory plays a central role in cryptography; as a matter of fact, any large finite group can potentially be a basis of a cryptographic system.

Permutations are ubiquitous in cryptography. In Sect. 5.3, we will revisit permutations and learn that the RSA encryption function is in fact a permutation.

3.1 Permutations

3.1.1 Composition of Mappings. The Group of Permutations of Degree n

Let A, B, and C be three sets. Suppose we have mappings $f : A \rightarrow B$ and $g : B \rightarrow C$. For any element $a \in A$, we can find its image $f(a) \in B$ under f and for that element of B we can find its image $g(f(a)) \in C$ under g. We have now implicitly defined a third mapping which maps $a \in A$ onto $g(f(a))$. We denote this mapping as $f \circ g$ and call the *composition* of mappings f and g. As a formula, it can be written as $(f \circ g)(a) = g(f(a))$.

© Springer Nature Switzerland AG 2020

A. Slinko, *Algebra for Applications*, Springer Undergraduate Mathematics Series,

https://doi.org/10.1007/978-3-030-44074-9_3

Important Note The convention we use runs contrary to that used in Calculus, where $(f \circ g)(x) = f(g(x))$ (i.e., first compute $g(x)$, then apply the function f to the result). This may cause some minor problems to students used to a different convention. The great advantage to write the composition the way we do is that it is the same convention as the one used in GAP.

One of the properties of composition of major importance is its compliance with the associative law.

Proposition 3.1.1 *Composition of mappings is associative, that is, given the sets A, B, C, D and mappings $f : A \rightarrow B$, $g : B \rightarrow C$ and $h : C \rightarrow D$, we have*

$$(f \circ g) \circ h = f \circ (g \circ h).$$

Proof Two mappings from A to D are equal when they assign exactly the same images in D to every element in A. Let us calculate the image of $a \in A$ first under the mapping $(f \circ g) \circ h$ and then under $f \circ (g \circ h)$:

$$((f \circ g) \circ h)(a) = h((f \circ g)(a)) = h(g(f(a))),$$
$$(f \circ (g \circ h))(a) = (g \circ h)(f(a)) = h(g(f(a))).$$

The image of a under both mappings is the same. Since $a \in A$ was arbitrary, the two mappings are equal. $\qquad\square$

Let A be any set. It is a well-known that if $f : A \rightarrow A$ is a function which is both one-to-one and onto then f is *invertible*, i.e., there exists a function $g : A \rightarrow A$ such that

$$g \circ f = f \circ g = \mathrm{id}, \tag{3.1}$$

where id is the identity mapping on A. In this case f and g are called *mutual inverses* and we use the notation $g = f^{-1}$ and $f = g^{-1}$ to express that. Equation (3.1) means that g maps $f(a)$ to a while f maps $g(a)$ to a, i.e., g undoes the work of f, and f undoes the work of g.

Example 3.1.1 Let $\mathbb{R}_+$ be the set of positive real numbers. Let $f : \mathbb{R}_+ \rightarrow \mathbb{R}$ and $g : \mathbb{R} \rightarrow \mathbb{R}_+$ be given as $f(x) = \ln x$ and $g(x) = e^x$. These are mutual inverses and hence both functions are invertible.

In what follows we assume that the set A is finite and consider mappings from A into itself. If A has n elements, for convenience, we assume that the elements of A are the numbers $1, 2, \ldots, n$ (the elements of any finite set can be labelled with the first few integers, so this does not restrict generality).

Definition 3.1.1 Let n be a positive integer. A *permutation of degree n* is a function

$$\pi: \{1, 2, \ldots, n\} \to \{1, 2, \ldots, n\},$$

which is one-to-one and onto.

Since a function is specified if we indicate what the image of each element is, we can specify a permutation π by listing each element together with its image, as

$$\pi = \begin{pmatrix} 1 & 2 & 3 & \cdots\cdots & n-1 & n \\ \pi(1) & \pi(2) & \pi(3) & \cdots\cdots & \pi(n-1) & \pi(n) \end{pmatrix}.$$

Given that π is one-to-one, no number is repeated in the second row of the array. Given that π is onto, each number from 1 to n appears somewhere in the second row. In other words, the second row is just a rearrangement of the first.[1]

Example 3.1.2 $\pi = \begin{pmatrix} 1\,2\,3\,4\,5\,6\,7 \\ 2\,5\,3\,1\,7\,6\,4 \end{pmatrix}$ is the permutation of degree 7 which maps 1 to 2, 2 to 5, 3 to 3, 4 to 1, 5 to 7, 6 to 6, and 7 to 4.

Example 3.1.3 The mapping $\sigma: \{1, 2, \ldots, 6\} \to \{1, 2, \ldots, 6\}$ given by $\sigma(i) = 3i \bmod 7$ is a permutation of degree 6. Indeed,

$$\sigma(1) = 3, \ \sigma(2) = 6, \ \sigma(3) = 2, \ \sigma(4) = 5, \ \sigma(5) = 1, \ \sigma(6) = 4,$$

and thus

$$\sigma = \begin{pmatrix} 1\,2\,3\,4\,5\,6 \\ 3\,6\,2\,5\,1\,4 \end{pmatrix}.$$

Theorem 3.1.1 *There are exactly $n!$ permutations of degree n.*

Proof Let us consider a permutation of degree n. It is completely determined by its bottom row. There are n ways to fill the first position of this row, $n-1$ ways to fill the second position (since we must not repeat the first entry), etc., leading to a total of $n \cdot (n-1) \cdot \ldots \cdot 2 \cdot 1 = n!$ different possibilities. $\qquad\square$

The composition of two permutations of degree n is again a permutation of degree n (see and do Exercise 1 after this section). Most of the time we will omit the symbol $\circ$ for function composition, and speak of the *product* $\pi\sigma$ of two permutations π and σ, meaning the composition $\pi \circ \sigma$.

---

[1] Clearly, in this case of finite sets, one-to-one implies onto and vice versa but this will no longer be true for infinite sets.

Example 3.1.4 Let

$$\sigma = \begin{pmatrix} 1\,2\,3\,4\,5\,6\,7\,8 \\ 2\,4\,5\,6\,1\,8\,3\,7 \end{pmatrix}, \qquad \pi = \begin{pmatrix} 1\,2\,3\,4\,5\,6\,7\,8 \\ 4\,6\,1\,3\,8\,5\,7\,2 \end{pmatrix}.$$

Then

$$\begin{aligned}
\sigma\pi &= \begin{pmatrix} 1\,2\,3\,4\,5\,6\,7\,8 \\ 2\,4\,5\,6\,1\,8\,3\,7 \end{pmatrix} \begin{pmatrix} 1\,2\,3\,4\,5\,6\,7\,8 \\ 4\,6\,1\,3\,8\,5\,7\,2 \end{pmatrix} \\
&= \begin{pmatrix} 1\,2\,3\,4\,5\,6\,7\,8 \\ 6\,3\,8\,5\,4\,2\,1\,7 \end{pmatrix},
\end{aligned}$$

and

$$\begin{aligned}
\pi\sigma &= \begin{pmatrix} 1\,2\,3\,4\,5\,6\,7\,8 \\ 4\,6\,1\,3\,8\,5\,7\,2 \end{pmatrix} \begin{pmatrix} 1\,2\,3\,4\,5\,6\,7\,8 \\ 2\,4\,5\,6\,1\,8\,3\,7 \end{pmatrix} \\
&= \begin{pmatrix} 1\,2\,3\,4\,5\,6\,7\,8 \\ 6\,8\,2\,5\,7\,1\,3\,4 \end{pmatrix}.
\end{aligned}$$

Explanation: the calculation of $\sigma\pi$ requires us to find

- the image of 1 when we apply *first σ, then π*, ($1 \xmapsto{\sigma} 2 \xmapsto{\pi} 6$, so write the 6 under the 1),
- the image of 2 when we apply *first σ, then π*, ($2 \xmapsto{\sigma} 4 \xmapsto{\pi} 3$, so write the 3 under the 2),
- etc.

    ⋮   All this is easily done at a glance and can be written down immediately; BUT
        be careful to start with the left-hand factor!

The calculation of $\pi\sigma$ requires us to find

- the image of 1 when we apply *first π, then σ*, ($1 \xmapsto{\pi} 4 \xmapsto{\sigma} 6$, so write the 6 under the 1)
- the image of 2 when we apply *first π, then σ*, ($2 \xmapsto{\pi} 6 \xmapsto{\sigma} 8$, so write the 8 under the 2)
- etc.

    ⋮   All this is easily done at a glance and can be written down immediately; BUT
        be careful to start with the left-hand factor again!

Important Note The example shows clearly that $\pi\sigma \neq \sigma\pi$, that is, the commutative law for permutations does not hold; so we have to be very careful about the order of the factors in a product of permutations. But the good news is that the composition of permutations is associative. This follows from Proposition 3.1.1.

We can also calculate the inverse of a permutation; for example, using the same π as above, we find

$$\pi^{-1} = \begin{pmatrix} 1\ 2\ 3\ 4\ 5\ 6\ 7\ 8 \\ 3\ 8\ 4\ 1\ 6\ 2\ 7\ 5 \end{pmatrix}.$$

Explanation: just read the array for π from the bottom up: since $\pi(1) = 4$, we must have $\pi^{-1}(4) = 1$, hence write 1 under the 4 in the array for π^{-1}, since $\pi(2) = 6$, we must have $\pi^{-1}(6) = 2$, hence write 2 under the 6 in the array for π^{-1}, etc. In this case we will indeed have $\pi\pi^{-1} = \text{id} = \pi^{-1}\pi$.

Similarly, we calculate

$$\sigma^{-1} = \begin{pmatrix} 1\ 2\ 3\ 4\ 5\ 6\ 7\ 8 \\ 5\ 1\ 7\ 2\ 3\ 4\ 8\ 6 \end{pmatrix}.$$

Simple algebra shows that the inverse of a product can be calculated from the product of the inverses (but note how the order is reversed!):

$$(\pi\sigma)^{-1} = \sigma^{-1}\pi^{-1}. \tag{3.2}$$

To justify this, we need only check if the product of $\pi\sigma$ and $\sigma^{-1}\pi^{-1}$ equals the identity, and this is pure algebra: it follows from the associative law that

$$(\pi\sigma)(\sigma^{-1}\pi^{-1}) = \pi(\sigma(\sigma^{-1}\pi^{-1})) = \pi((\sigma\sigma^{-1})\pi^{-1}) = \pi\pi^{-1} = \text{id}.$$

Definition 3.1.2 The set of all permutations of degree n, with the operation of composition is called the *symmetric group of degree n*, and is denoted by S_n.

We call S_n a *group* since the following axioms are satisfied:

1. S_n is associative, i.e., $(\pi\sigma)\tau = \pi(\sigma\tau)$ for all $\pi, \sigma, \tau \in S_n$;
2. S_n has an identity element id, i.e., $\pi\,\text{id} = \text{id}\,\pi = \pi$ for all $\pi \in S_n$;
3. every element $\pi \in S_n$ has an inverse π^{-1}, i.e., $\pi\pi^{-1} = \text{id} = \pi^{-1}\pi$.

In Sect. 1.6 we defined a commutative group. This group is not commutative as $\pi\sigma$ is not necessarily equal to $\sigma\pi$. The concept of a group was introduced into mathematics by Évariste Galois.[2]

---

[2] **Évariste Galois (1811–1832)**, a French mathematician who was the first to use the word "group" (French: groupe) as a technical term in mathematics to represent a group of permutations. While still in his teens, he was able to determine a necessary and sufficient condition for a polynomial to be solvable by radicals, thereby solving a long-standing problem. His work laid the foundations for Galois theory, a major branch of abstract algebra.

Exercises

1. In the following two cases calculate $f \circ g$ and $g \circ f$. Note that they are different in both cases and even their natural domains are different.

    (a) $f(x) = \sin x$ and $g(x) = 1/x$,
    (b) $f(x) = e^x$ and $g(x) = \sqrt{x}$.

2. Let R_θ be an anticlockwise rotation of the plane about the origin through an angle θ. Show that R_θ is invertible with the inverse $R_{2\pi-\theta}$.

3. Show that any reflection H of the plane in any line is invertible and the inverse of H is H itself.

4. Determine how many permutations of degree n act identically on a fixed set of k elements of $\{1, 2, \ldots, n\}$.

5. Show that the mapping $\sigma: \{1, 2, \ldots, 8\} \to \{1, 2, \ldots, 8\}$ given by $\sigma(i) = 5i$ mod 9 is a permutation by writing it down in the form of a table.

6. Let the mapping $\pi: \{1, 2, \ldots, 12\} \to \{1, 2, \ldots, 12\}$ be defined by $\pi(k) = 3k$ mod 13. Show that π is a permutation of S_{12}.

7. The mapping $\tau: \{1, 2, \ldots, 12\} \to \{1, 2, \ldots, 12\}$ is defined by $\tau(k) = k^2$ mod 13. Show that τ is not a permutation of S_{12} by showing that both one-to-one and onto properties are violated.

8. Calculate the inverse and all distinct powers of the permutations:

$$\rho = \begin{pmatrix} 1\,2\,3\,4\,5\,6 \\ 3\,4\,5\,6\,1\,2 \end{pmatrix}, \quad \tau = \begin{pmatrix} 1\,2\,3\,4\,5\,6 \\ 4\,6\,5\,1\,3\,2 \end{pmatrix}.$$

9. Let

$$\sigma = \begin{pmatrix} 1\,2\,3\,4\,5\,6\,7\,8\,9 \\ 2\,4\,5\,6\,1\,9\,8\,3\,7 \end{pmatrix}, \quad \gamma = \begin{pmatrix} 1\,2\,3\,4\,5\,6\,7\,8\,9 \\ 6\,2\,7\,9\,3\,8\,1\,4\,5 \end{pmatrix}.$$

    Calculate $(\sigma\gamma)^{-1}$ and check yourself with GAP.

10. Prove rigorously that the composition of two permutations of degree n is a permutation of degree n.

3.1.2 Block Permutation Cipher

A permutation π of order n can be used as a secret key in the following cryptosystem called *permutation cipher*. In this cryptosystem a plaintext and ciphertext are both over the same alphabet. Let $m = a_1 a_2 \ldots a_n$ be a message of fixed length n over alphabet A. Then the corresponding ciphertext is defined as

$$E(m, \pi) = a_{\pi(1)} a_{\pi(2)} \ldots a_{\pi(n)},$$

which means the symbols of the message are permuted in accord with the permutation π. If the message is longer than n, we split it into smaller segments of length n (it is always possible to add some junk letters to make the total length of the message divisible by n.)

Example 3.1.5 Suppose the secret key is

$$\pi = \begin{pmatrix} 1 & 2 & 3 & 4 & 5 & 6 & 7 & 8 & 9 & 10 & 11 & 12 & 13 & 14 & 15 & 16 \\ 2 & 12 & 3 & 16 & 4 & 10 & 9 & 15 & 7 & 8 & 6 & 5 & 14 & 1 & 13 & 11 \end{pmatrix}$$

and the message is

ALL ALL ARE GONE THE OLD FAMILIAR FACES

We split it into two submessages of length 16 each:

ALLALLAREGONETHE OLDFAMILIARFACES

and then apply π to both submessages:

LNLEAGEHARLLTAEO LFDSFAIEILMACOAR

The final message is then

LNLEAGEHARLLTAEOLFDSFAIEILMACOAR

The permutation cipher is difficult to break with the knowledge of ciphertext only. Indeed, the length of blocks is unknown, and even if known, the space of secret keys is very large: it has $n!$ possible permutations and $n!$ grows very fast. Even for reasonably small n like $n = 128$, the number of possible keys is astronomical. However, if one can guess even a fragment of the plaintext, it may become easy. To make guessing the plaintext difficult, a substitution cipher can be applied first. The combination of substitutions and permutations is called a *product cipher*. The product ciphers are not normally used on their own but they are an indispensable part of modern cryptography. For example, adopted on 23 November 1976 the Data Encryption Standard (DES) involved 16 rounds of substitutions and permutations.

The main steps of the DES algorithm are as follows:

- Partitioning of the text into 64-bit blocks;
- Initial permutation within each block;
- Breakdown of the blocks into two parts: left and right, named L and R;
- Permutation and substitution steps repeated 16 times (called rounds) on each part;
- Re-joining of the left and right parts then the inverse of the initial permutation.

DES is now considered to be insecure for many applications. In 1997, a call was launched for projects to develop an encryption algorithm in order to replace

DES. After an international competition, in 2001, a new block cipher Rijndael[3] was
selected as a replacement for DES; it is now referred to as the Advanced Encryption
Standard (AES).

3.1.3   Cycles and Cycle Decomposition

A permutation π of order n which "cyclically permutes" some of the numbers
$1, \ldots, n$ (and leaves all others fixed) is called a *cycle*.

For example, the permutation $\pi = \begin{pmatrix} 1\,2\,3\,4\,5\,6\,7 \\ 1\,5\,3\,7\,4\,6\,2 \end{pmatrix}$ is a cycle, because we have

$5 \xrightarrow{\pi} 4 \xrightarrow{\pi} 7 \xrightarrow{\pi} 2 \xrightarrow{\pi} 5$, and each of the other elements of $\{1, 2, 3, 4, 5, 6, 7\}$, namely
1,3,6, stay unchanged. To see this, we must of course chase elements around; the nice
cyclic structure is not immediately evident from our notation. We write $\pi = (5\,4\,7\,2)$,
meaning that all numbers not on the list are mapped to themselves, while the ones in
the bracket are mapped to the one listed to the right, except the rightmost one, which
goes back to the leftmost on the list.

Note Cycle notation is not unique, since there is no beginning or end to a circle.
We can write $\pi = (5\,4\,7\,2)$ and $\pi = (2\,5\,4\,7)$, as well as $\pi = (4\,7\,2\,5)$ and $\pi =
(7\,2\,5\,4)$—they all denote one and the same cycle.

We say that a cycle is of *length k* (or a *k-cycle*) if it moves k numbers. For example,
$(3\,6\,4\,9\,2)$ is a 5-cycle, $(3\,6)$ is a 2-cycle, $(1\,3\,2)$ is a 3-cycle. We note also that the
inverse of a cycle is again a cycle. For example $(1\,2\,3)^{-1} = (1\,3\,2)$ (or $(3\,2\,1)$ if
you prefer). Similarly, $(1\,2\,3\,4\,5)^{-1} = (1\,5\,4\,3\,2)$. To find the inverse of a cycle one
has to reverse the arrows. This leads us to the following.

Theorem 3.1.2 $(i_1\,i_2\,i_3\,\ldots\,i_k)^{-1} = (i_k\,i_{k-1}\,\ldots\,i_2\,i_1)$.

Not all permutations are cycles; for example, the permutation

$$\sigma = \begin{pmatrix} 1\,2\,3\ \ 4\ \ 5\,6\,7\,8\,9\ 10\ 11\ 12 \\ 4\,3\,2\ 11\ 8\,9\,5\,6\,7\ 10\ \ 1\ \ 12 \end{pmatrix} \tag{3.3}$$

is not a cycle (we have $1 \xmapsto{\sigma} 4 \xmapsto{\sigma} 11 \xmapsto{\sigma} 1$, but the other elements are not all fixed
(2 goes to 3, for example). Let us chase other elements. We find: $2 \xmapsto{\sigma} 3 \xmapsto{\sigma} 2$ and
$5 \xmapsto{\sigma} 8 \xmapsto{\sigma} 6 \xmapsto{\sigma} 9 \xmapsto{\sigma} 7 \xmapsto{\sigma} 5$. So in the permutation σ three cycles coexist peacefully.

---

[3] J. Daemen and V. Rijmen. The block cipher Rijndael, Smart Card research and Applications, LNCS
1820, Springer-Verlag, pp. 288–296.

Two cycles $(i_1\ i_2\ i_3\ \ldots\ i_k)$ and $(j_1\ j_2\ j_3\ \ldots\ j_m)$ are said to be *disjoint*, if the sets $\{i_1, i_2, \ldots, i_k\}$ and $\{j_1, j_2, \ldots, j_m\}$ have empty intersection. For instance, we may say that

$$(1\ 5\ 8) \quad \text{and} \quad (2\ 4\ 3\ 6\ 9)$$

are disjoint. Any two disjoint cycles σ and τ commute, i.e., $\sigma\tau = \tau\sigma$ (see Exercise 1). For example,

$$(1\ 2\ 3\ 4)(5\ 6\ 7) = (5\ 6\ 7)(1\ 2\ 3\ 4).$$

However, if we multiply any of the cycles which are not disjoint, we have to watch their order; for example: $(1\ 2)(1\ 3) = (1\ 2\ 3)$, whilst $(1\ 3)(1\ 2) = (1\ 3\ 2)$, and $(1\ 3\ 2) \neq (1\ 2\ 3)$.

The relationship between a cycle and the permutation group it belongs to is much like that between a prime and the natural numbers.

Theorem 3.1.3 *Every permutation can be written as a product of disjoint cycles. Moreover, any such representation is unique up to the order of the factors.*

Proof Let σ be a permutation of degree n. Take any element $i_1 \in \{1, 2, \ldots, n\}$ and start a cycle: $\sigma(i_1) = i_2$, $\sigma(i_2) = i_3$, etc. Suppose that $i_1, i_2, \ldots, i_k$ were all different and $\sigma(i_k) \in \{i_1, i_2, \ldots, i_k\}$ (this has to happen sooner or later since the set $\{1, 2, \ldots, n\}$ is finite). If $\sigma(i_k) = i_1$, we have a cycle. No other possibility can exist. If $\sigma(i_k) = i_\ell$ for $2 \le \ell \le k$, then $\sigma(i_{\ell-1}) = i_\ell = \sigma(i_k)$, which contradicts to σ being one-to-one. We observe then that $\sigma = (i_1\ i_2\ i_3\ \ldots\ i_k)\sigma'$, where σ' does not move any element of the set $\{i_1, i_2, \ldots, i_k\}$ and acts as σ on the complement of this set. So σ' fixes strictly more elements than σ does. This operation can be now applied to σ' and so on. It will terminate at some stage and at that moment σ will be represented as a product of disjoint cycles.   □

In particular, the permutation σ given in (3.3) can be represented as

$$\sigma = (1\ 4\ 11)(2\ 3)(5\ 8\ 6\ 9\ 7\ 5).$$

Exercises

1. Explain why any two disjoint cycles commute.
2. Let the mapping $\pi \colon \{1, 2, \ldots, 12\} \to \{1, 2, \ldots, 12\}$ be defined by $\pi(k) = 3k$ mod 13. This is a permutation, don't prove this. Find the decomposition of π into disjoint cycles.
3. Calculate the following product of permutations in S_5

$$(1\ 2)(1\ 3\ 5\ 2)^{-1}(4\ 3\ 5)(2\ 5).$$

and represent it as a product of disjoint cycles.

4. Let

$$\sigma = \begin{pmatrix} 1\ 2\ 3\ 4\ 5\ 6\ 7\ 8\ 9 \\ 9\ 8\ 7\ 6\ 5\ 3\ 1\ 4\ 2 \end{pmatrix}, \qquad \tau = \begin{pmatrix} 1\ 2\ 3\ 4\ 5\ 6\ 7\ 8\ 9 \\ 6\ 2\ 1\ 4\ 7\ 5\ 9\ 3\ 8 \end{pmatrix}.$$

Calculate $(\sigma\tau)^{-1}$ and represent the result as a product of disjoint cycles.

5°. Find the decomposition into disjoint cycles of

(a) $\sigma = \begin{pmatrix} 1\ \ 2\ 3\ 4\ 5\ 6\ \ 7\ \ 8\ 9\ 10\ 11\ 12 \\ 12\ 3\ 2\ 5\ 9\ 6\ 11\ 1\ 4\ \ 7\ \ 8\ 10 \end{pmatrix}$;

(b) $\tau = (1\ 4\ 3\ 2)^{-1}(3\ 4\ 5\ 6)(1\ 2\ 5\ 6)$.

6°. Show that the mapping σ given by $\sigma(i) = 5i \mod 11$ is a permutation from S_{10}. Write it down in the standard table view and as a product of disjoint cycles.

3.1.4   Orders of Permutations

An element of a group has an important characteristic—its *order*. Orders are very important for cryptography. Now we will define the order of a permutation, and show how the decomposition of this permutation into a product of disjoint cycles allows us to calculate its order.

It is clear that if a permutation τ is a cycle of length k, then $\tau^k = $ id, i.e., if this permutation is repeated k times, we will have the identity permutation as a result of this repeated action. Moreover, for no positive integer s smaller than k we will have $\tau^s = $ id. Also it is clear that if $\tau^m = $ id for some positive integer m, then k is a divisor of m. This observation motivates our next definition.

Definition 3.1.3 Let π be a permutation. The smallest positive integer i such that $\pi^i = $ id is called the *order* of π.

It is not immediately obvious that any permutation has order. We will see later that this is indeed the case.

Example 3.1.6 The order of the cycle $(3\ 2\ 6\ 4\ 1)$ is 5, as we noted before.

Example 3.1.7 Let us calculate the order of the permutation $\pi = (1\ 2)(3\ 4\ 5)$. We have:

$$\pi = (1\ 2)(3\ 4\ 5),$$
$$\pi^2 = (3\ 5\ 4),$$
$$\pi^3 = (1\ 2),$$
$$\pi^4 = (3\ 4\ 5),$$
$$\pi^5 = (1\ 2)(3\ 5\ 4),$$
$$\pi^6 = \text{id}.$$

So the order of σ is $2 \cdot 3 = 6$ (note that π has been given as a product of two disjoint cycles with relatively prime lengths).

We see that for those powers $k = 2, 3, 4, 5$ for which $(1\ 2)^k = $ id, we have $(3\ 5\ 4)^k \neq $ id and the other way around. This happens because the orders of $(1\ 2)$ and $(3\ 4\ 5)$ are relatively prime.

Example 3.1.8 The order of permutation $\rho = (1\ 2)(3\ 4\ 5\ 6)$ is four. To see this let us calculate

$$\rho = (1\ 2)(3\ 4\ 5\ 6),$$
$$\rho^2 = (3\ 5)(4\ 6),$$
$$\rho^3 = (1\ 2)(3\ 6\ 5\ 4),$$
$$\rho^4 = \text{id.}$$

So the order of σ is 4 (note that ρ has been given as a product of disjoint cycles but their lengths were not coprime).

More generally, this suggests that the order of a product of disjoint cycles equals the least common multiple of the lengths of those cycles. We will upgrade this suggestion into a theorem.

Theorem 3.1.4 *Let σ be a permutation and $\sigma = \tau_1 \tau_2 \cdots \tau_r$ be the decomposition of σ into a product of disjoint cycles. Let k be the order of σ and $k_1, k_2, \ldots, k_r$ be the orders (lengths) of $\tau_1, \tau_2, \ldots, \tau_r$, respectively. Then*

$$k = lcm\,(k_1, k_2, \ldots, k_r). \tag{3.4}$$

Proof We first notice that $\tau_i^m = $ id if and only if m is a multiple of k_i. Then, since the cycles τ_i are disjoint, we know that they commute and hence for $k = \text{lcm}\,(k_1, k_2, \ldots, k_r)$

$$\sigma^k = \tau_1^k \tau_2^k \ldots \tau_r^k = \text{id.}$$

So the order of σ is not greater than $\text{lcm}\,(k_1, k_2, \ldots, k_r)$.

Suppose now $\sigma^m = $ id for some m. Let us prove that m is a multiple of $\text{lcm}\,(k_1, k_2, \ldots, k_r)$. We have

$$\sigma^m = \tau_1^m \tau_2^m \ldots \tau_r^m = \text{id.}$$

The powers of cycles τ_1^m, τ_2^m, $\ldots$, τ_r^m act on disjoint sets of indices and, since $\sigma^m = $ id, it must be $\tau_1^m = \tau_2^m = \ldots = \tau_r^m = $ id. If, however, $\tau_s^m(i) = j$ with $i \neq j$, then the product $\tau_1^m \tau_2^m \ldots \tau_r^m$ cannot be equal to id because all permutations $\tau_1^m, \ldots, \tau_{s-1}^m, \tau_{s+1}^m, \ldots, \tau_r^m$ leave i and j invariant. Thus the order of σ is a multiple of each of the $k_1, k_2, \ldots, k_r$ and hence the multiple of the least common multiple of them. Thus the order of σ is not smaller than $\text{lcm}\,(k_1, k_2, \ldots, k_r)$. This proves the theorem. $\qquad\square$

Example 3.1.9 The order of

$$\sigma = (1\ 2\ 3\ 4)(5\ 6\ 7)(8\ 9)(10\ 11\ 12)(13\ 14\ 15\ 16\ 17)$$

is lcm$(4, 3, 2, 3, 5) = 60$. Before applying the formula (3.4) we must carefully check that the cycles are disjoint.

Example 3.1.10 To determine the order of an arbitrary permutation, first write it as product of disjoint cycles. For example, to determine the order of

$$\sigma = \begin{pmatrix} 1\ 2\ 3\ \ 4\ \ 5\ 6\ 7\ 8\ 9\ 10\ 11\ 12 \\ 4\ 3\ 2\ 11\ 8\ 9\ 5\ 6\ 7\ 10\ \ 1\ \ 12 \end{pmatrix}$$

we represent it as

$$\sigma = (1\ 4\ 11)(2\ 3)(5\ 8\ 6\ 9\ 7),$$

and therefore the order of σ is 30.

Exercises
1. Find the orders of the permutations

   (a) $\sigma = \begin{pmatrix} 1\ 2\ 3\ 4\ 5\ 6\ 7\ 8\ 9 \\ 5\ 3\ 6\ 7\ 1\ 2\ 8\ 9\ 4 \end{pmatrix}$,

   (b) $\tau = (1\ 2)(2\ 3\ 4)(4\ 5\ 6\ 7)(7\ 8\ 9\ 10\ 11)$.

2. There is an amusing legend about Flavius Josephus, a famous historian and mathematician who lived in the first century A.D. The story says that in the Jewish revolt against Rome, Josephus and 40 of his comrades were holding out against the Romans in a cave. With defeat imminent, they resolved that, like the rebels at Masada, they would rather die than be slaves to the Romans. They decided to arrange themselves in a circle. One man was designated as number one, and they proceeded clockwise around the circle of 41 men killing every third man. At first it is obvious whose turn it was to be killed. Initially, the men in positions $3, 6, 9, 12, \ldots, 39$ were killed. The next man to be killed was in position 1 and then in the position 5 (since the man in position 3 was slaughtered earlier), and so on.

   Josephus (according to the story) instantly figured out where he ought to stand in order to be the last man to go. When the time came, instead of killing himself, he surrendered to the Romans and lived to write his famous histories: "The Antiquities" and "The Jewish War".

   (a) Find the permutation σ (called the Josephus permutation) for which $\sigma(i)$ is the number of the man who was ith to be killed.

   (b) In which position did Josephus stand around the circle?

   (c) Find the cyclic structure of the Josephus permutation.

   (d) What is the order of the Josephus permutation?

   (e) Calculate σ^2 and σ^3.

3. The mapping $\pi(i) = 13i \mod 23$ is a permutation of S_{22} (do not prove this). Find the decomposition of π into a product of disjoint cycles and determine the order of this permutation.

4°. Without using GAP, find the disjoint cycle decomposition and the order of each of the following two permutations. Remember that permutations are read from left to right.

(a) $\sigma = \begin{pmatrix} 1 & 2 & 3 & 4 & 5 & 6 & 7 & 8 & 9 & 10 & 11 & 12 \\ 10 & 5 & 2 & 3 & 12 & 6 & 11 & 1 & 4 & 7 & 8 & 9 \end{pmatrix}$

(b) $\tau = (1, 2, 5, 6)^{-1} (3, 4, 5, 6) (1, 4, 3, 2)$.

5°. Generate in GAP permutation group S_4, then list its elements and then list their orders by giving commands:

```
gap> S4:=SymmetricGroup(4); List(S4); List(S4,Order);
```

6°. What is the largest possible order of a permutation in S_{42} that has exactly two cycles?

3.1.5 Analysis of Repeated Actions

In this section we consider one important application of permutations. Sometimes (and often in cryptography) a certain action is performed repeatedly and we are interested in the outcome that results after a number of repetitions.

As one particularly instructive example, we will analyse the so-called interlacing shuffle that card players often do with a deck of cards. Suppose that we have a deck of $2n$ cards (normally 52) and suppose that our cards were numbered from 1 to $2n$ and the original order of cards in the deck was

$$a_1 a_2 a_3 \ldots a_{2n-1} a_{2n}.$$

We split the deck into two halves which contain the cards $a_1, a_2, \ldots, a_n$ and $a_{n+1} a_{n+2}, \ldots, a_{2n}$, respectively. Then we interlace them as follows. We put the first card of the second pile first, then the first card of the first pile, then the second card of the second pile, then the second card of the first pile etc. This is called the *interlacing shuffle*. After this operation the order of the cards will be

$$a_{n+1} a_1 a_{n+2} a_2 \ldots a_{2n} a_n.$$

We put the permutation

$$\sigma_n = \begin{pmatrix} 1 & 2 & 3 & \ldots & n & n+1 & n+2 & \ldots & 2n \\ 2 & 4 & 6 & \ldots & 2n & 1 & 3 & \ldots & 2n-1 \end{pmatrix}$$

in correspondence to this shuffle. All it says is that the first card goes to the second position, the second card is moved to the fourth position, etc. We see that we can define this permutation by the formula:

$$\sigma_n(i) = 2i \bmod 2n + 1$$

and $\sigma_n(i)$ is the position of the ith card after the shuffle. What will happen after $2, 3, 4, \ldots$ shuffles? The resulting change will be characterised by the permutations $\sigma_n^2, \sigma_n^3, \sigma_n^4, \ldots,$ respectively.

Example 3.1.11 For $n = 4$

$$\sigma_4 = \begin{pmatrix} 1\ 2\ 3\ 4\ 5\ 6\ 7\ 8 \\ 2\ 4\ 6\ 8\ 1\ 3\ 5\ 7 \end{pmatrix} = \begin{pmatrix} 1\ 2\ 4\ 8\ 7\ 5 \end{pmatrix} \begin{pmatrix} 3\ 6 \end{pmatrix}$$

The order of σ_4 is 6.

Example 3.1.12 For $n = 5$

$$\sigma_5 = \begin{pmatrix} 1\ 2\ 3\ 4\ 5\ 6\ 7\ 8\ 9\ 10 \\ 2\ 4\ 6\ 8\ 10\ 1\ 3\ 5\ 7\ 9 \end{pmatrix} =$$
$$= \begin{pmatrix} 1\ 2\ 4\ 8\ 5\ 10\ 9\ 7\ 3\ 6 \end{pmatrix}.$$

Also $\sigma_5^{10} = $ id and 10 is the order of σ_5. Hence all cards will be back to their initial positions after 10 shuffles but not before.

Let us deal with the real thing that is the deck of card of 52 cards. We know that the interlacing shuffle is defined by the equation $\sigma_{26}(i) = 2i \bmod 53$. GAP helps us to investigate. We have:

```
gap> lastrow:=[1..52];;
gap> for i in [1..52] do
> lastrow[i]:=2*i mod 53;
> od;
gap> lastrow;
[ 2, 4, 6, 8, 10, 12, 14, 16, 18, 20, 22, 24, 26, 28, 30, 32, 34, 36, 38, 40,
  42, 44, 46, 48, 50, 52, 1, 3, 5, 7, 9, 11, 13, 15, 17, 19, 21, 23, 25, 27,
  29, 31, 33, 35, 37, 39, 41, 43, 45, 47, 49, 51 ]
gap> PermList(lastrow);
(1,2,4,8,16,32,11,22,44,35,17,34,15,30,7,14,28,3,6,12,24,48,43,33,13,26,52,51,
49,45,37,21,42,31,9,18,36,19,38,23,46,39,25,50,47,41,29,5,10,20,40,27)
gap> Order(last);
52
```

Thus the interlacing shuffle σ_{26} is a cycle of length 52 and has order 52.

Exercises
1. A shuffle of a deck of 15 cards is made as follows. The top card is put at the bottom, the deck is cut into three equal decks, the bottom third is switched with the middle third, and then the resulting bottom card is placed on the top. How many times must this shuffle be repeated to get the cards back in the initial order? Write down the permutation corresponding to this shuffle and find its decomposition into disjoint cycles.
2. Use GAP to determine the decomposition into disjoint cycles and the order of the interlacing shuffle σ_{52} for the deck of 104 cards which consists of two copies of ordinary decks with 52 cards in each.
3. On a circle there are n beetles. At a certain moment they start to move all at once and with the same speed (but maybe in different directions). When two beetles meet, both of them reverse their directions and continue to move with the same speed. Prove that there will be a moment when all beetles again occupy their initial positions. (Hint: Suppose one beetle makes the full circle in time t. Think about what will happen after time t when all beetles move.)
4°. Let σ_n be the permutation corresponding to the interlacing card shuffle of a deck of $2n$ cards.
   (a) What is the order of σ_n when $n = 2, 4, 8, 16, 32$?
   (b) What is the order of σ_n when $n = 1, 3, 7, 15, 31$?
   (c) Can you guess what is the order of σ when $n = 2^{10} = 1024$? and when $n = 2^{10} - 1 = 1023$?

3.1.6 Transpositions. Even and Odd

Cycles of length 2 are the simplest permutations, as they move only two elements. We define the following.

Definition 3.1.4 A cycle of length 2 is called a *transposition*.

It is intuitively plausible that any permutation is a product of transpositions (indeed, every arrangement of n objects can be obtained from a given starting position by making a sequence of swaps). We will observe, first, that a cycle of arbitrary length can be expressed as a product of transpositions. Then using Theorem 3.1.3 we will be able to express any permutation as product of transpositions. Here are some examples.

Example 3.1.13 $(1\ 2\ 3\ 4\ 5) = (1\ 2)(1\ 3)(1\ 4)(1\ 5)$ (just check that the left-hand side equals the right-hand side!).

We can express an arbitrary cycle as a product of transpositions in exactly the same way:

$$(i_1\ i_2\ \ldots\ i_r) = (i_1\ i_2)(i_1\ i_3)\ldots(i_1\ i_r). \tag{3.5}$$

As a result we have come to the following theorem.

Theorem 3.1.5 *Every permutation can be expressed as a product of transpositions.*

Proof To express any permutation σ as product of transpositions, first decompose σ into a product of disjoint cycles, then write the cycles as product of transpositions as in formula (3.5). $\qquad\square$

Example 3.1.14 Here is a decomposition of a familiar to us permutation:

$$\begin{pmatrix} 1\ 2\ 3\ \ 4\ \ 5\ 6\ 7\ 8\ 9\ 10\ 11 \\ 4\ 3\ 2\ 11\ 8\ 9\ 5\ 6\ 7\ 10\ \ 1 \end{pmatrix} = (1\ 4\ 11)(2\ 3)(5\ 8\ 6\ 9\ 7) =$$

$$(1\ 4)(1\ 11)(2\ 3)(5\ 8)(5\ 6)(5\ 9)(5\ 7).$$

Example 3.1.15 Note that there are many, many different ways to write a permutation as product of transpositions; for example, $(1\ 2\ 3\ 4\ 5)$ can be written in any of the following forms:

$$(1\ 2)(1\ 3)(1\ 4)(1\ 5) = (3\ 4)(3\ 5)(3\ 1)(3\ 2) = (3\ 4)(3\ 5)(2\ 3)(1\ 3)(2\ 3)(2\ 1)(3\ 1)(3\ 2).$$

(Don't ask how these products were found! The point is to check that all these products are equal, and to note that there is nothing unique about how one can write a permutation as product of transpositions.)

However, there is something in common in all decompositions of a given permutation into a product of transpositions. As we will see the number of such transpositions will be either always even or always odd.

Definition 3.1.5 A permutation is called *even* if it can be written as a product of an even number of transpositions. A permutation is called *odd* if it can be written as a product of an odd number of transpositions.

To make this definition meaningful we need to prove that there is no permutation which is at the same time even and odd—this justifies the use of the terminology. We will establish that by looking at the polynomial

$$f(x_1, x_2, \ldots, x_n) = \prod_{i<j}(x_i - x_j).$$

For example, for $n = 4$ we get a polynomial

$$f(x_1, x_2, x_3, x_4) = (x_1 - x_2)(x_1 - x_3)(x_1 - x_4)(x_2 - x_3)(x_2 - x_4)(x_3 - x_4).$$

It is clear that for $\pi = (i\ i+1)$ we have

$$f(x_{\pi(1)}, x_{\pi(2)}, \ldots, x_{\pi(n)}) = -f(x_1, x_2, \ldots, x_n) \tag{3.6}$$

since in all brackets but one will remain the same except $(x_i - x_{i+1})$. It will become $(x_{i+1} - x_i) = -(x_i - x_{i+1})$ so we will have one change of sign.

Arguing by induction we suppose that (3.6) is true for all permutations $\pi = (i\ j)$ for which $|j - i| < \ell$. Suppose now that $|j - i| = \ell$. Since

$$(i\ j) = (j{-}1\ j)(i\ j{-}1)(j{-}1\ j)$$

we conclude that (3.6) holds for the transposition $\pi = (i\ j)$ with $|j - i| = \ell$ too. Hence (3.6) holds for any product of an odd number of transpositions. It is now also clear that

$$f(x_{\pi(1)}, x_{\pi(2)}, \ldots, x_{\pi(n)}) = +f(x_1, x_2, \ldots, x_n) \tag{3.7}$$

whenever π is a product of an even number of transpositions. This implies that there is no permutation which is both even and odd.

Example 3.1.16 $(1\ 2\ 3\ 4)$ is an odd permutation, because $(1\ 2\ 3\ 4) = (1\ 2)(1\ 3)$ $(1\ 4)$. On the other hand the permutation $(1\ 2\ 3\ 4\ 5)$ is even, because $(1\ 2\ 3\ 4\ 5) = (1\ 2)(1\ 3)(1\ 4)(1\ 5)$.

Example 3.1.17 Since $id = (1\ 2)(1\ 2)$, the identity permutation is even.

Example 3.1.18 Let $\pi = \begin{pmatrix} 1\ 2\ 3\ 4\ 5\ 6\ 7\ 8\ 9 \\ 4\ 3\ 2\ 5\ 1\ 6\ 9\ 8\ 7 \end{pmatrix}$. Is π even or odd?

First we decompose π into a product of cycles, then use the result above:

$$\pi = (1\ 4\ 5)(2\ 3)(7\ 9) = (1\ 5)(1\ 4)(2\ 3)(7\ 9).$$

This shows that π is even.

Theorem 3.1.6 *A k-cycle is even if k is odd and odd if k is even.*

Proof Immediately follows from (3.5).   □

Definition 3.1.6 We say that two permutations have the *same parity* if they are both odd or both even and *different parity* if one of them is odd and another is even.

Theorem 3.1.7 *In any symmetric group S_n*

  (i) *The product of two even permutations is even.*
 (ii) *The product of two odd permutations is even.*
(iii) *The product of an even permutation and an odd one is odd.*
(iv) *A permutation and its inverse have the same parities.*

Proof Only the statements 4 needs a comment. It follows from (iii). Indeed, for any permutation π we have $\pi\pi^{-1} = \mathrm{id}$, and, since the identity permutation is even, by (iii), π and π^{-1} cannot have different parities. □

Theorem 3.1.8 *Exactly half of the elements of S_n are even and half of them are odd.*

Proof Denote by E the set of even permutations in S_n, and by O the set of odd permutations in S_n. If τ is any fixed transposition from S_n, we can establish a one-to-one correspondence between E and O as follows: for π in E we know that $\tau\pi$ belongs to O. Therefore we have a mapping $f : E \to O$ defined by $f(\pi) = \tau\pi$. The function f is one-to-one since $\tau\pi = \tau\sigma$ implies that $\pi = \sigma$; f is onto, because if κ is an odd permutation then $\tau\kappa$ is even, and $f(\tau\kappa) = \tau\tau\kappa = \kappa$. □

Corollary 3.1.1 *The number of even permutations in S_n is $\frac{n!}{2}$. The number of odd permutations in S_n is also $\frac{n!}{2}$.*

Corollary 3.1.2 *The set A_n of all even permutations of degree n is a group relative to the operation of composition called the* alternating group of degree n.

Example 3.1.19 We can have a look at the elements of S_4, listing all of them, and checking which of them are even, which of them are odd.

$$S_4 = \{\mathrm{id}, (1\ 2\ 3), (1\ 3\ 2), (1\ 2\ 4), (1\ 4\ 2), (2\ 3\ 4), (2\ 4\ 3),$$
$$(1\ 3\ 4), (1\ 4\ 3), (1\ 2)(3\ 4), (1\ 3)(2\ 4), (1\ 4)(2\ 3),$$
$$(1\ 2), (1\ 3), (1\ 4), (2\ 3), (2\ 4), (3\ 4), (1\ 2\ 3\ 4), (1\ 4\ 3\ 2),$$
$$(1\ 3\ 2\ 4), (1\ 4\ 2\ 3), (1\ 2\ 4\ 3), (1\ 3\ 4\ 2)\}.$$

The elements in the first two lines are even permutations, and the remaining elements are odd. We have

$$A_4 = \{\mathrm{id}, (1\ 2\ 3), (1\ 3\ 2), (1\ 2\ 4), (1\ 4\ 2), (2\ 3\ 4), (2\ 4\ 3),$$
$$(1\ 3\ 4), (1\ 4\ 3), (1\ 2)(3\ 4), (1\ 3)(2\ 4), (1\ 4)(2\ 3)\}.$$

Exercises

1. Write the permutations

$$(1\ 3\ 7)(5\ 8)(2\ 4\ 6\ 9), \quad (1\ 3\ 7)(5\ 7\ 8)(2\ 3\ 4\ 6\ 9)$$

   as a products of transpositions.
2. What would be the parity of the product of 11 odd permutations?
3. Let $\pi, \rho \in S_n$ be two permutations. Prove that π and $\rho^{-1}\pi\rho$ have the same parity.
4. Let $\pi, \rho \in S_n$ be two permutations. Prove that $\pi^{-1}\rho^{-1}\pi\rho$ is an even permutation.
5. Determine the parity of the permutation σ of order n such that $\sigma(i) = n + 1 - i$.
6°. Which of the following permutations in S_{10} are even? and which are odd?
   (a) $(1, 2)(3, 4, 5, 6, 7)$
   (b) $(1, 2)(3, 4, 5, 6, 7, 8)$
   (c) $(1, 2)(3, 4, 5)(6, 7, 8)$
   (d) $(1, 2)(3, 4, 5)(6, 7, 8, 9)$
   (e) $(1, 2, 3)(4, 5, 6, 7, 8, 9, 10)$.

3.1.7 Puzzle 15

We close this section with a few words about a game played with a simple toy. This game seems to have been invented in the 1870s by the famous puzzle-maker Sam Loyd. It caught on and became the rage in the United States in the 1870s, and finally led to a discussion by W. Johnson in the scholarly journal, the *American Journal of Mathematics*, in 1879. It is often called the *15-puzzle*.

Consider a toy made up of 16 squares, numbered from 1 to 15 inclusive and with the lower right-hand corner blank.

1	2	3	4
5	6	7	8
9	10	11	12
13	14	15	

The toy is constructed so that the squares can be slid vertically and horizontally, such moves being possible because of the presence of the blank square. Start with the position shown above and perform a sequence of slides in such a way that, at the end, the lower right-hand square is again blank. Call the new position *realisable*. The natural question is: How can we determine whether or not the given position is realisable?

What do we have here? After a sequence of slides we have shuffled about the numbers from 1 to 15; that is, we have effected a permutation of the numbers from 1 to 15. To ask which positions are realisable is merely to ask which permutations can be carried out. This is a permutation of S_{16} since the blank square also moves in the process. In other words, in S_{16}, the symmetric group of degree 16, which permutations can be reached via the toy? For instance, can the following position be realised?

13	4	12	15
1	14	9	6
8	3	2	7
10	5	11	

We will denote the empty square by the number 16. The position

a_1	a_2	a_3	a_4
a_5	a_6	a_7	a_8
a_9	a_{10}	a_{11}	a_{12}
a_{13}	a_{14}	a_{15}	a_{16}

will be then characterised by the permutation

$$\begin{pmatrix} 1 & 2 & \ldots & 16 \\ a_1 & a_2 & \ldots & a_{16} \end{pmatrix}.$$

Example 3.1.20 The position

1	3	5	7
9	11	13	15
2	4		6
8	10	12	14

will correspond to the permutation

$$\sigma = \begin{pmatrix} 1\ 2\ 3\ 4\ 5\ 6\ \ 7\ \ 8\ 9\ 10\ 11\ 12\ 13\ 14\ 15\ 16 \\ 1\ 3\ 5\ 7\ 9\ 11\ 13\ 15\ 2\ 4\ \ 16\ 6\ \ 8\ \ 10\ 12\ 14 \end{pmatrix}.$$

If we make a move pulling down the square 13, then the new position will be

1	3	5	7
9	11		15
2	4	13	6
8	10	12	14

and the new permutation will be

$$\begin{pmatrix} 1\ 2\ 3\ 4\ 5\ 6\ \ 7\ \ 8\ 9\ 10\ 11\ 12\ 13\ 14\ 15\ 16 \\ 1\ 3\ 5\ 7\ 9\ 11\ 16\ 15\ 2\ 4\ \ 13\ 6\ \ 8\ \ 10\ 12\ 14 \end{pmatrix} = \sigma\ (13\ 16).$$

We observe the rule how the permutation changes: when we swap the square with number i on it with the neighbouring empty square, the permutation is being multiplied on the right by the transposition $(i\ 16)$.

Theorem 3.1.9 *If a position characterised by the permutation σ can be transformed by legal moves to the initial position, then there exist transpositions $\tau_1, \tau_2, \ldots, \tau_m$ such that*

$$\mathrm{id} = \sigma\ \tau_1 \tau_2 \ldots \tau_m. \tag{3.8}$$

If the empty square was initially in the right bottom corner, then m is even and σ is even.

Proof Suppose that a position characterised by the permutation σ can be transformed by legal moves to the initial position. As we noted in Example 3.1.20, every legal move is equivalent to a multiplication by a transposition $(i\ 16)$ for some $i \in \{1, 2, \ldots, 15\}$. Since the initial position is characterised by the identity permutation, we see that (3.8) follows. It implies

$$\sigma = \tau_m \tau_{m-1} \ldots \tau_2 \tau_1$$

from which we see that the parity of σ is the same as the parity of m.

Let us colour the board in the chessboard pattern.

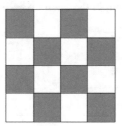

Every move changes the colour of the empty square. Thus if at the beginning and at the end the empty square was blank, then there was an even number of moves made. Therefore, if initially the right bottom corner was empty and we could transform this position to the initial position, then an even number of moves was made, m is even, and σ is also even. □

It can be shown that every position, with an even permutation σ, can be transformed to the initial position but no easy proof is known.

Exercises

1. Given the following two positions in 15-puzzle show that one of them is real-

14	10	13	12
6	11	9	8
7	3	5	1
4	15	2	

10	14	13	12
6	11	9	8
7	3	5	1
4	15	2	

isable and one is not without writing down the corresponding permutations and determining their parities.

2°. For each of the following arrangements of 15-puzzle, determine the parity of the corresponding permutation.

1	3	2	4
6	5	7	8
9	13	15	11
14	10	12	

13		5	3
9	2	7	10
1	15	14	8
12	11	6	4

Which one is realisable and which is not.

3°. In the 15-puzzle, suppose the initial state (on the left) is transformed by legal moves to the state on the right in the diagram below:

1	2	3	4
5	6	7	8
9	10	11	12
13	14	15	

12	6	3	1
15	2	4	14
13	8	11	9
5	10	7	

How many times must this transformation be repeated to return to the initial state?

3.2   General Groups

3.2.1   Definition of a Group. Examples

Surprisingly many objects in mathematics satisfy the same properties as symmetric groups defined in Definition 3.1.2. There is good reason to study all such objects simultaneously. For this purpose we introduce the concept of a general group.

Definition 3.2.1 A set G together with a binary operation $*$ is called a *group* if it satisfies the following three properties:

1. The operation $*$ is associative; i.e.,

$$(a * b) * c = a * (b * c) \qquad \text{for all } a, b, c \in G.$$

2. G contains an identity element; i.e., there exists an element $e \in G$ such that

$$e * g = g * e = g \qquad \text{for all } g \in G.$$

   (This element is often also denoted by 1, or, if the group operation is written as addition, it is usually denoted by 0.)
3. Every element of G possesses an inverse; i.e., given $g \in G$ there exists a unique element h in G such that

$$g * h = h * g = e \qquad (e \text{ is the identity element of } G).$$

   The element h is called the *inverse of g*, and denoted by g^{-1} (when the operation is written as addition, the inverse is usually denoted by $-g$).

We denote this group $(G, *)$, or simply G, when this invites no confusion. A group G in which the commutative law holds ($a * b = b * a$ for all $a, b \in G$) is called a *commutative group* or an *abelian group*.

In any group $(G, *)$ we have the familiar formula for the inverse of the product

$$(a * b)^{-1} = b^{-1} * a^{-1}$$

for all $a, b \in G$. This can be proved in the same way as can (3.2).

Example 3.2.1 We established in the previous sections that S_n is a group, the operation being multiplication of permutations (i.e., composition of functions). This group is not abelian.

Example 3.2.2 Here is an example where the group operation is written as addition: $\mathbb{Z}_n$ is an abelian group under addition $\oplus$ modulo n. This was established in Theorem 1.4.1.

Example 3.2.3 $\mathbb{Z}_n^*$ (the set of invertible elements in the ring $\mathbb{Z}_n$) is a group under multiplication modulo n. In particular, $\mathbb{Z}_8^* = \{1, 3, 5, 7\}$ with $3^{-1} = 3$, $5^{-1} = 5$, $7^{-1} = 7$.

When we talk about a group, it is important to be clear about the group operation; either it must be explicitly specified, or the group operation must be clear from the context and tacitly understood. The following are cases where there is a clear understanding of the operation, so it will often not be made explicit. Most important are:

- When we talk about the group $\mathbb{Z}_n$, we mean the set of integers modulo n *under addition modulo m*.
- When we talk about the group $\mathbb{Z}_n^*$, we mean the set of invertible elements in the ring $\mathbb{Z}_n$ *under multiplication modulo n*.

Normally, when making general statements about groups, we write the statements in multiplicative notation; but it is important to be able to apply them also in situations where the group operation is written as addition (some obvious modifications must be made).

Definition 3.2.2 Let G be a group and e be its identity element. The number of elements of G is called the *order* of G and denoted by $|G|$.

Example 3.2.4 Orders of several groups:

- S_n is a group of order $n!$.
- $\mathbb{Z}_n$ is a group of order n.
- $\mathbb{Z}_n^*$ is a group of order $\phi(n)$, where ϕ is Euler's totient function; for example, $|\mathbb{Z}_{12}^*| = 4$.
- $\mathbb{Z}$ is an infinite group.
- Positive integers $\mathbb{R}^+$ with the operation of usual multiplication of the reals is also an infinite group.

Exercises

1. Show that the division $a \star b = a : b$ is a binary operation on $\mathbb{R} \setminus \{0\}$. Show that it is not associative.
2. Show that $a \star b = a^b$ is a binary operation on the set $\mathbb{R}^+$ of positive real numbers. Show that it does not have a neutral element.
3. Let $\mathbb{C}_n$ be the set of all complex numbers satisfying the equation $z^n = 1$. Prove that this is an abelian group of order n.
4. Prove that the set $GL_n(\mathbb{R})$ of all invertible $n \times n$ matrices is a non-abelian group.
5. Prove that for arbitrary four elements g_1, g_2, g_3, g_4 of a group G (where operation is written as multiplication)

$$(g_1 g_2)(g_3 g_4) = (g_1(g_2 g_3))g_4.$$

List all possible arrangements of brackets on the product $g_1 g_2 g_3 g_4$ and show that the result will be always the same so that we can write

$$g_1 g_2 g_3 g_4$$

for all of them without confusion. Finally you may try to prove that a product

$$g_1 g_2 \cdots g_n$$

involving $n \geq 3$ elements is independent of the way in which these elements are combined and associated.

6°. Let G be the set of all points of the segment $[0, 1)$ of reals with the addition $a \oplus b = \{a + b\}$, where $\{x\}$ is the fractional part of x. Check that this is a group.

3.2.2 Powers, Multiples and Orders. Cyclic Groups

Definition 3.2.3 Let G be a group whose operation is written multiplicatively, g an element of G, e the identity element of G, and $n \in \mathbb{Z}$. We define

$$g^n = \begin{cases} \underbrace{gg \cdots g}_{n \text{ times}} & \text{if } n > 0, \\ e & \text{if } n = 0, \\ \underbrace{g^{-1}g^{-1} \cdots g^{-1}}_{|n| \text{ times}} & \text{if } n < 0. \end{cases}$$

Since we know that the product $g_1 g_2 \ldots g_n$ is independent of the way in which these elements are associated, it becomes clear that the usual law of exponents $g^i g^j = g^{i+j}$ holds (totally obvious in the case where both i and j are positive, and still trivial in all other cases). The set of all powers of $g \in G$, we denote by $< g >$.

Definition 3.2.4 Let G be a group whose operation is written additively, g an element of G, 0 the identity element of G, and $n \in \mathbb{Z}$. We define

$$ng = \begin{cases} \underbrace{g + g \cdots + g}_{n \text{ times}} & \text{if } n > 0, \\ 0 & \text{if } n = 0, \\ \underbrace{(-g) + (-g) + \cdots + (-g)}_{|n| \text{ times}} & \text{if } n < 0. \end{cases}$$

The usual law of multiples $mg + ng = (m + n)g$ also holds. The set of all multiples of $g \in G$, we also denote by $< g >$.

Definition 3.2.5 Any group G which consists of powers (multiples) of a single element g is called *cyclic*. This fact can be written as $G =< g >$. Element g in this case is called the *generator* of G.

We note that every cyclic group is abelian since $g^i g^j = g^j g^i$ and $mg + ng = ng + mg$.

Example 3.2.5 Several examples:

- S_n is NOT a cyclic group since it is not abelian.
- $\mathbb{Z}_n =< 1 >$ and is cyclic.
- $\mathbb{Z}_5^* =< 2 >$ and is cyclic. Check this by calculating all multiples of 2.
- $\mathbb{Z} =< 1 >$ is an infinite cyclic group.

Later (see, e.g., Exercise 1) we will see that abelian groups do not have to be cyclic.

Definition 3.2.6 Let G be a group and e be its identity element. Then the *order of g in G* is the least positive integer i such that $g^i = e$, if such an integer exists; otherwise we say that the order of g is infinite. It is denoted by ord (g).

We note that this definition is consistent with the definition of the order of a permutation given earlier.

In an additively written group G the order of $g \in G$ is the least positive integer m such that $mg = 0$, if such an integer exists; if no such integer exists, we say that the order of g is infinite.

Example 3.2.6 Confirm for yourself that:

- Each of the non-identity elements of $\mathbb{Z}_{12}^*$ have order 2;
- In $\mathbb{Z}_{12}$ the element 10 has order 6;
- Element 6 in the group $\mathbb{Z}$ has infinite order.

As we will see later, in a finite group G, the orders of its elements and the order of the group $|G|$ are closely related.

We start to establish this link with the following

Lemma 3.2.1 *If ord $(g) = n$, then $< g >= \{e, g, g^2, \ldots, g^{n-1}\}$, and all n powers of g in this set are distinct, i.e., $| < g > | = n$. Conversely, if $| < g > | = n$, then g is an element of order n.*

Proof Suppose ord $(g) = n$. Then $g^n = e$ and all powers of g belong to the set $\{e, g, g^2, \ldots, g^{n-1}\}$. Indeed, for any $k \in \mathbb{Z}$ we may divide k by n with remainder $k = qn + r$, where $0 \le r < n$. Then $g^k = g^{qn+r} = g^{qn} g^r = (g^n)^q g^r = g^r$, which belongs to $\{e, g, g^2, \ldots, g^{n-1}\}$. Hence $< g >= \{e, g, g^2, \ldots, g^{n-1}\}$, On the other hand, if any two powers in this set are equal, say $g^i = g^j$ with $i < j$, then $g^j = g^i g^{j-i} = g^i$ and $g^{j-i} = e$. This is a contradiction since $j - i < n$ and n is the order of g. Therefore, if the order of g is finite, the order of g is the same as the cardinality of the set $< g >$.

Suppose now that the cardinality of $< g >$ is n. Then we have only n distinct powers of g and there will exist two distinct integers k and m such that $g^k = g^m$. If we assume that $k > m$ then we will find that $g^{k-m} = e$, and g will have finite order. We have already proved that in this case the order of g and the size of $< g >$ coincide. Hence ord $(g) = n$. $\square$

In the following corollary a link between the two concepts of "order". It is often useful since we can decide whether a group is cyclic or not by looking at the orders of its elements.

Corollary 3.2.1 *For a group of order n to be cyclic, it is necessary and sufficient that it has an element of order n.*

Example 3.2.7 $\mathbb{Z}_8^*$ is NOT a cyclic group, because $|\mathbb{Z}_8^*| = \phi(8) = 4$ and there is no element of order 4 in this group (indeed, check that they all have order 2).

The following theorem is an important tool that allows to calculate orders of elements in $\mathbb{Z}_n$.

Theorem 3.2.1 *The order of* $i \in \mathbb{Z}_n$ *is* $\mathrm{ord}\,(i) = \dfrac{n}{gcd(i, n)}$.

Proof To see this, note that the group is written additively, so the order of i is the smallest positive integer k such that $ki \equiv 0 \mod n$. That is, ki is the smallest positive number which is a multiple of i as well as of n. This means that $ki = \mathrm{lcm}(i, n)$. Now solve this equation for k using (1.131):

$$k = \frac{\mathrm{lcm}(i, n)}{i} = \frac{in}{i\,gcd(i, n)} = \frac{n}{gcd(i, n)}.$$

This proves the theorem. □

Example 3.2.8 The order of 110 in $\mathbb{Z}_{121}$ is

$$\mathrm{ord}\,(110) = \frac{121}{gcd(121, 110)} = \frac{121}{11} = 11.$$

Exercises

1. Find the orders of elements 5, 1331, 594473 in $\mathbb{Z}_{16427202}$.
2. Find all elements of order 7 in $\mathbb{Z}_{84}$.
3. Find the order of $i = 41670852902912$ in the abelian group $\mathbb{Z}_n$, where $n = 563744998038700032$.
4. Show that $\mathbb{Z}_{12}^*$ is an abelian group which is not cyclic.
5. Show that the order of the interlacing shuffle σ_n (defined in Sect. 3.1.5) is equal to the order of 2 in $\mathbb{Z}_{2n+1}^*$.
6°. Find all elements of order 7 in $\mathbb{Z}_{105}$.
7°. Consider the function $\theta: \mathbb{Z}_{15} \to \mathbb{Z}_{15}$ given by $\theta(k) = 7k + 2 \pmod{15}$ for all $k \in \mathbb{Z}_{15}$.

   (a) Show that θ is a permutation, and find its order.
   (b) Find two permutations ψ and τ of orders 4 and 15 such that $\theta = \psi\tau$.

3.2.3   Isomorphism

A single group may have several very different presentations. To deal with this problem mathematics introduces the concept of isomorphism.

Definition 3.2.7 Let G and H be two groups with operations $*$ and $\circ$, respectively. An onto and one-to-one mapping $\sigma \colon G \to H$ is called an *isomorphism* if

$$\sigma(g_1 * g_2) = \sigma(g_1) \circ \sigma(g_2) \tag{3.9}$$

for all $g_1, g_2 \in G$.

What it says is that if we rename the elements of H appropriately and change the name for the operation in H, we will obtain the group G. If two groups G and H are isomorphic, we write $G \cong H$. Equation (3.9) written as

$$g_1 * g_2 = \sigma^{-1}(\sigma(g_1) \circ \sigma(g_2))$$

has also a computational interpretation. It says that instead of computing $g_1 * g_2$ of elements g_1 and g_2 in group G, one can compute $\sigma(g_1) \circ \sigma(g_2)$ for images $\sigma(g_1)$ and $\sigma(g_2)$ of these elements in H and take the preimage of the result.

Example 3.2.9 A classical, most known example of an isomorphism is the isomorphism of the group $\mathbb{R}$, which is the reals with the operation of addition, and the group $\mathbb{R}^+$, which is positive reals with the operation of multiplication. The isomorphism $\sigma \colon \mathbb{R} \to \mathbb{R}^+$ between these two groups is given by $\sigma(x) = e^x$. Indeed, the condition (3.9) is satisfied since

$$\sigma(x + y) = e^{x+y} = e^x e^y = \sigma(x)\sigma(y).$$

The famous slide rule—a commonly used calculation tool in science and engineering before electronic calculators became available—was based on this isomorphism.

Example 3.2.10 We claim $\mathbb{Z}_4 \cong \mathbb{Z}_5^*$. Let us look at their addition and multiplication tables, respectively.

$$
\begin{array}{c}
\mathbb{Z}_4 \\
\begin{array}{c|cccc}
\oplus & 0 & 1 & 2 & 3 \\
\hline
0 & 0 & 1 & 2 & 3 \\
1 & 1 & 2 & 3 & 0 \\
2 & 2 & 3 & 0 & 1 \\
3 & 3 & 0 & 1 & 2
\end{array}
\end{array}
\qquad
\begin{array}{c}
\mathbb{Z}_5^* \\
\begin{array}{c|cccc}
\odot & 1 & 2 & 3 & 4 \\
\hline
1 & 1 & 2 & 3 & 4 \\
2 & 2 & 4 & 1 & 3 \\
3 & 3 & 1 & 4 & 2 \\
4 & 4 & 3 & 2 & 1
\end{array}
\end{array}
$$

We may observe that the first table can be converted into the second one if we make the following substitution:

$$0 \to 1, \quad 1 \to 2, \quad 2 \to 4, \quad 3 \to 3$$

(check it right now). Therefore this mapping, let us call it σ, from $\mathbb{Z}_4$ to $\mathbb{Z}_5^*$ is an isomorphism. The mystery behind this mapping is clarified if we notice that we actually map

$$0 \rightarrow 2^0, \quad 1 \rightarrow 2^1, \quad 2 \rightarrow 2^2, \quad 3 \rightarrow 2^3.$$

Then the isomorphism property (3.9) follows from the formula $2^i \odot 2^j = 2^{i \oplus j}$.

Before continuing with the study of isomorphism we make a useful observation: in any group G the only element that satisfies $g^2 = g$ is the identity element. Indeed, multiplying this equation by g^{-1} we get $g = e$.

Proposition 3.2.1 *Let $(G, *)$ and $(H, \circ)$ be two groups and e be the identity element of G. Let $\sigma: G \rightarrow H$ be an isomorphism of these groups. Then $\sigma(e)$ is the identity of H.*

Proof Let $\sigma(e) = \epsilon$, where e is the identity element of G. Let us prove that ϵ is the identity element of H. We note that $\epsilon^2 = \sigma(e)^2 = \sigma(e^2) = \sigma(e) = \epsilon$. Any element in a group with this property must be the identity so ϵ is the identity of H. □

Theorem 3.2.2 *Let $(G, *)$ and $(H, \circ)$ be two groups and $\sigma: G \rightarrow H$ be an isomorphism. Then $\sigma^{-1}: H \rightarrow G$ is also an isomorphism.*

Proof We need to prove

$$\sigma^{-1}(h_1 \circ h_2) = \sigma^{-1}(h_1) * \sigma^{-1}(h_2) \tag{3.10}$$

for all $h_1, h_2 \in H$. For this reason we apply σ to both sides of this equation. As $\sigma\sigma^{-1} = \mathrm{id}_G$ and $\sigma^{-1}\sigma = \mathrm{id}_H$, and due to (3.9)

$$\sigma(\sigma^{-1}(h_1 \circ h_2)) = h_1 \circ h_2,$$
$$\sigma(\sigma^{-1}(h_1) * \sigma^{-1}(h_2)) = \sigma(\sigma^{-1}(h_1)) \circ \sigma(\sigma^{-1}(h_2)) = h_1 \circ h_2.$$

The result is the same both times. As σ is one-to-one (3.10) is proven. □

Any isomorphism preserves the orders of elements.

Theorem 3.2.3 *Let $\sigma: G \rightarrow H$ be an isomorphism and g be an element of G of finite order. Then $\mathrm{ord}\,(g) = \mathrm{ord}\,(\sigma(g))$.*

Proof By Proposition 3.2.1 $\sigma(e) = \epsilon$, where e is the identity element of G and ϵ is the identity of H. Suppose now $\mathrm{ord}\,(g) = n$. Then $g^n = e$. Let us now apply σ to both sides of this equation. We obtain $\sigma(g)^n = \sigma(g^n) = \sigma(e) = \epsilon$, from which we see that $\mathrm{ord}\,(\sigma(g)) \leq n$, i.e., $\mathrm{ord}\,(\sigma(g)) \leq \mathrm{ord}\,(g)$. Since σ^{-1} is also an isomorphism, which takes $\sigma(g)$ to g, we obtain $\mathrm{ord}\,(g) \leq \mathrm{ord}\,(\sigma(g))$. This proves the theorem. □

We now move on to one of the main theorems of this section. The theorem will, in particular, give us a tool for calculating orders of elements of cyclic groups which are also written multiplicatively.

Theorem 3.2.4 *Every cyclic group G of order n is isomorphic to $\mathbb{Z}_n$.*

Proof Since $G = <g>$ has cardinality n, by Lemma 3.2.1 we have ord $(g) = n$ and $G = \{g^0, g^1, g^2, \ldots, g^{n-1}\}$. We define $\sigma \colon \mathbb{Z}_n \to G$ by setting $\sigma(i) = g^i$. Then

$$\sigma(i \oplus j) = g^{i \oplus j} = g^{i+j} = g^i g^j = \sigma(i)\sigma(j),$$

where $\oplus$ is the addition modulo n. This checks (3.9) and proves that the mapping σ is indeed an isomorphism. $\square$

Now we can reap benefits of Theorem 3.2.4.

Corollary 3.2.2 *Let G be a multiplicative cyclic group and $G =< g >$, where g is an element of order n. Then*

$$ord\,(g^i) = \frac{n}{gcd(i, n)}. \tag{3.11}$$

Proof This now follows from the theorem we have just proved and Theorems 3.2.1 and 3.2.3. Indeed, the order of g^i in G must be the same as the order of i in $\mathbb{Z}_n$. $\square$

Exercises
1. Let $\sigma \colon G \to H$ be an isomorphism and g be an element of G. Prove that $\sigma(g^{-1}) = \sigma(g)^{-1}$.
2. Let $\mathbb{C}_n$ be the group of all complex numbers satisfying the equation $z^n = 1$. Prove that $\mathbb{C}_n \cong \mathbb{Z}_n$.
3. Prove that the multiplicative group of complex numbers $\mathbb{C}^*$ is isomorphic to the group of matrices

$$G = \left\{ \begin{bmatrix} a & -b \\ b & a \end{bmatrix} \mid a, b \in \mathbb{R} \right\}$$

under the usual multiplication of matrices.
4. Both groups $G_1 = \mathbb{Z}_{191}^*$ and $G_2 = \mathbb{Z}_{193}^*$ are cyclic (do not try to prove this). Which of these groups does contain elements of order 19? How many?
5. Knowing that 2 is a generating element for the cyclic group $\mathbb{Z}_{211}^*$, determine the order of 2^{150} in $\mathbb{Z}_{211}^*$.
6. 264 is a generator of $\mathbb{Z}_{271}^*$, i.e., the (multiplicative) order of 264 in $\mathbb{Z}_{271}$ is 270, as is shown by the following calculation:

```
gap> OrderMod(264,271);
270
```

Without GAP determine the multiplicative order of 264^{72} in $\mathbb{Z}_{271}^*$.

7°. Given that the multiplicative group $\mathbb{Z}_{14591}^*$ of field $\mathbb{Z}_{14591}$ is cyclic:

    (a) Give reasons why there exist elements of multiplicative order 1459 in $\mathbb{Z}_{14591}^*$ but not 1458;

    (b) Using GAP, find one element of order 1459 in this group.

8°. Are the groups $\mathbb{Z}_{12}^*$ and $\mathbb{Z}_5$ isomorphic? Give reasons.

9°. Let G be a group of even order. Consider the inversion function $\nu \colon G \to G$ given by $\nu(x) = x^{-1}$ for all $x \in G$.

    (a) Show that ν is a bijection.

    (b) Show that ν is an isomorphism if and only if G is abelian.

    (c) What are the elements of G with the property that $\nu(x) = x$ (i.e. "fixed" by ν)?

    (d) By pairing the elements not fixed by ν, show that if $|G|$ is even, then G has an even number of elements of order greater than 2, and at least one element of order 2.

    (e) What happens when the order of G is odd?

10°. Show that a group of order 6 is either commutative or isomorphic to S_3.

3.2.4 Subgroups

Definition 3.2.8 Let G be a group. We say that a subset H of G is a *subgroup of G* if it satisfies the following properties:

1. H contains the identity element of G.
2. H is closed under the group operation; i.e., if a and b belong to H, then ab also belongs to H.
3. H is closed under inverses; i.e., if a belongs to H then a^{-1} also belongs to H.

We write $H \leq G$ to denote that H is a subgroup of G. If G is any group, then $G \leq G$ and $\{e\} \leq G$. These are trivial examples. Let us consider a non-trivial one.

Firstly, we would like to introduce a construction which, given an element $g \in G$, will always give us a subgroup containing this element. Moreover, this subgroup will be the smallest subgroup with this property. This is familiar to us $< g >= \{g^i \mid i \in \mathbb{Z}\}$ which is the set of all powers of g.

Proposition 3.2.2 *Let G be a group, $g \in G$. Then $< g >$ is a subgroup of G. This is the smallest subgroup of G that contains g.*

Proof To decide whether or not $< g >$ is a subgroup, we must answer three questions:

- Does the identity e of G belong to $< g >$? The answer is YES, because $g^0 = e$ and $< g >$ consists of all powers of g.
- If $x, y \in < g >$, does xy also belong to $< g >$?

$x \in <g>$ means that $x = g^i$ for some integer i; similarly, $y = g^j$ for some integer j. Then $xy = g^i g^j = g^{i+j}$, which shows that xy is a power of g and therefore belongs to $<g>$.

- If $x \in <g>$, does x^{-1} also belong to $<g>$?
  $x \in <g>$ means that $x = g^i$ for some integer i; then $x^{-1} = g^{-i}$, i.e., x^{-1} is also a power of g and therefore belongs to $<g>$.

So $<g>$ is indeed a subgroup. It is the smallest subgroup containing $g \in G$ since any subgroup that contains g must also contain all powers of g. □

We will call $<g>$ the subgroup *generated* by $g \in G$.
Another example gives us a subgroup of a non-commutative group.

Example 3.2.11 Let us establish that the set of permutations $V = \{e, a, b, c\} \subset S_4$, where e is the identity permutation and

$$a = (1\ 2)(3\ 4), \quad b = (1\ 3)(2\ 4), \quad c = (1\ 4)(2\ 3),$$

is a subgroup of S_4. This statement makes the following claims:

1. The identity e belongs to V. This is obvious.
2. The product of two elements of V also belongs to V. We check:

$$ab = ba = c, \quad bc = cb = a, \quad ac = ca = b, \quad a^2 = b^2 = c^2 = e,$$

   and see that this is true.
3. V is closed under taking inverses. This is also true since $a^{-1} = a$, $b^{-1} = b$, $c^{-1} = c$.

We see that V is indeed a subgroup of S_4. This group is known as the *Klein four-group*.

Additional information about orders may be extracted using Lagrange's theorem. We will state and prove this theorem below, but first we need to introduce the cosets of a subgroup. Let G be a group, H a subgroup of G, and $g \in G$. The set $gH = \{gh \mid h \in H\}$ is called a *left coset* of H and the set $Hg = \{hg \mid h \in H\}$ is called a *right coset* of H.

Example 3.2.12 Let us consider $G = S_4$ and $H = V$, the Klein four-group which is a subgroup of S_4. Let $g = (12)$. Then the corresponding left coset consists of the permutations

$$(12)V = \{(12), (34), (1\ 4\ 2\ 3), (1\ 3\ 2\ 4)\}.$$

Indeed, $(12) = (12)\,e$, $(34) = (12)\,a$, $(1\ 4\ 2\ 3) = (12)\,b$, $(1\ 3\ 2\ 4) = (12)\,c$.

Proposition 3.2.3 *If H is finite, then $|gH| = |Hg| = |H|$ for any $g \in G$.*

Proof We need to prove that all elements gh are different, i.e., if $gh_1 = gh_2$, then $h_1 = h_2$. This is obvious since we can multiply both sides of the equation $gh_1 = gh_2$ by g^{-1} on the left. This proves $|gH| = |H|$. The proof of $|Hg| = |H|$ is similar. $\square$

We are now ready to state and prove Lagrange's theorem.

Theorem 3.2.5 (Lagrange's theorem) *Let G be a finite group, H a subgroup of G. Then the order of H is a divisor of the order of G.*

Proof Our proof relies on the decomposition of G into a disjoint union of left cosets of H, all of which have the same number of elements, namely $|H|$. Let us prove that such decomposition exists. All we need to show is that any two cosets are either disjoint or coincide.

Suppose the two cosets aH and bH have a nonzero intersection, i.e., $ah_1 = bh_2$ for some $h_1, h_2 \in H$. Then $b^{-1}a = h_2 h_1^{-1} \in H$. In this case any element $ah \in aH$ can be expressed as $b(b^{-1}a)h$, where $(b^{-1}a)h$ belongs to H. This proves $aH \subseteq bH$ and hence $aH = bH$ as both sets have the same cardinality. Hence these cosets must coincide. We obtain a partition of G into a number of disjoint cosets each of which has cardinality $|H|$. If k is the number of cosets in the partition, then in total G has $k|H|$ elements. This proves the theorem. $\square$

Corollary 3.2.3 *The order of an element g of a finite group G is a divisor of the order of G. In particular, $g^{|G|} = 1$.*

Proof Just note that by Lemma 3.2.1 the order of an element $g \in G$ equals the order of the subgroup $<g>$ of G. Then Lagrange's theorem implies that the order of g is a divisor of $|G|$. Let ord $(g) = m$, $|G| = n$, and $n = mk$ for some integer k. Then $g^n = g^{mk} = (g^m)^k = 1^k = 1$. $\square$

Example 3.2.13 Find the order of the element $2 \in \mathbb{Z}_{17}^*$.

A naive approach is to calculate *all* powers of 2, until one such power is found to be the identity. We have a more economical way to find the order: since $\mathbb{Z}_{17}^*$ has 16 elements, it is sufficient to calculate all the powers 2^i where i is a divisor of 16 until the result equals 1. We know that $2^{16} \mod 17 = 1$ and we need to calculate only $2^2 \mod 17$, $2^4 \mod 17$, and $2^8 \mod 17$. Our calculations will terminate when we find that $2^8 = 1 \mod 17$; the order of 2 in $\mathbb{Z}_{17}^*$ is therefore 8.

Example 3.2.14 Find out if 2 is a generator of the group $\mathbb{Z}_{13}^*$.

The question asks: Is the order of 2 in $\mathbb{Z}_{13}^*$ equal to 12? Now we see that it is not necessary to calculate each of the powers of 2, but only those powers 2^i where i is a divisor of 12 (which is $\phi(13)$ as the order of the (multiplicative) group $\mathbb{Z}_n^*$ is $\phi(n)$). So we calculate $2^2 \mod 13$, $2^3 \mod 13$, $2^4 \mod 13$, and $2^6 \mod 13$. If none of them turn out to be 1, then we can be sure that the order of 2 in $\mathbb{Z}_{13}^*$ is 12, and that 2 is a generator of the group $\mathbb{Z}_{13}^*$ (which is therefore cyclic). It turns out that $2^k \mod 13 \neq 1$ for $k = 2, 3, 4, 6$ and 2 therefore is indeed a generator of $\mathbb{Z}_{13}^*$.

Exercises

1. Let $SL_n(\mathbb{R})$ be the set of all real matrices with determinant 1. Prove that this is a subgroup of $GL_n(\mathbb{R})$.
2. Let m, n be positive integers and let m be a divisor of n. Prove that $\mathbb{C}_m$ is a subgroup of $\mathbb{C}_n$.
3. Prove that a cyclic group G of order n has exactly $\phi(n)$ generators, i.e., elements $g \in G$ such that $G = <g>$.
4. Let G be a finite group with $|G|$ even. Prove that it contains an element of order 2.
5. Prove that any finite subgroup of the multiplicative group $\mathbb{C}^*$ of the field $\mathbb{C}$ of complex numbers is cyclic.
6°. Is the group $\mathbb{Z}_8^*$ isomorphic to $\mathbb{Z}_3$? or to $\mathbb{Z}_4$? or to V_4?
7°. (a) Can a group of order 12 have a subgroup of order 4?
   (b) Does every group of order 12 have a subgroup of order 6?
   (c) Can a group of order 12 have a subgroup of order 9?
8°. Let A, B, C, and D be the (additive) cyclic subgroups of $\mathbb{Z}_{20}$ generated by 2, 4, 5, and 10. Find the orders of these subgroups, and the sizes of their pairwise intersections $A \cap B$, $A \cap C$, $A \cap D$, $B \cap C$, $B \cap D$ and $C \cap D$. Is there anything special that you can see happening every time?
9°. Find all the subgroups of A_4.
10°. Find two different right cosets Hx and Hy for the subgroup $H = \langle (1, 2, 3, 4) \rangle$ in the symmetric group S_4.

3.3   The Abelian Group of an Elliptic Curve

During the last 20 years, the theory of elliptic curves over finite fields has been found to be of great value to cryptography. As methods of factorisation of integers are getting better and computers are getting more powerful, to maintain the same level of security the prime numbers p and q in RSA have to be chosen bigger and bigger, which slows calculations down. The idea of using elliptic curves over finite fields belong to Neal Koblitz [11] and Victor Miller [16] who in 1985 independently proposed cryptosystems based on groups of points of elliptic curves. By now their security has been thoroughly tested and in 2009 the National Security Agency of the USA stated that "Elliptic Curve Cryptography provides greater security and more efficient performance than the first generation public key techniques (RSA and

Diffie–Hellman) now in use". Some researchers also see elliptic curves as the source of cryptosystems of the next generation. Certicom www.certicom.com is the first company that markets security products using elliptic curve cryptography.

3.3.1 Elliptic Curves. The Group of Points of an Elliptic Curve

Elliptic curves are not ellipses and do not look like them. They received their name due to their similarities with denominators of elliptic integrals that arise in calculations of the arc length of ellipses.

Definition 3.3.1 Let F be a field, and a, b be scalars in F such that the cubic $X^3 + aX + b$ has no multiple roots. An *elliptic curve* E over a field F is the set of solutions $(X, Y) \in F^2$ to the equation

$$Y^2 = X^3 + aX + b, \tag{3.12}$$

plus a "point at infinity" denoted by ∞.

When F is the field of real numbers the condition on the cubic can be expressed in terms of a and b. Let r_1, r_2, r_3 be the roots (maybe complex) of $X^3 + aX + b$, taken together with their multiplicities, such that

$$X^3 + aX + b = (X - r_1)(X - r_2)(X - r_3). \tag{3.13}$$

Then it is possible to check that

$$d = (r_1 - r_2)^2 (r_1 - r_3)^2 (r_2 - r_3)^2 = -(4a^3 + 27b^2). \tag{3.14}$$

This real number is called the *discriminant* of the cubic, and the cubic has no multiple roots if and only if this discriminant is nonzero, i.e.,

$$d = -(4a^3 + 27b^2) \neq 0. \tag{3.15}$$

This condition also guarantees the absence of multiple roots over an arbitrary field F.

Example 3.3.1 The equation

$$Y^2 = X^3 + 3X + 4$$

defines an elliptic curve over $\mathbb{Z}_7$ since the discriminant $d = -(4a^3 + 27b^2) = 6 \neq 0$. The point $(5, 2) \in \mathbb{Z}_7^2$ belongs to this curve since $2^2 \equiv 5^3 + 3 \cdot 5 + 4 \bmod 7$ with both sides being equal to 4.

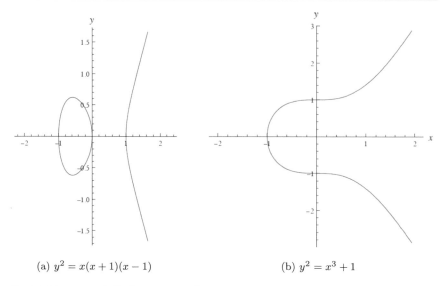

(a) $y^2 = x(x + 1)(x - 1)$ (b) $y^2 = x^3 + 1$

Fig. 3.1 Two types of elliptic curves over $\mathbb{R}$

When $F = \mathbb{R}$ is a field of reals, the graph of an elliptic curve can have two different forms depending on whether the cubic on the right-hand side of (3.12) has one or three real roots (see Fig. 3.1).

Jacobi[4] (1835) was the first to suggest using the group law on a cubic curve. In this section we will introduce the addition law for points of the elliptic curve (3.12), so that it will become an abelian group. We will do this first for elliptic curves over the familiar field of real numbers. These curves have the advantage that they can be represented graphically.

Definition 3.3.2 Let E be an elliptic curve over $\mathbb{R}$ and $P = (x, y) \in E$. Then we define $-P$ as the point $(x, -y)$, which is symmetric to P about x-axis. It is clear that $(x, -y) \in E$ whenever $(x, y) \in E$.

Definition 3.3.3 Let E be an elliptic curve over $\mathbb{R}$ and $P, Q \in E$.

(a) Suppose that $P \neq Q$ and that the line PQ is not parallel to the y-axis. Then PQ intersects E at the third point R (will be shown). Then we define $P + Q$ as $-R$ (see picture below).
(b) Suppose that $P = Q$ and the tangent line to the curve at P is not parallel to the y-axis. Further, suppose that the tangent line to the curve at P intersects E at the third point R. Then we define $2P = P + P = -R$. (If the tangent line has a "double tangency" at P, i.e., P is a point of inflection, then R is taken to be P.)

---

[4]**Carl Gustav Jacob Jacobi (1804–1851)** was a German mathematician, who made fundamental contributions to elliptic functions, dynamics, differential equations, and number theory.

(c) Suppose that $P \neq Q$ and PQ is parallel to the y-axis. Then we define $P + Q = \infty$.

(d) Suppose that $P = Q$ and the tangent line to the curve at P is parallel to the y-axis. Then we define $2P = P + P = \infty$.

(e) For every $P \in E$ (including $P = \infty$) we define $P + \infty = P$.

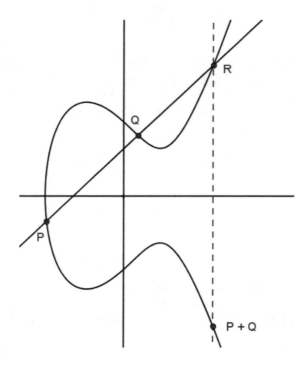

Theorem 3.3.1 *The elliptic curve E over $\mathbb{R}$ relative to this addition is an (infinite) abelian group. If $P = (x_1, y_1)$ and $Q = (x_2, y_2)$ are two points of E, then $P + Q = (x_3, y_3)$, where*

1. in case (a)

$$x_3 = \left(\frac{y_2 - y_1}{x_2 - x_1} \right)^2 - x_1 - x_2, \qquad (3.16)$$

$$y_3 = -y_1 + \left(\frac{y_2 - y_1}{x_2 - x_1} \right) (x_1 - x_3). \qquad (3.17)$$

2. in case (b)

$$x_3 = \left(\frac{3x_1^2 + a}{2y_1}\right)^2 - 2x_1,$$ (3.18)

$$y_3 = -y_1 + \left(\frac{3x_1^2 + a}{2y_1}\right)(x_1 - x_3).$$ (3.19)

Proof First, we have to prove that the addition is defined for every pair of (not necessary distinct) points of E. Suppose we are in case (a), which means $x_1 \neq x_2$. Then we have to show that the third point R on the line PQ exists. The equation of this line is $y = mx + c$, where $m = \frac{y_2 - y_1}{x_2 - x_1}$ and $c = y_1 - mx_1$. A point $(x, mx + c)$ of this line lies on the curve if and only if $(mx + c)^2 = x^3 + ax + b$ or

$$x^3 - m^2 x^2 + (a - 2mc)x - (c^2 - b) = 0.$$ (3.20)

Since we have already two real roots of this polynomial x_1 and x_2, we will have the third one as well. Dividing the left-hand side of (3.20) by $(x - x_1)(x - x_2)$ will give the factorisation

$$x^3 - m^2 x^2 + (a - 2mc)x - (c^2 - b) = (x - x_1)(x - x_2)(x - x_3),$$

where x_3 is this third root. Knowing x_1 and x_2, the easiest way to find x_3 is to notice that $x_1 + x_2 + x_3 = m^2$, and express the third root as $x_3 = m^2 - x_1 - x_2$. Since $m = \frac{y_2 - y_1}{x_2 - x_1}$, this is exactly (3.16). Now we can also calculate y_3 as follows

$$y_3 = -(mx_3 + c) = -mx_3 - (y_1 - mx_1) = -y_1 + m(x_1 - x_3)$$

(remember (x_3, y_3) represents $-R$, hence the minus). This will give us (3.17).

Case (b) is similar, except that m can now be calculated as the derivative dy/dx at P. Implicit differentiation of (3.12) gives us

$$2y\frac{dy}{dx} = 3x^2 + a,$$

or $dy/dx = (3x^2 + a)/2y$. Hence $m = (3x_1^2 + a)/2y_1$. (We note that $y_1 \neq 0$ in this case.) This implies (3.18) and (3.19). $\square$

It helps to visualise the point at infinity ∞ as located far up the y-axis. Then it becomes the third point of intersection of any vertical line with the curve. Then (c), (d), and (e) of Definition 3.3.2 will implement the same set of rules as (a) and (b), for the case when the point at infinity is involved.

We deduced formulae (3.16)–(3.19) for the real field $\mathbb{R}$ but they make sense for any field. Of course we have to remove references to parallel lines and interpret the addition rule in terms of coordinates only.

Definition 3.3.4 Let F be a field and let E be the set of pairs $(x, y) \in F^2$ satisfying (3.12) plus a special symbol ∞. Then for any $(x_1, y_1), (x_2, y_2) \in E$ we define:

(a) If $x_1 \neq x_2$, then $(x_1, y_1) + (x_2, y_2) = (x_3, y_3)$, where x_3, y_3 are defined by formulae (3.16) and (3.17).
(b) If $y_1 \neq 0$, then $(x_1, y_1) + (x_1, y_1) = (x_3, y_3)$, where x_3, y_3 are defined by formulae (3.18) and (3.19).
(c) $(x, y) + (x, -y) = \infty$ for all $(x, y) \in E$ (including the case $y = 0$).
(d) $(x, y) + \infty = \infty + (x, y) = (x, y)$ for all $(x, y) \in E$.
(e) $\infty + \infty = \infty$.

Theorem 3.3.2 *For any field F and for any elliptic curve*

$$Y^2 = X^3 + aX + b, \qquad a, b \in F,$$

the set E with the operation of addition defined in Definition 3.3.4 is an abelian group.

Proof It is easy to check that the identity element is ∞ and the inverse of $P = (x, y)$ is $-P = (x, -y)$. So two axioms of a group are obviously satisfied. It is not easy to prove that the addition, so defined, is associative. We omit this proof since it is a tedious calculation. □

Example 3.3.2 Suppose $F = \mathbb{Z}_{11}$ and the curve is given by the equation $Y^2 = X^3 + 7$. Then $P = (5, 0)$ and $Q = (3, 10)$ belong to the curve. We have

$$P + Q = (6, 5), \qquad 2Q = (3, 1), \qquad 2P = \infty.$$

Indeed, if $P + Q = (x_3, y_3)$, then $m = \frac{y_2 - y_1}{x_2 - x_1} = \frac{10}{-2} = \frac{-1}{-2} = \frac{1}{2} = 6$ and

$$x_3 = m^2 - x_1 - x_2 = 3 - 5 - 3 = 6,$$
$$y_3 = -y_1 + m(x_1 - x_3) = 0 + 6(-1) = 5.$$

So $P + Q = (6, 5)$. Calculating $2Q = (x_4, y_4)$, we get $m = \frac{3 \cdot 3^2 + 0}{9} = 3$ and

$$x_4 = m^2 - 2x_1 = 9 - 2 \cdot 3 = 3,$$
$$y_4 = -y_1 + m(x_1 - x_4) = -10 + 3 \cdot 0 = 1.$$

So $2Q = (3, 1)$. The last equation $2P = \infty$ follows straight from the definition (part (c) of Definition 3.3.4).

The calculations in the last exercise can be done with GAP. The program has to read the files `elliptic.gd` and `elliptic.gi` first (given in appendix). Then the command `EllipticCurveGroup(a,b,p);` calculates the points of the elliptic curve $Y^2 = aX + b \bmod p$. As you see below GAP uses the multiplicative notation for operations of elliptic curves:

```
Read("/.../elliptic.gd");
Read("/.../elliptic.gi");
gap> G:=EllipticCurveGroup(0,7,11);
EllipticCurveGroup(0,7,11)
gap> points:=AsList(G);
[ ( 2, 2 ), ( 2, 9 ), ( 3, 1 ), ( 3, 10 ), ( 4, 4 ), ( 4, 7 ), ( 5, 0 ),
  ( 6, 5 ), ( 6, 6 ), ( 7, 3 ), ( 7, 8 ), infinity ]
gap> P:=points[7];
( 5, 0 )
gap> Q:=points[4];
( 3, 10 )
gap> P*Q;
( 6, 5 )
gap> Q^2;
( 3, 1 )
gap> P^2;
infinity
```

Exercises

1. Which of the following equations define an elliptic curve over $\mathbb{Z}_{13}$:

$$Y^2 = X^3 + 4X + 11, \qquad Y^2 = X^3 + 6X + 11?$$

2. Prove that from Eq. (3.13) it follows that

$$r_1 + r_2 + r_3 = 0, \quad r_1 r_2 + r_1 r_3 + r_2 r_3 = a, \quad r_1 r_2 r_3 = -b. \tag{3.21}$$

3. Prove (3.14) using (3.21).
4. Consider elliptic curve E given by $Y^2 = X^3 + X - 1 \bmod 7$.
   (a) Check that $(1, 1)$, $(2, 3)$, $(3, 1)$, $(4, 2)$, $(6, 2)$ are the points on E;
   (b) Find another six points on this curve;
   (c) Calculate $-(2, 3)$, $2(4, 2)$, $(1, 1) + (3, 1)$;
   (d) Use GAP to show that E has 11 points in total.
5. Let $F = \mathbb{Z}_{13}$ and let the elliptic curve E be given by the equation $Y^2 = X^3 + 5X + 1$.
   (a) Using GAP list all the elements of the abelian group E of this elliptic curve. Hence find the order of the abelian group E.
   (b) Find (manually) the order of $P = (0, 1)$ in E. Is E cyclic?
6. Using GAP generate the elliptic curve $Y^2 = X^3 + 7X + 11$ in $\mathbb{Z}_{46301}$. Determine its order and check that it is cyclic.
7°. Let E be the abelian group of the elliptic curve $Y^2 = X^3 + 1234X + 17$ over $\mathbb{Z}_{346111}$. (GAP will take about a minute to generate this group using the command EllipticCurveGroup. Be patient. Do not try to display this group on your screen.)
   (a) Check that $Q = (283468, 291812)$ is a point on this curve.
   (b) Choose a random point on E, let it be called P (use command Random(G)).

(c) Use P as a reference point to input $Q = (283468, 291812)$ as a point of E. (See Sect. 9.4.2 to see how this can be done.)

(d) What is the order of Q in E?

(e) Calculate the order of G and decide whether Q is a generator of G.

8°. The elliptic curve E over $\mathbb{Z}_{11}$ is given by $Y^2 = X^3 + X + 4$. Let G be the group of points on this elliptic curve.

(a) Check that points $P = (0, 9)$ and $Q = (2, 5)$ are on the curve.

(b) Calculate the sum $P + Q$ of these two points in the group G.

(c) Bob wants to represent messages 8 and 9 as x-coordinates of points on E. In which case will he succeed?

9°. On the basis of the following GAP calculation

```
gap> G:=EllipticCurveGroup(11,2,17);
EllipticCurveGroup(11,2,17)
gap> Size(G);
13
```

proves that the group of points of the elliptic curve $Y^2 = X^3 + 11X + 2$ over $\mathbb{Z}_{17}$ is cyclic.

10°. Using trial and error, find the smallest integer x which is greater than or equal to 2010 and for which there exist a positive integer y such that (x, y) is a point on the elliptic curve $y^2 = x^3 + 111x + 328 \mod 3001$. (Use command RootMod(x,p) for extracting a square root of x mod p (it will give you one of them).)

3.3.2   Quadratic Residues and Hasse's Theorem

Definition 3.3.5 Let F be a finite field. An element $h \in F^*$ is called a *quadratic residue* if there exists another element $g \in F$ such that $g^2 = h$. Otherwise, it is called a *quadratic non-residue*.

Theorem 3.3.3 *If $F = \mathbb{Z}_p$ for $p > 2$, then exactly half of all nonzero elements of the field $\mathbb{Z}_p^*$ are quadratic residues.*

Proof Since p is odd, $p - 1$ is even. Then all nonzero elements of $\mathbb{Z}_p$ can be split into pairs,

$$\mathbb{Z}_p \setminus \{0\} = \{\pm 1, \pm 2, \ldots, \pm(p-1)/2\}.$$

Since $i^2 = (-i)^2$, each pair gives us only one quadratic residue, hence we cannot have more than $(p-1)/2$ quadratic residues. On the other hand, if we have $x^2 = y^2$, then $x^2 - y^2 = (x - y)(x + y) = 0$. Due to the absence of zero divisors, we have $x = \pm y$. Therefore we have exactly $(p-1)/2$ nonzero quadratic residues. □

Example 3.3.3 In $\mathbb{Z}_7$ we have $1^2 = 6^2 = 1$, $2^2 = 5^2 = 4$, and $3^2 = 4^2 = 2$ so the set of nonzero quadratic residues is $\{1, 2, 4\}$.

Determining whether or not a particular element a of $\mathbb{Z}_p^*$ is a quadratic residue or non-residue has a great importance for applications of elliptic curves. Even more important are the algorithms for finding a square root of a, if it exists. The first question can be efficiently solved by using the following criterion.

Theorem 3.3.4 (Euler's criterion) *Let p be an odd prime and $a \in \mathbb{Z}_p^*$. Then*

$$a^{\frac{p-1}{2}} = \begin{cases} 1 & \text{if } a \text{ is a quadratic residue} \\ -1 & \text{if } a \text{ is a quadratic non-residue} \end{cases}$$

Proof Since by Fermat's little theorem $\left(a^{\frac{p-1}{2}}\right)^2 = a^{p-1} = 1$, we conclude that $a^{\frac{p-1}{2}} \in \{-1, 1\}$. If a is a quadratic residue with $b^2 = a$, then by Fermat's little theorem $a^{\frac{p-1}{2}} = b^{p-1} = 1$. For the converse see Exercise 5. □

The importance of this criterion is that we can use the Square and Multiply algorithm to raise a to the power of $\frac{p-1}{2}$ and thus check if a is a quadratic residue or not. By Theorem 2.3.2 the Square and Multiply algorithm has linear complexity, hence this is an easy problem to solve. It is somewhat more difficult to find a square root of an element of $\mathbb{Z}_p$, given that it is a quadratic residue. Reasonably fast polynomial time algorithms exist—most notably Tonelli–Shanks algorithm[5]— however it is not fully deterministic as it requires finding at least one quadratic non-residue. This necessary quadratic non-residue is easy to find using trial and error method with the average expected number of trials being only 2. No fully deterministic polynomial time algorithm is known.

GAP uses Tonelli–Shanks algorithm to extract square roots in finite fields. For example, the following calculation shows that 12 is a quadratic non-residue in $\mathbb{Z}_{103}$ and 13 is a quadratic residue in this field:

```
gap> RootMod(12,103);
fail
gap> RootMod(13,103);
61
```

Let p be a large prime. Let us try to estimate the number of points on the elliptic curve $Y^2 = f(X)$ over $\mathbb{Z}_p$, where $f(X)$ is a cubic. For a solution with the first coordinate X to exist, it is necessary and sufficient that $f(X)$ is a quadratic residue. It is plausible to suggest that $f(X)$ will be a quadratic residue for approximately half of all points $X \in \mathbb{Z}_p$. On the other hand, if $f(X)$ is a nonzero quadratic residue, then the equation $Y^2 = f(X)$ will have two solutions with X as the first coordinate. Hence it is reasonable to expect that the number of points on the curve will be approximately

---

[5]Daniel Shanks. Five Number Theoretic Algorithms. Proceedings of the Second Manitoba Conference on Numerical Mathematics, pp. 51–70, 1973.

$\frac{p}{2} \cdot p + 1 = p + 1$ (p plus the point at infinity). Hasse[6] (1930) gave the exact bound, which we give here without a proof:

Theorem 3.3.5 (Hasse's theorem) *Suppose E is an elliptic curve over $\mathbb{Z}_p$ and let N be the number of points on E. Then*

$$p + 1 - 2\sqrt{p} \leq N \leq p + 1 + 2\sqrt{p}. \tag{3.22}$$

It was also shown that for any p and N satisfying (3.22) there exists a curve over $\mathbb{Z}_p$ having exactly N points.

As we have already seen, cryptography works with large objects with which it is difficult to calculate. Large elliptic curves are of great interest to it. Hasse's theorem says that to have a large curve we need a large field. This can be achieved in two ways. First is to have a large prime p. The second is to keep p small but to try to build a new large field F, as extension of $\mathbb{Z}_p$. As we will see later, for every n there is a field containing exactly $q = p^n$ elements. There is a more general version of Theorem 3.3.5 which also often goes by the name of "Hasse's Theorem".

Theorem 3.3.6 *Suppose E is an elliptic curve over a field F containing q elements and let N be the number of points on E. Then*

$$q + 1 - 2\sqrt{q} \leq N \leq q + 1 + 2\sqrt{q}. \tag{3.23}$$

For cryptographic purposes, it is not uncommon to use elliptic curves over fields of 2^{150} or more elements. It is worth noting that for $n \geq 20$, it is infeasible to list all points on the elliptic curve over a field of 2^n elements.

Despite the fact that each curve has quite a few points, there does not exist a deterministic algorithm which will produce, in less than exponential time, a point on a given curve $Y^2 = f(X)$. In particular, it is difficult to find X such that $f(X)$ is a quadratic residue. In practice, fast probabilistic methods are used.

Example 3.3.4 Let $F = \mathbb{Z}_5$. Consider the curve $Y^2 = X^3 + 2$. Let us list all the points on this curve and calculate the addition table for the corresponding abelian group E. The quadratic residues of $\mathbb{Z}_5$ are $1 = 1^2 = 4^2$ and $4 = 2^2 = 3^2$. We shall

---

[6]**Helmut Hasse (1898–1979)** was a German mathematician who worked on algebraic number theory, and was known for many fundamental contributions. The period, when Hasse's most important discoveries were made, was a very difficult time for German mathematics. When the Nazis came to power in 1933, a great number of mathematicians with Jewish ancestry were forced to resign and many of them left the country. Hasse did not compromise his mathematics for political reasons, he struggled against Nazi functionaries who tried (sometimes successfully) to subvert mathematics to political doctrine. On the other hand, he made no secret of his strong nationalistic views and his approval of many of Hitler's policies.

list all possibilities for x and see what y can in each case be:

$$x = 0 \implies y^2 = 2, \text{ no solution}$$
$$x = 1 \implies y^2 = 3, \text{ no solution}$$
$$x = 2 \implies y^2 = 0 \implies y = 0$$
$$x = 3 \implies y^2 = 4 \implies y = 2, 3$$
$$x = 4 \implies y^2 = 1 \implies y = 1, 4$$

Hence we can list all the points of E. We have $E = \{\infty, (2, 0), (3, 2), (3, 3), (4, 1), (4, 4)\}$. Let us calculate the addition table.

$+$	∞	$(2, 0)$	$(3, 2)$	$(3, 3)$	$(4, 1)$	$(4, 4)$
∞	∞	$(2, 0)$	$(3, 2)$	$(3, 3)$	$(4, 1)$	$(4, 4)$
$(2, 0)$	$(2, 0)$	∞	$(4, 1)$	$(4, 4)$		
$(3, 2)$	$(3, 2)$	$(4, 1)$	$(3, 3)$	∞		
$(3, 3)$	$(3, 3)$	$(4, 4)$	∞	$(3, 2)$		
$(4, 1)$	$(4, 1)$					∞
$(4, 4)$	$(4, 4)$				∞	$(3,2)$

We see that $2 \cdot (2, 0) = \infty$, hence ord $((2, 0)) = 2$. Also $3 \cdot (3, 2) = 3 \cdot (3, 3) = \infty$, while $2 \cdot (3, 2) \neq \infty$ and $2 \cdot (3, 3) \neq \infty$, hence ord $((3, 2)) = $ ord $((3, 3)) = 3$.

Exercises

1. Fill the remaining empty slots of the table above and find the orders of $(4, 1)$ and $(4, 4)$.
2. Find all quadratic residues of the field $\mathbb{Z}_{17}$.
3. Use Hasse's theorem to estimate the number of points on an elliptic curve over $\mathbb{Z}_{2011}$.
4. Prove that

   (a) the product of two quadratic residues and the inverse of a quadratic residues are quadratic residues;
   (b) the product of a quadratic residue and a quadratic non-residue is a quadratic non-residue;
   (c) the product of two quadratic non-residues is a quadratic residue.
5. Prove that, if a is a quadratic non-residue, then $a^{\frac{p-1}{2}} = -1$. (Use Wilson's theorem, which is Exercise 7, Sect. 1.4.)
6. Use trial and error method to find a quadratic non-residue in $\mathbb{Z}_p$, where

$$p = 3593340859686228310419601885980436610653887726959079837.$$

7°. For each of the two elements 96 and 97 of the field $\mathbb{Z}_{331}$, determine whether or not it is a quadratic residue or a quadratic non-residue. (Use the Euler's criterion formulated in Theorem 3.3.4.)

8°. Is there an elliptic curve E over $\mathbb{Z}_5$ such that
   (a) E contains exactly 11 points (including the point at infinity ∞)?
   (b) E contains exactly 10 points (including the point at infinity ∞)? If the answer is positive, find such a curve and list all of its points, if it is negative, prove it.

9°. Given an elliptic curve E over the finite field $\mathbb{Z}_{101}$. Find the number of points on this curve if it is known that there is a point of order 116 on this curve.

3.3.3 Calculating Large Multiples Efficiently

For calculating multiples efficiently the same rules apply as to calculating powers. Below we give a complete analogue of the Square and Multiply algorithm.

Theorem 3.3.7 *Given $P \in E$, for any positive integer N it is possible to calculate $N \cdot P$ using no more than $2\lfloor \log_2 N \rfloor$ additions.*

Proof We assume N is already written in binary (otherwise we need another $\lfloor \log_2 N \rfloor$ divisions to convert N into binary representation):

$$N = 2^{m_0} + 2^{m_1} + \cdots + 2^{m_s},$$

where $m_0 = \lfloor \log_2 N \rfloor$ and $m_0 > m_1 > \cdots > m_s$. We can find all multiples $2^{m_i} \cdot P$, $i = 1, 2, \ldots, s$ by successive doubling in m_0 additions:

$$2^1 \cdot P = P + P,$$
$$2^2 \cdot P = 2^1 \cdot P + 2^1 \cdot P,$$
$$\cdots$$
$$2^{m_0} \cdot P = 2^{m_0-1} \cdot P + 2^{m_0-1} \cdot P$$

Now to calculate

$$N \cdot P = (2^{m_0} + 2^{m_1} + \cdots + 2^{m_s}) \cdot P = 2^{m_0} \cdot P + 2^{m_1} \cdot P + \cdots + 2^{m_s} \cdot P$$

we need no more than m_0 extra additions. In total no more than $2m_0 = 2\lfloor \log_2 N \rfloor$. Since $n = \lfloor \log_2 N \rfloor$ is the length of the input, the complexity function $f(n)$ is at most linear in n or $f(n) = \Theta(n)$. $\square$

The algorithm presented here can be called *Double and Add algorithm*. It has linear complexity. Up to an isomorphism, this is the same algorithm as Square and Multiply.

We see that it is an easy task to calculate multiples of any point P on an elliptic curve. That is it is easy to calculate $N \cdot P$ given an integer N and a point P on the curve. However there are no easy ways to calculate N given $N \cdot P$ and P. So the function $N \mapsto N \cdot P$ is a one way function and it has been recognised by now that it has a great significance for cryptography. This branch of cryptography is called *Elliptic Curve Cryptography (ECC)*. It was proposed in 1985 by Victor Miller and Neil Koblitz as a mechanism for implementing public key cryptography alternative to RSA. We will show one of the cryptosystems of ECC in the next section.

Exercises

1. If GAP uses the Double and Add algorithm to compute large multiples of points of elliptic curves, how many additions will GAP perform when calculating $1729 \cdot P$ for some point P?

3.4   Applications to Cryptography

3.4.1   Encoding Plaintext

This is not so straightforward as for RSA to encode a message as a point of the given elliptic curve. To illustrate the difficulties we may face here, it is enough to say that there is no known polynomial time algorithm for finding a single point on that curve. This problem has not been fully resolved yet. However there are fast probabilistic methods which work for most messages, however for small proportion of them these methods fail to produce a point. The probability of such an unwanted event can be managed and made arbitrarily small.

The following method was suggested by Koblitz (1985). Suppose that we have an elliptic curve over $\mathbb{Z}_p$ given by the equation $Y^2 = X^3 + aX + b$. We may assume that our message is already represented by a number m. We will try to embed this number in the x-coordinate of the point $P = (X, Y) \in E$. Of course, we would like to make $X = m$ but this is not always possible since $f(m) = m^3 + am + b$ is a quadratic residue only in about 50% of cases. A failure rate of $1/2$ is, of course, unacceptable.

Suppose that a failure rate of 2^{-k} is acceptable for some sufficiently large positive integer k. Then, for each of the numbers $m_i = km + i$, where $0 \le i < k$, we check if $f(m_i)$ is a quadratic residue. If $f(m_i)$ is a quadratic residue, then we can find a point $P = (X, Y) \in E$, for which $X = m_i$ (using, for example, the GAP command RootMod($f(m_i)$,p); to find a matching Y). This will be the plaintext. The message m can always be recovered as $m = \lfloor X/k \rfloor$. We should choose p sufficiently large so that $(m + 1)k < p$ for any message m. Since we now have k numbers m_i that represent the message, the probability that for none of them $f(m_i)$ is a quadratic residue will be less than 2^{-k}.

If $k = 10$, then this means that we can add another junk digit to m (it will be placed in the rightmost position) in order to get a point on the curve. This junk digit will be discarded at the receiving end. If $k = 100$, then we can add two junk digits. This is already sufficient for practical purposes as 2^{-100} is very small.

Suppose we have chosen prime number $p = 17487707$ and we would like to represent the message "HAPPY NEW YEAR" using points of elliptic curve $Y^2 = X^3 + 123X + 456 \mod p$. Let us encode letters as follows: A = 11, B = 12, ..., Z = 36 and suppose we view the failure rate 2^{-10} as acceptable. Since our chosen prime has 8 digits, we can make messages 6 digits long and still have a possibility to add one junk digit. This means we have split our message into blocks with three letters in each:

$$[\text{HAP, PYN, EWY, EAR}] \rightarrow [x_1, x_2, x_3, x_4] = [181126, 263524, 153335, 151128].$$

The message will be encoded as

$$[(1811261, 11301481), (2635241, 14638357), (1533350, 13487258), (1511282, 9580769)].$$

Calculating this, we initially added a junk digit zero to every x_i and tried to find a matching y_i. If we failed, we would change the last digit to 1, and, in the case of another failure to 2, etc. We see that x_3 was a quadratic residue straightaway, x_1 and x_2 needed the second attempt with the last digit 1 and x_4 needed three attempts with the last digits 0,1,2.

Exercises

1. Use trial and error method to find a quadratic residue r and a quadratic non-residue n in $\mathbb{Z}_p$, where

   $$p = 359334085968622831041960188598043661065388726959079837.$$

   Find an element $s \in \mathbb{Z}_p$ such that $r = s^2$ in $\mathbb{Z}_p$.
2. Represent message **CHRISTMAS** using the points of elliptic curve $Y^2 = X^3 + 123X + 456 \mod 17487707$. (Note that you do not have to generate the whole group of points for this curve, which would be time consuming.)
3°. Encode message **CRYPTO** as a sequence of two points of elliptic curve $Y^2 = X^3 + 111x + 1111 \mod 65231563$. Use the method with the failure rate $1/2^{10}$. Use the following numerical encodings for the letters of English alphabet:

   $$A = 11, \ B = 12, \ \ldots, \ Z = 36.$$

   (Note that it is not necessary to generate the full group of points for this question which may be beyond GAP.)

3.4.2 Additive Diffie–Hellman Key Exchange and the ElGamal Cryptosystem

The exponential Diffie–Hellman key exchange can be easily adapted for elliptic curves. Suppose that E is a publicly known elliptic curve over $\mathbb{Z}_p$. Alice and Bob, through an open channel, agree upon a point $Q \in E$. Alice chooses a secret positive integer k_A (her private multiplier) and sends $k_A \cdot Q$ to Bob. Bob chooses a secret positive integer k_B (his private multiplier) and sends $k_B \cdot Q$ to Alice. Bob then calculates $P = k_B \cdot (k_A \cdot Q) = k_A k_B \cdot Q$, and Alice calculates $P = k_A \cdot (k_B \cdot Q) = k_A k_B \cdot Q$. They now both know the point P which they can use as the key for a conventional secret key cryptosystem. An eavesdropper wanting to spy on Alice, and Bob would face the following task called the Diffie–Hellman problem for elliptic curves.

Diffie–Hellman Problem Given Q, $k_A \cdot Q$, $k_B \cdot Q$ (but not k_A or k_B), find $P = k_A k_B \cdot Q$. No polynomial time algorithms are known for this problem.

Elgamal[7] (1985) modified the Diffie–Hellman idea to adapt it for message transmission (see [2], p. 287). It starts as above with Alice and Bob publicly announcing Q and exchanging $k_B \cdot Q$ and $k_A \cdot Q$, which play the role of their public keys. Alternatively you may think that there is a public domain run by a trusted authority where Q is stored and that any new entrant, say Cathy, chooses her private multiplier k_C and publishes her public key $k_C \cdot Q$ there.

Suppose that messages can be interpreted as points of an elliptic curve E in an agreed upon way, and that Bob wants to send Alice a message $M \in E$. He chooses a secret random integer s (for each message a distinct random number should be generated)), reads Alice's public key $k_A \cdot Q$ from the public domain and sends Alice the pair of points $C_1 = s \cdot Q$ and $C_2 = M + s \cdot (k_A \cdot Q)$. On the receiving end, Alice, using her private multiplier k_A, can calculate the plaintext as $M = C_2 - k_A \cdot C_1$. Nobody else can do this without knowing Alice's private multiplier k_A.

Exercises

1. Alice and Bob are setting up the ElGamal elliptic curve cryptosystem for private communication. They've chosen a point $Q = (88134, 77186)$ on the elliptic curve E given by $Y^2 = X^3 + 12345$ over $\mathbb{Z}_{95701}$. They've chosen their private multipliers $k_A = 373$ and $k_B = 5191$ and published the points $Q_A = (27015, 92968)$ and $Q_B = (55035, 17248)$, respectively. They agreed

---

[7]**Taher Elgamal** (born 18 August 1955) is an Egyptian-born American cryptographer. In 1985, Elgamal published a paper titled "A Public Key Cryptosystem and a Signature Scheme Based on Discrete Logarithms" in which he proposed the design of the ElGamal discrete logarithm cryptosystem and of the ElGamal signature scheme. He is also recognised as the "father of SSL", which is a protocol for transmitting private documents via the Internet that is now the industry standard for Internet security and ecommerce.

to cut the messages into two-letter segments and encode the letters as A = 11, B = 12, ..., Z = 36, space = 41, ' = 42, . = 43, , = 44, and ? = 45. They also agreed that, for each point (x, y), only the first four digits of x are meaningful (so that they can add additional junk digits to their messages, if needed, to obtain a point on the curve).

(a) Alice got the message:

```
[ [ ( 87720, 6007  ), ( 59870, 82101 ) ], [ ( 34994, 7432  ), ( 36333, 86213 ) ],
  [ ( 50702, 2643  ), ( 33440, 56603 ) ], [ ( 34778, 12017 ), ( 81577, 501   ) ],
  [ ( 93385, 52237 ), ( 38536, 21346 ) ], [ ( 63482, 12110 ), ( 70599, 87781 ) ],
  [ ( 16312, 46508 ), ( 62735, 69061 ) ], [ ( 64937, 58445 ), ( 41541, 36985 ) ],
  [ ( 40290, 45534 ), ( 11077, 77207 ) ], [ ( 64001, 62429 ), ( 32755, 18973 ) ],
  [ ( 81332, 47042 ), ( 35413, 9688  ) ], [ ( 5345, 68939  ), ( 475, 53184    ) ] ] ]
```

Help her to decrypt this message.

(b) She suspects that the sender of the message was Bob. Show how Alice may reply to this message and how Bob will decrypt it.

2°. Consider the elliptic curve variant of the Diffie–Hellman key exchange protocol with the elliptic curve $E : y2 = x^3 + 3x + 4 \mod 17$ and the point $P = (1, 5)$. Let Alice's choice of integer be $k_a = 3$ and let Bob's choice be $k_b = 4$. Finish the protocol and show your steps.

Fields

4

Oh field of battle, field of dying,
Who sank on you with glory here?

Ruslan and Liudmila. Alexander Pushkin (1799–1837)

4.1   Introduction to Fields

In Sect. 1.4 we defined a field and proved that, for any prime p, the set of integers $\mathbb{Z}_p = \{0, 1, 2, \ldots, p - 1\}$ with the operations:

$$a \oplus b = a + b \bmod p,$$
$$a \odot b = ab \bmod p$$

is a field. This field has cardinality p. So far, these are the only finite fields we have learned. In this chapter we prove that a finite field must have cardinality p^n for some prime p and positive integer n, i.e., its cardinality may be only a power of a prime. Such fields exist and we lay the grounds for the construction of such fields in Chap. 5. In this chapter we also prove a very important result that the multiplicative group of any finite field is cyclic. This makes it possible to define discrete logarithms—special functions on finite fields that are difficult to compute, and widely used in cryptography. We show that the Elgamal cryptosystem can also be based on the multiplicative group of a large finite field.

© Springer Nature Switzerland AG 2020
A. Slinko, *Algebra for Applications*, Springer Undergraduate Mathematics Series,
https://doi.org/10.1007/978-3-030-44074-9_4

4.1.1 Examples and Elementary Properties of Fields

We recap that an algebraic system consisting of a set F set equipped with two operations addition $+$ and multiplication $\cdot$ is called a *field* if the following nine axioms are satisfied:

F1. The addition is commutative, $a + b = b + a$, for all $a, b \in F$.
F2. The addition is associative, $a + (b + c) = (a + b) + c$, for all $a, b, c \in F$.
F3. There exists a unique element 0 such that $a + 0 = 0 + a = a$, for all $a \in F$.
F4. For every element $a \in F$ there exists a unique element $-a$ such that $a + (-a) = (-a) + a = 0$, for all $a \in F$.
F5. The multiplication is commutative, $a \cdot b = b \cdot a$, for all $a, b \in F$.
F6. The multiplication is associative, $a \cdot (b \cdot c) = (a \cdot b) \cdot c$, for all $a, b, c \in F$.
F7. There exists a unique element $1 \in F$ such that $a \cdot 1 = 1 \cdot a = a$, for all nonzero $a \in F$.
F8. The distributive law holds, that is, $a \cdot (b + c) = a \cdot b + a \cdot c$, for all $a, b, c \in F$.
F9. For every nonzero $a \in F$ there is a unique element $a^{-1} \in F$ such that $a \cdot a^{-1} = a^{-1} \cdot a = 1$.

Here and later, for any field F, the set of nonzero elements will be denoted as F^*. We note that axioms F1–F4 mean that F is an abelian group relative to the addition and axioms F5–F7 mean that F^* is also an abelian group but relative to the multiplication. Axioms F1–F8 mean that F is a commutative ring relative to the two operations. And only the last axiom is specific for fields only.

The examples of infinite fields are numerous. The most important are the fields of rational numbers $\mathbb{Q}$, real numbers $\mathbb{R}$ and complex numbers $\mathbb{C}$.

Definition 4.1.1 Let F be a field and G be a subset of F. Sometimes G is also a field relative to the same operations of addition and multiplication as in F. If so, we say that G is a *subfield* of F.

Example 4.1.1 $\mathbb{Q}$ is a subfield of $\mathbb{R}$, and $\mathbb{R}$ is a subfield of $\mathbb{C}$.

Three basic properties of fields are stated in the following theorem. The second one is called *absence of divisors of zero* and the third *solvability of linear equations*. We saw these properties hold for $\mathbb{Z}_p$ but now we would like to prove them for arbitrary fields.

Theorem 4.1.1 *Let F be a field. Then for any two elements $a, b \in F$*

(i) $a0 = 0$ *for all $a \in F$;*
(ii) $ab = 0$ *if and only if $a = 0$ or $b = 0$ (or both);*
(iii) *if $a \neq 0$, the equation $ax = b$ has a unique solution $x = a^{-1}b$ in F.*

Proof (i) Firstly we need to prove that $0 \cdot a = 0$ for all $a \in F$. We have

$$0 \cdot a \overset{F3}{=} (0+0) \cdot a \overset{F8}{=} 0 \cdot a + 0 \cdot a,$$

Adding $-(0 \cdot a)$ to both sides we get

$$0 = -(0 \cdot a) + (0 \cdot a + 0 \cdot a) \overset{F2}{=} (-(0 \cdot a) + 0 \cdot a) + 0 \cdot a \overset{F4}{=} 0 + 0 \cdot a \overset{F3}{=} 0 \cdot a.$$

Hence $0 \cdot a = 0$. This proves also the "if" part of (ii).

(ii) Suppose now that $ab = 0$ and either $a \neq 0$ or $b \neq 0$. Without loss of generality we assume the former. Then by F9 we know that a^{-1} exist. We have now

$$0 = a^{-1} \cdot 0 = a^{-1}(ab) \overset{F6}{=} (a^{-1}a)b \overset{F9}{=} 1 \cdot b \overset{F7}{=} b.$$

So $b = 0$ which proves the "only if" part of (ii).

Let us prove (iii). Since $a \neq 0$, we know a^{-1} exists. Suppose that the equation $ax = b$ has a solution. Then multiplying both sides by a^{-1} we get $a^{-1}(ax) = a^{-1}b$. As in the proof of (i) we calculate that the left-hand side of this equation is x. So $x = a^{-1}b$. It is also easy to check that $x = a^{-1}b$ is indeed a solution of $ax = b$. $\square$

A very important technique is enlarging a given field to obtain a larger field with some given property. After learning a few basic facts about polynomials we discuss how to make such extensions.

Exercises

1. Prove that the set of all non-negative rational numbers $\mathbb{Q}^+$ is NOT a field.
2. Prove that the set of all integers $\mathbb{Z}$ is NOT a field.
3. Prove that the set of all real numbers $\mathbb{Q}(\sqrt{2})$ of the form $x + y\sqrt{2}$, where x and y are in $\mathbb{Q}$ is a field.
4. Consider $\mathbb{Q}(\sqrt{3})$, which is defined similarly to the field from the previous exercise. Find the inverse element for $2 - \sqrt{3}$ and solve the equation

$$(2 - \sqrt{3})x = 1 + \sqrt{3}.$$

5. Solve the system of linear equations

$$3x + y + 4z = 1$$
$$x + 2y + z = 2$$
$$4x + y + 4z = 4$$

with coefficients in Z_5.

6°. Prove that the following matrices over $\mathbb{Z}_2$ relative to usual addition and multiplication of matrices

$$\begin{bmatrix} 0 & 0 \\ 0 & 0 \end{bmatrix}, \quad \begin{bmatrix} 1 & 0 \\ 0 & 1 \end{bmatrix}, \quad \begin{bmatrix} 0 & 1 \\ 1 & 1 \end{bmatrix}, \quad \begin{bmatrix} 1 & 1 \\ 1 & 0 \end{bmatrix}.$$

form a field consisting of 4 elements. Show that its multiplicative group is cyclic.

4.1.2 Vector Spaces

The reader familiar with linear algebra may well skip this section.

Definition 4.1.2 Suppose that the following objects are given:

VS1. a field F of scalars;
VS2. a set V of objects, called vectors;
VS3. a rule (or operation) called vector addition, which associates with each pair of
vectors $\mathbf{u}$, $\mathbf{v}$ in V a vector $\mathbf{u} + \mathbf{v}$ in V, called the sum of $\mathbf{u}$ and $\mathbf{v}$, in such a way
that

   (a) Addition is commutative, $\mathbf{u} + \mathbf{v} = \mathbf{v} + \mathbf{u}$;
   (b) Addition is associative, $\mathbf{u} + (\mathbf{v} + \mathbf{w}) = (\mathbf{u} + \mathbf{v}) + \mathbf{w}$;
   (c) There exists a unique vector $\mathbf{0}$ in V, called the zero vector, such that ·
        $\mathbf{u} + \mathbf{0} = \mathbf{u}$ for all $\mathbf{u}$ in V;
   (d) For each vector $\mathbf{u}$ in V there is a unique vector $-\mathbf{u}$ in V such that $\mathbf{u} +$
        $(-\mathbf{u}) = \mathbf{0}$;

VS4. a rule (or operation) called scalar multiplication, which associates with each
scalar a in F and vector $\mathbf{u}$ in V a vector $a\mathbf{u}$ in V, called the product of a and
$\mathbf{u}$, in such a way that

   (a) $1\mathbf{u} = \mathbf{u}$ for all $\mathbf{u}$ in V;
   (b) $a_1(a_2\mathbf{u}) = (a_1a_2)\mathbf{u}$;
   (c) $a(\mathbf{u} + \mathbf{v}) = a\mathbf{u} + a\mathbf{v}$;
   (d) $(a_1 + a_2)\mathbf{u} = a_1\mathbf{u} + a_2\mathbf{u}$.

Then we call V a *vector space over the field F*.

Example 4.1.2 Where F is a field, F^n is the set of n-tuples whose entries are scalars
from F. It is a vector space over F relative to the following addition and scalar

multiplication:

$$
\begin{bmatrix} a_1 \\ a_2 \\ \vdots \\ a_n \end{bmatrix} + \begin{bmatrix} b_1 \\ b_2 \\ \vdots \\ b_n \end{bmatrix} = \begin{bmatrix} a_1 + b_1 \\ a_2 + b_2 \\ \vdots \\ a_n + b_n \end{bmatrix}, \qquad k \begin{bmatrix} a_1 \\ a_2 \\ \vdots \\ a_n \end{bmatrix} = \begin{bmatrix} ka_1 \\ ka_2 \\ \vdots \\ ka_n \end{bmatrix}.
$$

In particular, $\mathbb{R}^n$, $\mathbb{C}^n$ and $\mathbb{Z}_p^n$ are vector spaces over the fields $\mathbb{R}$, $\mathbb{C}$ and $\mathbb{Z}_p$, respectively.

Example 4.1.3 Let $F_{m \times n}$ be the set of $m \times n$ matrices whose entries are scalars from a field F. It is a vector space over F relative to matrix addition and scalar multiplication. The sets of all $m \times n$ matrices $\mathbb{R}_{m \times n}$, $\mathbb{C}_{m \times n}$ and $(\mathbb{Z}_p)_{m \times n}$ with entries from $\mathbb{R}$, $\mathbb{C}$ and $\mathbb{Z}_p$ are vector spaces over the fields $\mathbb{R}$, $\mathbb{C}$ and $\mathbb{Z}_p$, respectively.

Example 4.1.4 Let F be a field, and $F_n[x]$ be the set of all polynomials of degree at most n whose coefficients are scalars from F. It is a vector space over F relative to the addition of polynomials and scalar multiplication. The sets of all polynomials $\mathbb{R}_n[x]$, $\mathbb{C}_n[x]$ and $(\mathbb{Z}_p)_n[x]$ of degree at most n with coefficients from $\mathbb{R}$, $\mathbb{C}$ and $\mathbb{Z}_p$ are vector spaces over the fields $\mathbb{R}$, $\mathbb{C}$ and $\mathbb{Z}_p$, respectively.

Example 4.1.5 Let F be a field, $F[x]$ be the set of all polynomials (without restriction on their degrees), whose coefficients are scalars from F. It is a vector space over F relative to addition of polynomials and scalar multiplication. The sets of all polynomials $\mathbb{R}[x]$, $\mathbb{C}[x]$ and $\mathbb{Z}_p[x]$ with coefficients from $\mathbb{R}$, $\mathbb{C}$ and $\mathbb{Z}_p$ are vector spaces over the fields $\mathbb{R}$, $\mathbb{C}$ and $\mathbb{Z}_p$, respectively.

The most interesting example for us is given in the following theorem.

Theorem 4.1.2 *Let F be a subfield of a field G. Then G is a vector space over F relative to the following operations. The addition of elements of G is the operation of addition in the field G. The scalar multiplication of elements of G by elements of F is performed as multiplication in the field G.*

Proof Check that the vector space axioms for G all follow from the field axioms. □

Example 4.1.6 The field of complex numbers $\mathbb{C}$ is a vector space over the reals $\mathbb{R}$ which is a subfield of $\mathbb{C}$. Both $\mathbb{C}$ and $\mathbb{R}$ are vector spaces over the rationals $\mathbb{Q}$.

The axioms of a vector space have many useful consequences. The two most important once are as follows:

Proposition 4.1.1 *For any element* **v** *of a vector space V we have*

$$0 \cdot \mathbf{v} = \mathbf{0}, \quad (-1) \cdot \mathbf{v} = -\mathbf{v}.$$

Proof We will prove only the first one, the second is an exercise. We will use VS4 (d) for this. We have

$$0 \cdot \mathbf{v} = (0 + 0) \cdot \mathbf{v} = 0 \cdot \mathbf{v} + 0 \cdot \mathbf{v}.$$

If we denote $\mathbf{x} = 0 \cdot \mathbf{v}$, then we will have $\mathbf{x} = \mathbf{x} + \mathbf{x}$ in the group $< V, + >$. Adding $-\mathbf{x}$ on both sides we will get $\mathbf{0} = \mathbf{x}$. $\qquad\square$

Definition 4.1.3 Let V be a vector space over the field F and $\mathbf{v}_1, \ldots, \mathbf{v}_k$ be arbitrary vectors in V. Then the set of all possible linear combinations $a_1 \mathbf{v}_1 + a_1 \mathbf{v}_2 + \cdots + a_k \mathbf{v}_k$ with coefficients $a_1, \ldots, a_k$ in F, is called the *span* of $\mathbf{v}_1, \ldots, \mathbf{v}_k$ and denoted $\text{span}\{\mathbf{v}_1, \ldots, \mathbf{v}_k\}$.

Definition 4.1.4 Let V be a vector space over the field F. The space V is said to be *finite dimensional*, if there exists a finite number of vectors $\mathbf{v}_1, \mathbf{v}_2, \ldots, \mathbf{v}_k$, which span V, that is $V = \text{span}\{\mathbf{v}_1, \mathbf{v}_2, \ldots, \mathbf{v}_k\}$.

Example 4.1.7 The space of polynomials $F_n[x]$ is finite dimensional as the set of monomials $\{1, x, x^2, \ldots, x^n\}$ spans it. The space of polynomials $F[x]$ is infinite dimensional.

Proof We will concentrate only on the second part of this example (for the first see exercise below). Suppose $F[x]$ is finite dimensional and there exist polynomials $f_1, f_2, \ldots, f_n$ such that

$$F[x] = \text{span}\{f_1, f_2, \ldots, f_n\}.$$

Let us choose a positive integer N such that $N > \deg(f_i)$ for all $i = 1, \ldots, n$. As $\{f_1, f_2, \ldots, f_n\}$ spans $F[x]$ we can find scalars $a_1, a_2, \ldots, a_n$ such that $x^N = a_1 f_1 + a_2 f_2 + \cdots + a_n f_n$. Then the polynomial

$$G(x) = x^N - a_1 f_1(x) - a_2 f_2(x) - \ldots - a_n f_n(x)$$

is a polynomial of degree N which is identically zero. This is a contradiction since $G(x)$ cannot have more than N roots. (When $F = \mathbb{R}$, this result is well known. For an arbitrary field this will be proved in Proposition 5.1.3.) $\qquad\square$

Definition 4.1.5 Let V be a vector space over the field F. A subset $\{\mathbf{v}_1, \mathbf{v}_2, \ldots, \mathbf{v}_k\}$ of V is said to be *linearly dependent* if there exist scalars $a_1, a_2, \ldots, a_k$ in F, *not all of which are* 0, such that

$$a_1 \mathbf{v}_1 + a_2 \mathbf{v}_2 + \cdots + a_k \mathbf{v}_k = 0.$$

A set which is not linearly dependent is called *linearly independent*.

Example 4.1.8 Let $F_{m \times n}$ be the space of all $m \times n$ matrix with entries from F. Let E_{ij} be the matrix whose (ij)-entry is 1 and all other entries are 0. Such a matrix is called a *matrix unit*. The set of all n^2 matrix units is linearly independent.

Example 4.1.9 The set of monomials $\{1, x, x^2, \ldots, x^n\}$ is linearly independent in $F_n[x]$.

Definition 4.1.6 Let V be a vector space. A *basis* for V is a linearly independent set of vectors which spans V.

Theorem 4.1.3 *Let V be a finite-dimensional vector space. Then every spanning subset of V can be reduced to a basis.*

Proof Suppose $V = \mathrm{span}\{v_1, v_2, \ldots, v_k\}$ but $\{v_1, v_2, \ldots, v_k\}$ is not linearly independent. Then

$$a_1 v_1 + a_2 v_2 + \cdots + a_k v_k = 0$$

and at least one coefficient is nonzero. Without loss of generality we may assume that $a_k \neq 0$. Then

$$v_k = -(a_k^{-1} a_1) v_1 - \ldots - (a_k^{-1} a_{k-1}) v_{k-1}$$

and now every linear combination of $v_1, v_2, \ldots, v_k$ can be written as a linear combination of $v_1, v_2, \ldots, v_{k-1}$. Indeed,

$$b_1 v_1 + b_2 v_2 + \cdots + b_k v_k = (b_1 - a_k^{-1} a_1 b_k) v_1 + \ldots + (b_{k-1} - a_k^{-1} a_{k-1} b_k) v_{k-1}.$$

This implies that $V = \mathrm{span}\{v_1, v_2, \ldots, v_k\} = \mathrm{span}\{v_1, v_2, \ldots, v_{k-1}\}$. We continue this process until the remaining system of vectors is linearly independent. Then we will have arrived at a basis for V. $\square$

Proposition 4.1.2 *Let $\{v_1, v_2, \ldots, v_n\}$ be a basis for a finite-dimensional vector space V over a field F and $v \in V$. Then there exist a unique n-tuple $(a_1, a_2, \ldots, a_n)$ of elements of F such that*

$$v = a_1 v_1 + a_2 v_2 + \cdots + a_n v_n. \tag{4.1}$$

Proof The fact that there is at least one such n-tuple follows from the fact that $\{v_1, v_2, \ldots, v_n\}$ spans V. Suppose there were two different ones:

$$v = a_1 v_1 + a_2 v_2 + \cdots + a_n v_n = b_1 v_1 + b_2 v_2 + \cdots + b_n v_n.$$

Then

$$(a_1 - b_1) v_1 + \ldots + (a_n - b_n) v_n = 0,$$

which contradicts to $\{v_1, v_2, \ldots, v_n\}$ being linearly independent. $\square$

Lemma 4.1.1 *Let F be a finite field with q elements. Suppose $\{\mathbf{v}_1, \mathbf{v}_2, \ldots, \mathbf{v}_n\}$ is a basis for V over F. Then V contains q^n elements.*

Proof Every element $\mathbf{v}$ of V can be written in a unique way as a linear combination (4.1). Each coefficient a_i appearing in this linear combination may take any one of q values. The total number of such linear combinations will therefore be q^n. This is how many elements V has. □

In the case when F is finite, it is now clear that all bases are equinumerous, i.e., contain the same number of vectors. In general it is also true.

Definition 4.1.7 Let V be a finite-dimensional vector space over a field F. The *dimension* for V is the number of vectors in any basis of V. It is denoted as $\dim_F V$.

Exercises

1. Check that F^n satisfies all axioms of a vector space observing how these axioms follow from the axioms of a field.
2. Justify the statement in Example 4.1.8.
3. Justify the statement in Example 4.1.9.
4. Prove that the set of symmetric $n \times n$ matrices $S = \{A \in F_{n \times n} \mid A^T = A\}$ is a vector space over F. Find its dimension over F.
5. Let V be the set of positive real numbers with the addition

$$\mathbf{u} \oplus \mathbf{v} := \mathbf{uv},$$

   i.e., the new addition is the former multiplication. Also for any real number $a \in R$ and any $\mathbf{u} \in V$ we define the scalar multiplication

$$a \odot \mathbf{u} := \mathbf{u}^a.$$

   Prove that $< V, \oplus, \odot >$ is a vector space over $\mathbb{R}$.
6°. Let $V = < V, +, \cdot >$ be a vector space over the field of complex numbers $\mathbb{C}$. Let us define a new operation $\alpha \circ v := \bar{\alpha} \cdot v$, where $\alpha \in \mathbb{C}$ and $v \in V$. Prove that $V = < V, +, \circ >$ is also a vector space over $\mathbb{C}$.

4.1.3 Cardinality of a Finite Field

Theorem 4.1.4 *Any finite field F contains one of the fields $\mathbb{Z}_p$ for a certain prime p. In this case F is a vector space over $\mathbb{Z}_p$ and it contains p^n elements, where $n = \dim_{\mathbb{Z}_p} F$.*

Proof Let m be a positive integer. Consider the element $m \cdot 1$ of F which is obtained by adding m ones, that is $m \cdot 1 = 1 + \ldots + 1$ (m times). When $m = 1, 2, \ldots,$ we obtain the sequence

$$1, \; 2 \cdot 1, \; 3 \cdot 1, \; \ldots, m \cdot 1, \ldots$$

The following is clear from the ring axioms: for any positive integers a, b

$$a \cdot 1 + b \cdot 1 = (a + b) \cdot 1, \tag{4.2}$$
$$(a \cdot 1) \cdot (b \cdot 1) = (ab) \cdot 1. \tag{4.3}$$

Since F is finite, we get $m_1 \cdot 1 = m_2 \cdot 1$ or assuming that $m_1 < m_2$, we get $(m_2 - m_1) \cdot 1 = 0$. Let p be the minimal positive integer for which $p \cdot 1 = 0$. Then p is prime. If not, and $p = ab$ for $a < p$ and $b < p$, then $a \cdot 1 \neq 0$ and $b \cdot 1 \neq 0$ but $(a \cdot 1) \cdot (b \cdot 1) = (ab) \cdot 1 = p \cdot 1 = 0$. This is a contradiction since F, being a field, by Theorem 4.1.1 contains no zero divisors.

Now, since $p \cdot 1 = 0$, the Eqs. (4.2) and (4.3) become

$$a \cdot 1 + b \cdot 1 = (a \oplus b) \cdot 1,$$
$$(a \cdot 1) \cdot (b \cdot 1) = (a \odot b) \cdot 1,$$

where $\oplus$ and $\odot$ are the addition and multiplication modulo p. We can now recognise that the set $\{0, 1, 2 \cdot 1, \ldots, (p - 1) \cdot 1\}$ together with the operations of addition and multiplication in F is in fact $\mathbb{Z}_p$. By Theorem 4.1.2 F is a vector space over $\mathbb{Z}_p$. Moreover, F is finite dimensional over $\mathbb{Z}_p$ since it is finite. Let $n = \dim_{\mathbb{Z}_p} F$. By Lemma 4.1.1, there are exactly p^n elements of F. $\qquad \square$

The theorem we have proved states that the cardinality of any finite field is a power of a prime. The converse is also true.

Definition 4.1.8 If $p \cdot 1 = 0$ in a field F for some prime p, then this prime p is said to be *the characteristic* of F. If such prime does not exist, the field F is said to have characteristic 0.

Theorem 4.1.5 *For any prime p and any positive integer n there exists a field of cardinality p^n. This field is unique up to an isomorphism.*

Proof We will show how to construct the fields of cardinality p^n in the next chapter. The uniqueness, however, is beyond the scope of this book. $\qquad \square$

The unique field of cardinality p^n is denoted $GF(p^n)$ and is called the *Galois field* of p^n elements.[1]

---

[1] See Sect. 3.1.3 for a brief historic note about Évariste Galois.

Exercises

1. Let $n_1 = 449873499879757801$ and $n_2 = 449873475733618561$. Find out if there are fields $GF(n_1)$ and $GF(n_2)$. In case $GF(n_i)$ exists for $i = 1$ or $i = 2$, identify the prime number p such that $\mathbb{Z}_p \subseteq GF(n_i)$ and determine the dimension of $GF(n_i)$ over $\mathbb{Z}_p$.

2. Let F be a finite field of q elements. Prove that all its elements are roots of the equation $x^q - x = 0$. (Hint: Consider the multiplicative group $< F^*, \cdot >$ of this field and use Corollary 3.2.3).

3°. Prove that in a field of characteristic p the following identity holds for every positive integer m:

$$(x + y)^{p^m} = x^{p^m} + y^{p^m}.$$

4°. In the field $GF(p^n)$ the mapping $x \mapsto x^p$ is an automorphism of this field (i.e., isomorphism onto itself).

4.2   The Multiplicative Group of a Finite Field is Cyclic

In any field F the set F^* of all nonzero elements play a very important role. Axiom F9 states that all elements of F^* are invertible. Moreover, this axiom, together with axioms F5–F7 imply that F^* relative to the operation of multiplication is a commutative group. This group is called the *multiplicative group* of F. Our goal for the rest of this chapter is to prove that in any finite field F the multiplicative group of F is cyclic.

We will concentrate our attention on orders of elements in F^*. Eventually, we will find that there is always an element in F^* whose order is exactly the cardinality of this group, thus proving that F^* is cyclic.

We now look at the field $\mathbb{Z}_7$ to get an intuition of what is to come. In this case $\mathbb{Z}_7^* = \{1, 2, 3, 4, 5, 6\}$. Let us calculate the powers of each element:

Powers of 1: $1, 1^2 = 1$.

Powers of 2: $2, 2^2 = 4, 2^3 = 1$; so there are 3 elements in $\mathbb{Z}_7$ which are powers of 2.

Powers of 3: $3, 3^2 = 2, 3^3 = 6, 3^4 = 4, 3^5 = 5, 3^6 = 1$; so all nonzero elements are powers of 3.

Powers of 4: $4, 4^2 = 2, 4^3 = 1$, so there are three distinct powers of 4.

Powers of 5: $5, 5^2 = 4, 5^3 = 6, 5^4 = 2, 5^5 = 3, 5^6 = 1$, so all nonzero elements are powers of 5.

Powers of 6: $6, 6^2 = 1$, so there are two powers.

We summarise our experience: the element 1 has order 1, the elements 2 and 4 have order 3, the elements 3 and 5 have order 6, and the element 6 has order 2. Hence $\mathbb{Z}_7^* = < 3 > = < 5 >$, it is cyclic and has two generators 3 and 5.

4.2.1 Lemmas on Orders of Elements

Lemma 4.2.1 *Let G be a group and g be an element of G of finite order. Then* $ord\,(g) = ord\,(g^{-1})$.

Proof Since $(g^k)^{-1} = (g^{-1})^k$ it follows that $g^k = 1$ implies $(g^{-1})^k = 1$ and the other way around. Therefore the orders of g and of g^{-1} are the same. □

Lemma 4.2.2 *Every element of a finite group has finite order. Moreover, in a finite group the order of any element is a divisor of the total number of elements in the group.*

Proof Let G be a finite group containing g. Then by Proposition 3.2.1 ord $(g) = |<g>|$ which is a divisor of $|G|$ by Lagrange's theorem. □

Lemma 4.2.3 *Let G be a group and g be an element of G of finite order. Suppose that* $g^n = 1$. *Then ord* $(g)|n$, *i.e., ord* (g) *is a divisor of n.*

Proof Let ord $(g) = m$. Suppose $n = qm + r$, where $0 \le r < m$, and suppose that $r \ne 0$. Then $1 = g^n = g^{qm+r} = (g^m)^q \cdot g^r = g^r$ which contradicts to the minimality of m. □

Equation (3.11) will play a crucial role in the proof of our next lemma. To recap, Eq. (3.11) says that for any element $g \in G$ and positive integer i

$$ord\,(g^i) = \frac{ord\,(g)}{\gcd(i, ord\,(g))}. \qquad (4.4)$$

Lemma 4.2.4 *If g is an element of a group G and ord* $(g) = ki$, *where k and i are positive integers, then ord* $(g^i) = k$.

Proof Indeed by (4.4) we have

$$ord\,(g^i) = \frac{ord\,(g)}{\gcd(i, ord\,(g))} = \frac{ki}{i} = k.$$ □

Lemma 4.2.5 *Let G be a commutative group, and a and b be two elements of G that have orders m and n, respectively. Suppose that gcd(m, n) = 1. Then ord* $(ab) = mn$.

Proof Since $(ab)^{mn} = a^{mn}b^{mn} = 1$, we know by Lemma 4.2.3 that ord $(ab)|mn$. Suppose that for some k the equality $(ab)^k = 1$ holds. Then $(ab)^k = a^k b^k = 1$ and $a^k = (b^k)^{-1}$. Let $c = a^k = (b^k)^{-1}$. Then $c^m = (a^k)^m = (a^m)^k = 1$ and $c^n = ((b^k)^{-1})^n = ((b^n)^k)^{-1} = 1$. As $1 = \gcd(m, n) = um + vn$ for some integers u and v, we may write $c = c^{um+vn} = c^{um} \cdot c^{vn} = (c^m)^u \cdot (c^n)^v = 1$. Thus $a^k = b^k = 1$

and by Lemma 4.2.3 we have $m|k$ and $n|k$. This implies $mn|k$, because m and n are relatively prime. If $k = \text{ord}\,(ab)$, we get $mn|\text{ord}\,(ab)$ and together with $\text{ord}\,(ab)|mn$ we get $\text{ord}\,(ab) = mn$. $\square$

Corollary 4.2.1 *Let G be a commutative group and let $a_1, a_2, \ldots, a_k \in G$ be elements of finite order such that $\text{ord}\,(a_i) = p_i^{\alpha_i}$ and all primes $p_1, p_2, \ldots, p_k$ are distinct. Then*

$$\text{ord}\,(a_1 a_2 \ldots a_k) = p_1^{\alpha_1} p_2^{\alpha_2} \ldots p_k^{\alpha_k}.$$

Proof Follows immediately from Lemma 4.2.5. We have to apply it $k-1$ times. $\square$

Example 4.2.1 Let a, b, c be elements of a commutative group G with orders

$$\text{ord}\,(a) = 5^3 \cdot 17, \quad \text{ord}\,(b) = 7^2 \cdot 5, \quad \text{ord}\,(c) = 17^2 \cdot 7.$$

Let us show how to use these elements to construct an element $g \in G$ such that $\text{ord}\,(g) = m$ and $a^m = b^m = c^m = 1$.

We claim that m can be taken as $\text{lcm}(\text{ord}\,(a), \text{ord}\,(b), \text{ord}\,(c)) = 5^3 \cdot 7^2 \cdot 17^2$ and $g = a^{17}b^5c^7$. Indeed, by Lemma 4.2.4 we have

$$\text{ord}\,(a^{17}) = 5^3, \quad \text{ord}\,(b^5) = 7^2, \quad \text{ord}\,(c^7) = 17^2,$$

and by Corollary 4.2.1 we get $\text{ord}\,(g) = 5^3 \cdot 7^2 \cdot 17^2$. If $m = 5^3 \cdot 7^2 \cdot 17^2$, then $\text{ord}\,(a)|m$, $\text{ord}\,(b)|m$, $\text{ord}\,(c)|m$, which implies $a^m = b^m = c^m = 1$.

Exercises

1. Let G be an abelian group of order 105 containing elements of orders 3, 5 and 7. Prove that it is cyclic.
2. Let g, h, k be elements of a finite abelian group G of orders 183618, 131726, 127308, respectively. Use g, h, k to construct an element of G of order 1018264646281, i.e., express an element of this order using g, h, k.
3°. Let g, h, k, ℓ be elements of a finite abelian group G of orders

$$\text{ord}\,(g) = 114900555878827339711577456221410444 11,$$
$$\text{ord}\,(h) = 636429004384549388734738941328412410 58592519601,$$
$$\text{ord}\,(k) = 3810716361,$$
$$\text{ord}\,(\ell) = 1885175497551,$$

respectively. Use $g, h, k.\ell$ to construct an element e of G of order

$$36250854905823179045637822474202751020997236353450271790483684 2272670521.$$

Express it in terms of g, h, k, ℓ.

4.2.2   Proof of the Main Theorem

Theorem 4.2.1 *Let G be a finite commutative group. Then there exists an element* $g \in G$ *such that* $\mathrm{ord}\,(g) = m \leq |G|$ *and* $x^m = 1$ *for all* $x \in G$.

Proof Let us consider the set of integers $I = \{\mathrm{ord}\,(g) \mid g \in G\}$ and let $p_1, p_2, \ldots, p_n$ be the set of all primes that occur in the prime factorisations of integers from I. For each such prime p_i let us choose the element g_i such that $\mathrm{ord}\,(g_i) = p_i^{\alpha_i} q_i$, where $\gcd(p_i, q_i) = 1$ and the integer α_i is maximal among all elements of G. (Note that the same element might correspond to several primes, i.e., among $g_1, g_2, \ldots, g_n$ not all elements may be distinct.) Then by Lemma 4.2.4 for the element $h_i = g_i^{q_i}$ we have $\mathrm{ord}\,(h_i) = p_i^{\alpha_i}$. Set $g = h_1 h_2 \ldots h_n$. Then, by Corollary 4.2.1,

$$m = \mathrm{ord}\,(g) = p_1^{\alpha_1} p_2^{\alpha_2} \cdots p_n^{\alpha_n},$$

and it is also clear that the order of every element in G divides m, thus $x^m = 1$ for all $x \in G$. Moreover, $m \leq |G|$ by Lemma 4.2.2.   □

Theorem 4.2.2 *Let F be a finite field consisting of q elements. Then there exists an element* $g \in F^*$ *such that* $\mathrm{ord}\,(g) = |F^*| = q - 1$, *i.e.,* $F^* = <g>$.

Proof It is sufficient to prove that there exists an element $g \in F^*$ of order $q - 1$. By Theorem 4.2.1 there exists an element of order $m \leq q - 1$ such that $x^m = 1$ for all $x \in F^*$. In the next chapter we will prove that a polynomial of degree n over any field has no more than n roots in that field. The polynomial $x^m - 1$ can be considered as a polynomial from $F[x]$; it has degree m and $q - 1$ roots in F. Since $q - 1 \geq m$, this is possible only if $m = q - 1$. The theorem is proved.   □

Definition 4.2.1 Let F be a finite field consisting of q elements. Then an element $g \in F^*$ of order $q - 1$ is called a *primitive element* of F.

Corollary 4.2.2 *Every finite field has a primitive element.*

Corollary 4.2.3 *Let F be a finite field consisting of q elements. Then* $\mathrm{ord}\,(a)$ *divides* $q - 1$ *for every element* $a \in F^*$.

Proof Let g be a primitive element of F. Then $\mathrm{ord}\,(g) = q - 1$ and $a = g^i$ for some $1 \leq i < q - 1$. Then by Lemma 4.2.4,

$$\mathrm{ord}\,(a) = \mathrm{ord}\,(g^i) = q - 1/\gcd(i, q - 1),$$

which is a divisor of $q - 1$.   □

Theorem 4.2.3 *For each prime p and positive integer n there is a unique, up to isomorphism, finite field $GF(p^n)$ that consists of p^n elements. Its elements are the roots of the polynomial $f(x) = x^{p^n} - x$.*

Proof We cannot prove the first part of the statement, i.e., the existence of $F = GF(p^n)$ but we can prove the second. Suppose F does exist and g is a primitive element. Then every nonzero element a of F lies in F^*, which is a cyclic group of order $p^n - 1$ with generator g. By Corollary 4.2.3 ord (a) is a divisor of $p^n - 1$, hence, $a^{p^n-1} = 1$. It follows that $a^{p^n} = a$ for all $a \in F$, including 0, which proves the second part of the theorem.   □

The idea behind the proof of the existence of $GF(p^n)$ is as follows. Firstly we construct an extension $\mathbb{Z}_p \subset K$ such that every polynomial with coefficients in $\mathbb{Z}_p$ has a root in K. Then the polynomial $f(x) = x^{p^n} - x$ will have p^n roots in K and we have to check that $f(x)$ does not have multiple roots. These p^n distinct roots will then be a field $GF(p^n)$.

From our considerations it follows that, if $m|n$, then $GF(p^m)$ is a subfield of $GF(p^n)$. Indeed, any root of the equation $x^{p^m} = x$ will also be a root of the equation $x^{p^n} = x$ (see Exercise 3 that follows).

Exercises

1. Consider the field $\mathbb{Z}_p$, where $p = 192837481$ is prime. Do there exist elements in $\mathbb{Z}_p^*$ of orders 11561 and 58380?
2. Let p be a prime and m, n be positive integers. Then $p^m - 1$ divides $p^n - 1$ if and only if m divides n.
3. $GF(p^m)$ is a subfield of $GF(p^n)$ if and only if $m|n$.
4°. How many primitive elements are there in the field $\mathbb{Z}_{2017}$?
5°. Find generators of the multiplicative groups of fields (a) $\mathbb{Z}_{31}$ and (b) $\mathbb{Z}_{1237}$ (one for each field). In each case, how many other generators are there?

4.2.3   Proof of Euler's Criterion

Now we can give a more natural proof of Euler's criterion (Theorem 3.3.4) based on Theorem 4.2.2. Let g be a primitive element of $\mathbb{Z}_p$, where p is an odd prime. Then ord $(g) = p - 1$ and $p - 1$ is even. Suppose $p - 1 = 2^s t$, where t is an odd integer. As g is a generator of the cyclic group $\mathbb{Z}_p^*$ by Lemma 4.2.3 we have $g^n = 1$ if and only if ord $(g)|n$. Suppose a is not a quadratic residue in $\mathbb{Z}_p$. Then by Theorem 4.2.4 we have $a = g^k$, where k is odd. But then $k \cdot \frac{p-1}{2} = 2^{s-1} tk$, where tk is odd, so this number is not a multiple of $p - 1 = 2^s t$. Hence $a^{\frac{p-1}{2}} = g^{\frac{k(p-1)}{2}} \neq 1$ and then $a^{\frac{p-1}{2}} = -1$.

4.2.4   Discrete Logarithms

Definition 4.2.2 Let F be a finite field consisting of q elements and let $g \in F^*$ be a primitive element of F. Then the equation $g^x = h$ has a unique solution modulo $q - 1$ which is called the *discrete logarithm* of h to base g, denoted $\log_g(h)$.

Example 4.2.2 As was computed in the previous section

$$\mathbb{Z}_7^* = \{3^1 = 3, 3^2 = 2, 3^3 = 6, 3^4 = 4, 3^5 = 5, 3^6 = 1\}.$$

Thus 3 is a primitive element of $\mathbb{Z}_7$ and $\log_3(3) = 1$, $\log_3(2) = 2$, $\log_3(6) = 3$, $\log_3(4) = 4$, $\log_3(5) = 5$, $\log_3(1) = 6$.

Example 4.2.3 For example, $g = 3$ is a primitive element of $\mathbb{Z}_{19}$ as seen from the table featuring powers of 3:

n	1	2	3	4	5	6	7	8	9	10	11	12	13	14	15	16	17	18
3^n	3	9	8	5	15	7	2	6	18	16	10	11	14	4	12	17	13	1

Therefore the table of logarithms to base 3 will be:

n	1	2	3	4	5	6	7	8	9	10	11	12	13	14	15	16	17	18
$\log_3(n)$	18	7	1	14	4	8	6	3	2	11	12	15	17	13	5	10	16	9

Computing discrete logarithms in a finite field is believed to be computationally difficult. So we can now add the following problem to our list of *apparently hard* problems in number theory.

Discrete Logarithm Problem   Given a prime p, a generator g of $\mathbb{Z}_p^*$ and an element $h \in \mathbb{Z}_p^*$, find the integer x such that $g^x = h$ and $0 \le x \le p - 2$.

We recap that a nonzero element h of a finite field F is a quadratic residue if there exists another element $g \in F$ such that $g^2 = h$.

Theorem 4.2.4 *Let F be a finite field. Let $g \in F^*$ be any primitive element of F. Then an element $h \in F^*$ is a quadratic residue if and only if $\log_g(h)$ is even.*

Proof If $\log_g(h) = 2k$ is even, then $h = g^{2k} = (g^k)^2$, thus h is a quadratic residue. The reverse is clearly also true. Indeed, if h is a quadratic residue, then $h = (h_1)^2$ for some $h_1 \in F^*$. Since $h_1 = g^k$ for some k, we get $h = (g^k)^2 = g^{2k}$ and $\log_g(h) = 2k$ is even.   □

Exercises

1. How many primitive elements are there in the field $\mathbb{Z}_{1237}$?
2. Let $F = \mathbb{Z}_{17}$.

   (a) Decide whether 2 or 3 is a primitive element of F. Denote the one which is primitive by g.
   (b) Compute the table of powers of g in F and the table of discrete logs to base g.

3. Let g be a primitive element in a finite field F consisting of q elements. Prove that

$$\log_g(ab) \equiv \log_g(a) + \log_g(b) \mod q-1.$$

4°. (a) By trial and error, using the GAP command `OrderMod(m,n);` which determines the order of m in the multiplicative group $\mathbb{Z}_m^*$ of $\mathbb{Z}_m$, find a primitive element g of $\mathbb{Z}_{23}$ with the smallest numerical value.
   (b) Generate $\mathbb{Z}_{23}$ using the command `F:=GF(23);` and list its elements using the command `AsList(F);` note that GAP lists all nonzero elements of $F = GF(23)$ as powers of g.
   (c) Using the command `Int(a);` which gives the numerical value of $a \in F$ fill the table of discrete logarithms to base g.

5°. Characterise all primes $p \neq 2$ for which $-1 = p - 1$ is a quadratic residue.

4.3   Elgamal Cryptosystem Revisited

In Chap. 2 we studied a public-key cryptosystem whose security is based on the complexity of factoring integers. Here we present a cryptosystem whose security is based on the complexity of calculating discrete logarithms. It is based on the Diffie–Hellman key exchange agreement. It was invented by Taher Elgamal in 1985. The Elgamal algorithm is used in the free GNU Privacy Guard software, recent versions of PGP and other cryptosystems.

In a public domain, a large prime p and a primitive element α of $\mathbb{Z}_p$ are displayed. Each participant of the group, who wants to send or receive encrypted messages, creates their private and public keys. Alice, for example, selects a secret integer k_A and calculates α^{k_A} which she places in the public domain as her public key. Bob selects a secret integer k_B and calculates α^{k_B} which he places in a public domain as his public key. Now they can exchange messages.

Suppose, for example, that Bob wants to send a message m to Alice. We will assume that m is an integer such that $0 \leq m < p$. (If m is larger, he breaks it into several block as usual.) He chooses another secret random integer s and computes $c_1 = \alpha^s$ in $\mathbb{Z}_p$. He also takes Alice's public key α^{k_A} from the public domain and calculates $c_2 = m \cdot (\alpha^{k_A})^s$. He sends this pair (c_1, c_2) of elements of $\mathbb{Z}_p$ to Alice so this is the cyphertext. On the receiving end Alice uses her private key k_A to calculate

m as follows: $m = c_2 \cdot ((c_1)^{k_A})^{-1}$. For the evil eavesdropper Eve to figure out k_A she must solve a discrete logarithm problem, which is difficult.

Exercises

1. Alice and Bob agreed to use Elgamal cryptosystem based on the multiplicative group of field $\mathbb{Z}_p$ for $p = 53$. They also agreed to use 2 as the primitive element of $\mathbb{Z}_p$. Since p is small their messages consist of a single letter which is encoded as

$$A = 11, \ B = 12, \ \ldots, \ Z = 36.$$

Bob's public key is 32 and Alice sent him the message $(30, 42)$. Which letter did Alice send to Bob in this message?

2. Alice and Bob have set up the multiplicative Elgamal cryptosystem for private communication. They have chosen an element $g = 123456789$ in the multiplicative group of the field Z_p, where $p = 123456789987654353003$. They have chosen their private exponents $k_A = 373$ and $k_B = 5191$ and published the elements $g_A = 52808579942366933355$ and $g_B = 39318628345168608817$, respectively. They agreed to cut the messages into ten-letter segments and encode the letters as $A = 11, B = 12, \ldots, Z = 36$, space $= 41$, ' $= 42$, . $= 43$, , $= 44$, ? $= 45$. Bob got the following message from Alice:

```
[ [ 83025882561049910713,   66740266984208729661 ],
  [ 117087132399404660932,  44242256035307267278 ],
  [ 67508282043396028407,   77559274822593376192 ],
  [ 60938739831689454113,   14528504156719159785 ],
  [ 5059840044561914427,    59498668430421643612 ],
  [ 92232942954165956522,  105988641027327945219 ],
  [ 97102226574752360229,   46166643538418294423 ] ]
```

Help him to decrypt it.

3°. Alice and Bob have set up the multiplicative Elgamal cryptosystem for private communication. They agreed to cut the messages into ten-letter segments and encode the letters as $A = 11$, $B = 12$, ..., $Z = 36$, "," $= 37$, "." $= 38$, "!" $= 39$, space $= 40$.

Moreover, they agreed to use Elgamal cryptosystem based on the multiplicative group of the field $\mathbb{Z}_p$ for $p := 112211293740262525327$ with the primitive element $g = 3$. Bob's public key is $g^{k_B} = 2014$.

(a) Alice sent Bob the following message:

```
[ [ 109247194023665333196, 117625321671373665658702154840467 1132096 ],
  [ 106327406037146532225, 176112048708970133065563314862583 4924019 ],
  [ 7466307393140907153, 326143913360704463304844857246247116890 ],
  [ 9229617509201388951, 375800226407505562526391070077 14521998 ] ]
```

Show how Eve can decrypt it using command `LogMod(h,g,p);` which calculates $\log_g(h)$ in $\mathbb{Z}_p$.

4°. Alice and Bob agreed to use Diffie–Hellman secret key exchange to come up with a secret key for their secret-key cryptosystem. They openly agreed on the prime

$$p = 10014088944206281414040434711571$$

and primitive element $\alpha = 13$ of $\mathbb{Z}_p$. Alice has decided on her private key by choosing $k_A = 123456789$. She also got a message 9263920439873227653264 2490482 from Bob. Which message should she send to Bob and how she calculates the shared secret key?

Polynomials

5

> *A polynomial walks into a bar and asks for a drink. The barman declines: "We don't cater for functions."*
>
> *An old math joke.*

This chapter is about polynomials and their use. After learning the basics we discuss the concept of "secret sharing" and learn Shamir's secret sharing scheme, which relies on the properties of polynomials over large finite fields. Then, after proving some further results on polynomials, we give a construction of a finite field whose cardinality p^n is a power of a prime p. The field constructed will be an extension of $\mathbb{Z}_p$ and in this context we discuss minimal annihilating polynomials which we will need later for error-correcting codes.

5.1 The Ring of Polynomials

5.1.1 Introduction to Polynomials

Let F be a field. Any expression

$$f(x) = \sum_{i=0}^{k} a_i x^i, \qquad a_i \in F, \tag{5.1}$$

where k is an arbitrary positive integer, is called a *polynomial* over F. The set of all polynomials over F is denoted by $F[x]$. For $k = 0$ there is no distinction between the scalar a_0 and the polynomial $f(x) = a_0$. Thus we assume that $F \subset F[x]$. The

© Springer Nature Switzerland AG 2020

A. Slinko, *Algebra for Applications*, Springer Undergraduate Mathematics Series,

https://doi.org/10.1007/978-3-030-44074-9_5

zero polynomial 0 is a very special one. Any other polynomial $f(x) \neq 0$ we can write in the form (5.1) with $a_k \neq 0$ and define its degree as follows.

Definition 5.1.1 Given a <u>nonzero</u> polynomial $f(x) = \sum_{i=0}^{k} a_i x^i$, with $a_k \neq 0$, the number k is said to be the *degree* of $f(x)$ and will be denoted $\deg(f)$. Note that $\deg(f)$ is undefined if $f = 0$. Colloquially speaking, the degree of $f(x)$ is the highest power of x which appears.

Definition 5.1.2 Let

$$f(x) = \sum_{i=0}^{k} a_i x^i, \qquad g(x) = \sum_{i=0}^{m} b_i x^i$$

be two polynomials of degree $\deg(f) = k$ and $\deg(g) = m$, respectively. We say that these two polynomials are equal, and write $f(x) = g(x)$, if $k = m$ and $a_i = b_i$ for all $i = 0, 1, 2, \ldots, k$.

The addition and the multiplication in the field induces the corresponding operations over polynomials. Let

$$f(x) = \sum_{i=0}^{k} a_i x^i, \qquad g(x) = \sum_{i=0}^{m} b_i x^i$$

be two polynomials and assume that $\deg(f) = k \geq m = \deg(g)$. Then we define

$$f(x) + g(x) := \sum_{i=0}^{k} (a_i + b_i) x^i,$$

where for $i > \deg(g)$ we assume that $b_i = 0$. The multiplication is defined in such a way so that $x^i \cdot x^j = x^{i+j}$ is true. The only way to do this is to set

$$f(x)g(x) := \sum_{i=0}^{k+m} \left(\sum_{j=0}^{i} a_j b_{i-j} \right) x^i.$$

The same convention works also here: $a_p = 0$, when $p > \deg(f)$, and $b_q = 0$, when $q > \deg(g)$.

By defining these two operations we obtained an algebraic object which is called the polynomial ring over F; it is also denoted as $F[x]$.

Example 5.1.1 Let $f(x) = x^2 + x + 1$ and $g(x) = x^3 + x + 1$ be two polynomials from $\mathbb{Z}_2[x]$. Then

$$f(x) + g(x) = x^3 + x^2, \quad f(x)g(x) = x^5 + x^4 + 1.$$

(Some training in handling these operations is desirable. Try several examples your-self.)

We observe that

Proposition 5.1.1 *For any two nonzero polynomials* $f, g \in F[x]$

1. $deg\,(f + g) \leq \max(deg\,(f), deg\,(g));$
2. $deg\,(fg) = deg\,(f) + deg\,(g)$ *and, in particular,* $F[x]$ *has no zero divisors.*

Division with remainder is also possible.

Theorem 5.1.1 (Division Algorithm) *Given polynomials* $f(x)$ *and* $g(x)$ *in* $F[x]$ *with* $g(x) \neq 0$, *there exist a "quotient"* $q(x) \in F[x]$ *and a "remainder"* $r(x) \in F[x]$ *such that*

$$f(x) = g(x)q(x) + r(x)$$

and either $r(x) = 0$ *or* $deg\,(r) < deg\,(g)$. *Moreover, the quotient and the remainder are uniquely defined.*

Proof Let

$$f(x) = \sum_{i=0}^{k} a_i x^i, \qquad g(x) = \sum_{i=0}^{m} b_i x^i$$

be two polynomials with $deg\,(f) = k$ and $deg\,(g) = m$. Then there are two cases to consider:

Case 1. If $k < m$, then we can set $q(x) = 0$ and $r(x) = f(x)$.

Case 2. If $k \geq m$, we can define

$$f_1(x) = f(x) - b_m^{-1} a_k x^{k-m} g(x) = f(x) - g(x)q_1(x),$$

where $q_1(x) = b_m^{-1} a_k x^{k-m}$. This polynomial $f_1(x)$ will be of smaller degree than f, since $f(x)$ and $q_1(x)g(x)$ have the same degree m and the same leading coefficient a_m. By induction hypothesis,

$$f_1(x) = g(x)q_2(x) + r(x).$$

with either $r(x) = 0$ or $deg\,(r) < deg\,(g)$. Therefore

$$f(x) = g(x)(q_1(x) + q_2(x)) + r(x).$$

Now suppose that

$$f(x) = g(x)q_1(x) + r_1(x) = g(x)q_2(x) + r_2(x).$$

Then

$$g(x)(q_1(x) - q_2(x)) = r_2(x) - r_1(x).$$

This cannot happen for $r_1(x) \neq r_2(x)$ since the degree of the right-hand side is smaller than the degree of the left-hand side. Thus $r_2(x) - r_1(x) = 0$. This can happen only when $q_1(x) - q_2(x) = 0$, since $F[x]$ has no zero divisors. □

The quotient and the remainder can be computed by the following "polynomial long division" process, commonly taught in high school. For example, let us consider polynomials $f(x) = x^4 + x^3 + x^2 + x + 1$ and $g(x) = x^2 + 1$ from $\mathbb{Z}_2[x]$. Then

$$
\begin{array}{r}
x^2 \quad + x \qquad\qquad\qquad \\
x^2 + 1 \enclose{longdiv}{x^4 \quad + x^3 \quad + x^2 \quad + x \quad + 1} \\
\underline{x^4 \qquad\qquad + x^2 \qquad\qquad} \\
x^3 \qquad\quad + x \quad + 1 \\
\underline{x^3 \qquad\qquad + x \qquad} \\
+ 1
\end{array}
$$

encodes a division with remainder of polynomial $f(x)$ by $g(x)$. It shows that the quotient $q(x)$ and the remainder $r(x)$ are

$$q(x) = x^2 + x, \qquad r(x) = 1,$$

that is

$$x^4 + x^3 + x^2 + x + 1 = (x^2 + x)(x^2 + 1) + 1.$$

We say that a polynomial $f(x)$ is divisible by $g(x)$ if $f(x) = q(x)g(x)$, i.e., when the remainder is zero.

A polynomial (5.1) defines a function $f : F \to F$ with

$$f(\alpha) = \sum_{i=0}^{k} a_i \alpha^i.$$

It is straightforward to check that this function satisfies the following conditions:

$$(f + g)(\alpha) = f(\alpha) + g(\alpha), \qquad (fg)(\alpha) = f(\alpha)g(\alpha).$$

In analysis this function is always identified with the polynomial itself. However, working over a finite field we cannot do this. Indeed, $1^2 + 1 = 0$ and $0^2 + 0 = 0$. So the polynomial $f(x) = x^2 + x$ over $\mathbb{Z}_2$ is nonzero but the function associated with it is the zero function.

Definition 5.1.3 An element $\alpha \in F$ is called a *root*[1] of $f(x)$ if $f(\alpha) = 0$.

Proposition 5.1.2 *An element $\alpha \in F$ is a root of a polynomial $f(x)$ if and only if $f(x) = g(x)(x - \alpha)$, i.e., $f(x)$ is divisible by $x - \alpha$.*

Proof Let us divide $f(x)$ by $(x - \alpha)$ with remainder:

$$f(x) = q(x)(x - \alpha) + r,$$

where $r \in F$ is the remainder and $q(x)$ is the quotient. Substituting α in this equation we get $0 = 0 + r$, whence $r = 0$ and $f(x)$ is divisible by $(x - \alpha)$ and $q(x)$ can be taken as $g(x)$. Conversely, if $f(x) = g(x)(x - \alpha)$, then $f(\alpha) = g(\alpha) \cdot 0 = 0$.   □

Definition 5.1.4 We say that $\alpha \in F$ is a root of a polynomial $f(x)$ of *multiplicity k* if $f(x)$ is divisible by $(x - \alpha)^k$ but not divisible by $(x - \alpha)^{k+1}$.

Proposition 5.1.3 *A polynomial*

$$f(x) = \sum_{i=0}^{k} a_i x^i, \qquad a_i \in F. \tag{5.2}$$

of degree k cannot have more than k roots in the field F, where each root is counted as many times as its multiplicity.

Proof Suppose that α is a root of $f(x)$ of multiplicity m. Then

$$f(x) = (x - \alpha)^m g(x), \qquad \deg(g) = k - m.$$

If there are no other roots we are done. If $\beta \neq \alpha$ is also a root of $f(x)$, then it is a root of $g(x)$. Indeed, substituting β in this equation we get

$$0 = f(\beta) = (\beta - \alpha)^m g(\beta).$$

Since in any field there are no divisors of zero and $\beta - \alpha \neq 0$, we conclude that $g(\beta) = 0$. By induction hypothesis there are no more than $k - m$ roots in $g(x)$. Hence $f(x)$ has at most $m + (k - m) = k$ roots.   □

---

[1] A purist would talk about a zero of the polynomial $f(x)$ but a root of the equation $f(x) = 0$. We are not making this distinction.

Exercises

1. Consider the following polynomials in $\mathbb{Z}_7$:

$$f(x) = 5x^4 + x^2 + 3x + 4, \quad g(x) = 3x^2 + 2x + 1.$$

Find the quotient and the remainder of $f(x)$ on division by $g(x)$.
2. Find the roots of $f(x) = x^4 + 2x^3 + 2x^2 + 2x + 1$ in $\mathbb{Z}_5[x]$. Hence find a factorisation of $f(x)$ into linear factors.
3°. Use GAP to find the roots of $f(x) = x^4 + 5x^2 + 4x + 5$ in $\mathbb{Z}_7[x]$. Hence find a factorisation of $f(x)$ into linear factors.

5.1.2 Lagrange's Interpolation

Sometimes we need to reconstruct a polynomial knowing a number of values of this polynomial.

Proposition 5.1.4 *Let $\alpha_0, \alpha_1, \ldots, \alpha_k$ be distinct elements of F and $\beta_0, \beta_1, \ldots, \beta_k$ be arbitrary elements of F. Then there exists no more than one polynomial $f(x)$ of degree at most k such that $f(\alpha_i) = \beta_i$ for $i = 0, 1, \ldots, k$.*

Proof Suppose that two distinct polynomials $f(x) = \sum_{i=0}^{k} a_i x^i$ and $g(x) = \sum_{i=0}^{k} b_i x^i$ satisfy $f(\alpha_i) = \beta_i$ and $g(\alpha_i) = \beta_i$ for all $i = 0, 1, 2, \ldots, k$. Then the polynomial $h(x) = f(x) - g(x)$ is not zero, and its degree is not greater than k. Also

$$h(\alpha_i) = f(\alpha_i) - g(\alpha_i) = \beta_i - \beta_i = 0,$$

and $h(x)$ has at least $k + 1$ distinct roots $\alpha_0, \alpha_1, \ldots, \alpha_k$. However, by Proposition 5.1.3 this is impossible. □

Theorem 5.1.2 *Let $\alpha_0, \alpha_1, \ldots, \alpha_k$ be distinct elements of F and $\beta_0, \beta_1, \ldots, \beta_k$ be arbitrary elements of F. Then there exists a unique polynomial*

$$f(x) = \sum_{i=0}^{k} \beta_i \frac{(x - \alpha_0) \ldots (x - \alpha_{i-1})(x - \alpha_{i+1}) \ldots (x - \alpha_k)}{(\alpha_i - \alpha_0) \ldots (\alpha_i - \alpha_{i-1})(\alpha_i - \alpha_{i+1}) \ldots (\alpha_i - \alpha_k)} \qquad (5.3)$$

of degree at most k such that $f(\alpha_i) = \beta_i$ for $i = 0, 1, \ldots, k$.

Proof The polynomial (5.3) was constructed as follows. We first constructed polynomials $g_i(x)$ of degree k such that $g_i(\alpha_i) = 1$ and $g_i(\alpha_j) = 0$ for $i \neq j$. These polynomials are:

$$g_i(x) = \frac{(x - \alpha_0) \ldots (x - \alpha_{i-1})(x - \alpha_{i+1}) \ldots (x - \alpha_k)}{(\alpha_i - \alpha_0) \ldots (\alpha_i - \alpha_{i-1})(\alpha_i - \alpha_{i+1}) \ldots (\alpha_i - \alpha_k)}.$$

Then the desired polynomial was constructed as $f(x) = \sum_{i=0}^{k} \beta_i g_i(x)$. We immediately see that $f(\alpha_i) = \beta_i$, as required. This polynomial is unique because of Proposition 5.1.4. □

Example 5.1.2 As an example, let us construct a polynomial $f(x)$ of degree at most 2 over $F = \mathbb{Z}_5$ with the properties: $f(1) = 2$, $f(2) = 4$, $f(3) = 4$. We apply Theorem 5.1.2 to the case $F = \mathbb{Z}_5, k = 2, \alpha_0 = 1, \alpha_1 = 2, \alpha_2 = 3, \beta_0 = 2, \beta_1 = 4, \beta_2 = 4$.

The formula tells us that

$$f(x) = 2\frac{(x-2)(x-3)}{(1-2)(1-3)} + 4\frac{(x-1)(x-3)}{(2-1)(2-3)} + 4\frac{(x-1)(x-2)}{(3-1)(3-2)}$$

is the desired polynomial. If we want to know the coefficients of this polynomial, we have to expand all the expressions, bearing in mind that all the arithmetic is in $\mathbb{Z}_5$:

$$f(x) = 2\frac{x^2+1}{4\cdot 3} + 4\frac{x^2+x+3}{1\cdot 4} + 4\frac{x^2+2x+2}{2\cdot 1} =$$

$$x^2 + 1 + (x^2 + x + 3) + 2(x^2 + 2x + 2) = 4x^2 + 3.$$

(You can easily check that indeed $f(1) = 2$, $f(2) = 4$, $f(3) = 4$. Do it!)

Note: a simple alternative to using the formula is to calculate the coefficients of the desired polynomial as the unique solution of a system of linear equations: if $f(x) = ax^2 + bx + c$ and $f(1) = 2$, $f(2) = 4$, $f(3) = 4$, we have the system

$$\begin{cases} a + b + c = 2, \\ 4a + 2b + c = 4, \\ 4a + 3b + c = 4. \end{cases}$$

The usual method of Gaussian elimination (all arithmetic in $\mathbb{Z}_5$) leads to $a = 4, b = 0, c = 3$, confirming the result obtained by the previous method. Another way to solve this system of linear equation is of course to calculate the inverse of the matrix of this system and multiply it by the column on the right-hand side.

Corollary 5.1.1 *Let us consider the class of polynomials*

$$f(x) = \sum_{i=0}^{k} a_i x^i, \quad a_i \in F,$$

with an arbitrary but fixed $a_0 \in F$. Let $\alpha_1, \ldots, \alpha_k$ be distinct nonzero elements of F and $\beta_1, \ldots, \beta_k$ be arbitrary elements of F. Then there exists a unique polynomial $f(x)$ of degree at most k in this class such that $f(\alpha_i) = \beta_i$ for $i = 1, 2, \ldots, k$.

Proof Since $a_0 = f(0)$ it is enough to set $\alpha_0 = 0$ and $\beta_0 = a_0$ and apply Theorem 5.1.2. □

In the next chapter we will look at one particular application of Lagrange's interpolation to cryptography, namely to secret sharing.

Exercises

1. Use Lagrange's interpolation to find $f(x) = \sum_{i=0}^{2} a_i x^i \in \mathbb{Z}_7[x]$ with $f(1) = f(2) = 1$ and $f(3) = 2$.
2. Find the constant term of the polynomial $f(x)$ of degree no greater than 2 with coefficients in $\mathbb{Z}_7$ such that $f(1) = 3$, $f(3) = 2$, $f(4) = 1$.
3. Find the constant term of the polynomial $f(x)$ of degree at most 3 in $\mathbb{Z}_7$ such that

$$f(1) = 3, \ \ f(2) = 2, \ \ f(3) = 2, \ \ f(5) = 1.$$

4. Use GAP to find a polynomial $f(x) \in \mathbb{Z}_{13}[x]$ of degree at most 3 such that

$$f(1) = 5, \ \ f(2) = 7, \ \ f(3) = 0, \ \ f(5) = 3.$$

5. Let F be a finite field. Show that for every function $f : F \rightarrow F$ of F to itself there exist a polynomial $g(x) \in F[x]$ such that $f(a) = g(a)$ for all $a \in F$.

5.1.3   Factoring Polynomials

Definition 5.1.5 Any polynomial

$$f(x) = \sum_{i=0}^{k} a_i x^i, \qquad a_i \in F. \tag{5.4}$$

where k is an arbitrary positive integer and $a_k = 1$ is called a *monic* polynomial of degree k over F.

Example 5.1.3 The polynomial $f(x) = 5x^2 + x^5 - 1$ is a monic polynomial of degree 5. The polynomial $g(x) = x^2 + 2x^5 - 1$ is not monic.

Definition 5.1.6 A polynomial $f(x)$ from $F[x]$ is said to be *reducible over F* if there exist two polynomials $f_1(x)$ and $f_2(x)$ from $F[x]$, each of degree greater than or equal 1, such that $f(x) = f_1(x)f_2(x)$. Otherwise $f(x)$ is said to be *irreducible over F*.

Example 5.1.4 The polynomial $f(x) = x^2 + 1$ is irreducible over $\mathbb{R}$ and reducible over $\mathbb{C}$ since $f(x) = (x - i)(x + i)$. The polynomial $g(x) = x^2 - 2$ is irreducible over $\mathbb{Q}$ and reducible over $\mathbb{R}$. The polynomial $h_1(x) = x^2 + 2 \in \mathbb{Z}_5[x]$ is irreducible over $\mathbb{Z}_5$, and $h_2(x) = x^2 + 2 \in \mathbb{Z}_{11}[x]$ is reducible over $\mathbb{Z}_{11}$ since $x^2 + 2 = (x + 3)(x + 8)$.

We see that the reducibility or irreducibility of a given polynomial depends heavily on the field under consideration. We will be especially interested in irreducible polynomials over $\mathbb{Z}_2$. Of course, both linear polynomials x and $x + 1$ are irreducible. Since x^2, $(x + 1)^2 = x^2 + 1$, $x(x + 1) = x^2 + x$ are reducible, the only irreducible polynomial of degree 2 is $x^2 + x + 1$. There are eight polynomials of degree 3:

$$f_1(x) = x^3,$$
$$f_2(x) = x^3 + 1,$$
$$f_3(x) = x^3 + x + 1,$$
$$f_4(x) = x^3 + x,$$
$$f_5(x) = x^3 + x^2,$$
$$f_6(x) = x^3 + x^2 + 1,$$
$$f_7(x) = x^3 + x^2 + x,$$
$$f_8(x) = x^3 + x^2 + x + 1.$$

To check them for irreducibility, the following proposition is useful.

Proposition 5.1.5 *A polynomial $f(x) \in F[x]$ of degree 3 is irreducible over F if and only if it has no roots in F.*

Proof If $f(x)$ is irreducible clearly it has no linear factors, nor by Proposition 5.1.2 does it have any roots in F. Conversely, suppose that $f(x)$ has no roots in F. If it is reducible, then $f(x) = g(x)h(x)$, where either $g(x)$ or $h(x)$ has degree 1, and polynomial of degree 1 always has a root in F. This gives us a contradiction to Proposition 5.1.2.   □

Returning to our list, we know that any reducible polynomial $f(x)$ of degree 3 has a root in $\mathbb{Z}_2$, i.e., either $f(0) = 0$ or $f(1) = 0$. Six out of the eight polynomials in the table have roots in $\mathbb{Z}_2$ and only $f_3(x) = x^3 + x + 1$ and $f_6(x) = x^3 + x^2 + 1$ do not have roots, hence, irreducible.

Theorem 5.1.3 *If a polynomial $f(x) \in F[x]$ of degree n is not divisible by any irreducible polynomial over F of degree not greater than $\lfloor \frac{n}{2} \rfloor$, then it is irreducible over F.*

Proof If $f(x)$ is reducible over F, then $f(x) = g(x)h(x)$, where $g(x), h(x) \in F[x]$ both have degrees at least one. Then at least one of them will have degree not greater than $\lfloor \frac{n}{2} \rfloor$. Any of its irreducible factors will have degree not greater than $\lfloor \frac{n}{2} \rfloor$. Hence, if there are no irreducible polynomials over F of degree not greater than $\lfloor \frac{n}{2} \rfloor$ that divide $f(x)$, it must be irreducible over F.   □

Example 5.1.5 Let us determine whether or not $f(x) = x^5 + x^4 + 1$ is irreducible over $\mathbb{Z}_2$. We check that $f(0) = f(1) = 1$, that is $f(x)$ has no roots in $\mathbb{Z}_2$. But does this imply its irreducibility? Not at all. The absence of roots means the absence of linear factors. However it is possible now that a polynomial of degree five has no linear root but is reducible by having one quadratic irreducible factor and another one of degree three. We have now to check that there are no quadratic irreducible factors. The only possible irreducible quadratic factor is $x^2 + x + 1$, so we have to divide $f(x)$ by $x^2 + x + 1$ and calculate the remainder. We find that $f(x) = (x^2 + x + 1)(x^3 + x + 1)$. Hence $f(x)$ is reducible.

Irreducible polynomials play a similar role to that played by prime numbers. The following theorem can be proved using the same ideas as for integers.

Theorem 5.1.4 *Any polynomial $f(x)$ from $F[x]$ of degree no less than 1 can be uniquely represented as a product*

$$f(x) = c p_1(x)^{\alpha_1} p_2(x)^{\alpha_2} \ldots p_k(x)^{\alpha_k}$$

where $p_1(x), p_2(x), \ldots, p_k(x) \in F[x]$ are monic irreducible (over F) polynomials, c is a nonzero constant, and $\alpha_1, \alpha_2, \ldots, \alpha_k$ are positive integers. This representation is unique apart from the order of $p_1(x), p_2(x), \ldots, p_k(x)$.

Exercises

1. Let F be a field and let $f(x) \in F[x]$. True or false:

   (a) If $f(x)$ has a root in F then $f(x)$ is reducible in $F[x]$.
   (b) If $f(x)$ is reducible in $F[x]$ then $f(x)$ has a root in F.

2. Find all irreducible quadratic polynomials in $\mathbb{Z}_3[x]$.
3. Explain why checking irreducibility is much easier for cubic (degree 3) polynomials than for quartic (degree 4) polynomials.
4. Which of the following polynomials are irreducible in $\mathbb{Z}_3[x]$:

   (i) $f(x) = x^3 + 2x + 2$,
   (ii) $g(x) = x^4 + 2x^3 + 2x + 1$,
   (iii) $h(x) = x^4 + x^3 + x^2 + x + 1$?

5. Represent $f(x) = x^5 + x + 1 \in \mathbb{Z}_2[x]$ as a product of irreducible polynomials.
6. Show that $f(x) = x^5 + x^2 + 1 \in \mathbb{Z}_2[x]$ is an irreducible polynomial over $\mathbb{Z}_2$.

5.1.4 Greatest Common Divisor and Least Common Multiple

Definition 5.1.7 Let F be a field and $f(x), g(x)$ be two polynomials from $F[x]$. A monic polynomial $d(x) \in F[x]$ is called the *greatest common divisor* of $f(x)$ and $g(x)$ iff:

(a) $d(x)$ divides both $f(x)$ and $g(x)$, and
(b) $d(x)$ is of maximal degree with the above property.

The greatest common divisor of $f(x)$ and $g(x)$ is denoted $\gcd(f(x), g(x))$ or $\gcd(f, g)(x)$. Its uniqueness follows from the following

Theorem 5.1.5 (The Euclidean Algorithm) *Let f and g be two polynomials. We use the division algorithm several times to find:*

$$
\begin{aligned}
f &= q_1 g + r_1, & deg\,(r_1) &< deg\,(g), \\
g &= q_2 r_1 + r_2, & deg\,(r_2) &< deg\,(r_1), \\
r_1 &= q_3 r_2 + r_3, & deg\,(r_3) &< deg\,(r_2), \\
&\;\;\vdots \\
r_{s-2} &= q_s r_{s-1} + r_s, & deg\,(r_s) &< deg\,(r_{s-1}), \\
r_{s-1} &= q_{s+1} r_s.
\end{aligned}
$$

Then all common divisors of f and g are also divisors of r_s. Moreover, r_s divides both f and g. Thus $r_s = \gcd(f, g)$.

The extended Euclidean algorithm also holds.

Theorem 5.1.6 (The Extended Euclidean algorithm) *Let f and g be two polynomials. Let us form the following matrix with two rows R_1, R_2, and three columns C_1, C_2, C_3:*

$$
(C_1\, C_2\, C_3) = \begin{pmatrix} f & 1 & 0 \\ g & 0 & 1 \end{pmatrix}.
$$

In accordance with the Euclidean algorithm above, we perform elementary row operations $R_3 := R_1 - q_1 R_2$, $R_4 := R_2 - q_2 R_3$, $\ldots$, each time creating a new row, so as to obtain:

$$
(C_1'\, C_2'\, C_3') = \begin{pmatrix} f & 1 & 0 \\ g & 0 & 1 \\ r_1 & 1 & -q_1 \\ r_2 & -q_2 & 1 + q_1 q_2 \\ & \vdots & \\ r_s & m & n \end{pmatrix}.
$$

Then $gcd(f, g)(x) = r_s(x) = f(x)m(x) + g(x)n(x)$.

Proof The proof is exactly the same as for numbers.   □

Example 5.1.6 Let $f(x) = x^4 + x^3 + x^2 + 1$ and $g(x) = x^4 + x^2 + x + 1$ are from $\mathbb{Z}_2[x]$. We write:

$$x^4 + x^3 + x^2 + 1 = (x^4 + x^2 + x + 1) \cdot 1 + (x^3 + x)$$
$$x^4 + x^2 + x + 1 = (x^3 + x) \cdot x + (x + 1)$$
$$x^3 + x = (x + 1) \cdot (x^2 + x).$$

So $gcd(f, g)(x) = x + 1$.

The extended Euclidean algorithm gives

$x^4 + x^3 + x^2 + 1$	1	0
$x^4 + x^2 + x + 1$	0	1
$x^3 + x$	1	1
$x + 1$	x	$x + 1$

Hence $gcd(f, g)(x) = x + 1 = f(x)x + g(x)(x + 1)$.

Definition 5.1.8 Two polynomials $f(x), g(x) \in F[x]$ are said to be *coprime (relatively prime)* if $gcd(f, g)(x) = 1$.

Corollary 5.1.2 *Two polynomials $f(x), g(x) \in F[x]$ are coprime if and only if there exist polynomials $m(x), n(x) \in F[x]$ such that*

$$1 = f(x)m(x) + g(x)n(x).$$

Definition 5.1.9 Let F be a field and $f(x), g(x)$ be two polynomials from $F[x]$. A monic polynomial $m(x) \in F[x]$ is called *the least common multiple* of $f(x)$ and $g(x)$ if:

(a) $m(x)$ is a multiple of both $f(x)$ and $g(x)$;
(b) $m(x)$ is of minimal degree with the above property

It is denoted $lcm(f(x), g(x))$ or $lcm(f, g)(x)$.

All the usual properties of the least common multiple are satisfied. For example, as for the integers, we can prove:

Theorem 5.1.7 *Let $f(x)$ and $g(x)$ be two monic polynomials in $F[x]$. Then*

$$lcm(f(x), g(x)) \cdot gcd(f(x), g(x)) = f(x)g(x).$$

Example 5.1.7 Let $f(x) = x^4 + x^3 + x^2 + 1$ and $g(x) = x^4 + x^2 + x + 1$ be two polynomials in $\mathbb{Z}_2[x]$. We know that $gcd(f, g)(x) = x + 1$. Hence $lcm(f, g)(x) = \frac{f(x)g(x)}{x+1} = x^7 + x^6 + x^5 + x^4 + x^2 + 1$.

Exercises

1. Find the greatest common divisor $d(x)$ of the polynomials $f(x) = x^7 + 1$ and $g(x) = x^3 + x^2 + x + 1$ in $\mathbb{Z}_2[x]$ and represent it in the form $d(x) = f(x)m(x) + g(x)n(x)$.
2. Let $f(x) = x^5 + x^4 + 1$ and $g(x) = x^5 + x + 1$ in $\mathbb{Z}_2[x]$. Calculate by hand and check with GAP:

   (a) $gcd(f(x), g(x))$ and its representation in the form $a(x)f(x) + b(x)g(x)$,
   (b) $lcm(f(x), g(x))$.

   Check that $gcd(f(x), g(x))lcm(f(x), g(x)) = f(x)g(x)$.
3. Let $f(x) = a_0 + a_1x + \ldots + a_nx^n$ be a polynomial from $F[x]$, where F is any field. We define the derivative of $f(x)$ by the formula:

$$f'(x) = a_1 + 2a_2x + \ldots + na_nx^{n-1}.$$

   (a) Check that the product rule holds for such a derivative.
   (b) Prove that any multiple root of $f(x)$ is also a root of $gcd(f(x), f'(x))$.
   (c) Let p be a prime. Prove that the polynomial $f(x) = x^{p^n} - x$ does not have multiple roots in any field F of characteristic p.

5.2 Finite Fields

5.2.1 Polynomials Modulo $m(x)$

Let $m(x)$ be a polynomial over F of degree n. Let us consider the set

$$F[x]/(m(x)) = \{f(x) \mid f = 0 \text{ or } \deg(f) < n\}$$

of all polynomials of degree lower than n. This is exactly the set of all possible remainders on division by $m(x)$. Clearly $F[x]/(m(x))$ is an n dimensional vector space over F spanned by the monomials $1, x, \ldots, x^{n-1}$.

Let $f(x)$ be a polynomial from $F[x]$ and $r(x)$ be its remainder on division by $m(x)$. We denote

$$r(x) = f(x) \mod m(x).$$

We will also write $f(x) \equiv g(x) \mod m(x)$ if $f(x) \mod m(x) = g(x) \mod m(x)$. Note that $f(x) \mod m(x)$ belongs to $F[x]/(m(x))$ for all $f(x) \in F[x]$.

Let us now convert $F[x]/(m(x))$ into a ring[2] by introducing the following addition and multiplication:

$$f(x) \oplus g(x) := (f + g)(x) \mod m(x), \tag{5.5}$$

$$f(x) \odot g(x) := fg(x) \mod m(x). \tag{5.6}$$

Note that the "new" addition is not really new as it coincides with the old one. But we do indeed get a new multiplication. All properties of a commutative ring for $F[x]/(m(x))$ can be easily verified.

Example 5.2.1 Let us consider the ring $\mathbb{R}[x]/(x^2 + 1)$. Since $\deg(x^2 + 1) = 2$, this is two-dimensional space over the reals with basis $\{1, x\}$. The addition is

$$(a \cdot 1 + bx) \oplus (c \cdot 1 + dx) = (a + c) \cdot 1 + (b + d)x,$$

and the multiplication

$$(a \cdot 1 + bx) \odot (c \cdot 1 + dx) = ac \cdot 1 + (ad + bc)x + bdx^2$$
$$\equiv (ac - bd) \cdot 1 + (ad + bc)x.$$

One must be able to recognise the complex numbers (with x playing the role of i). In mathematical language the ring $\mathbb{R}[x]/(x^2 + 1)$ is said to be isomorphic to $\mathbb{C}$.

As in the case of the integers, and by using the same approach, we can prove

Theorem 5.2.1 $F[x]/(m(x))$ *is a field if and only if* $m(x)$ *is irreducible over* F.

Proof Suppose $m(x)$ is of degree n and is irreducible over F. Then we need to show that every nonzero polynomial $f(x) \in F[x]/(m(x))$ is invertible. We know that $\deg(f) < n$. Since $m(x)$ is irreducible we have $\gcd(f, m) = 1$ and by the extended Euclidean logrithm we can find $a(x), b(x) \in F[x]$ such that $a(x)f(x) + b(x)m(x) = 1$. Let us divide $a(x)$ by $m(x)$ with remainder: $a(x) = q(x)m(x) + r(x)$ and substitute into the previous equation. We will obtain

$$r(x)f(x) + (q(x)f(x) + b(x))m(x) = 1.$$

This means that $r(x) \odot f(x) = 1$ in $F[x]/(m(x))$, thus $f(x)$ is invertible and $r(x)$ is its inverse.

On the other hand, if $m(x)$ is not irreducible, we can write $m(x) = n(x)k(x)$, which will lead to $n(x) \odot k(x) = 0$ in $F[x]/(m(x))$. Then, having divisors of zero, by Lemma 1.4.2 $F[x]/(m(x))$ cannot be a field.   □

---

[2]Those familiar with basics of abstract algebra will recognise the quotient-ring of $F[x]$ by the principal ideal generated by $m(x)$.

From now on, we will not use the special symbols $\oplus$ and $\odot$ to denote the operations in $F[x]/(m(x))$; this will invite no confusion.

Example 5.2.2 Prove that $K = \mathbb{Z}_2[x]/(x^4 + x + 1)$ is a field, and determine how many elements it has. Then find $(x^3 + x^2)^{-1}$.

Solution To prove that K is a field we must prove that $m(x) = x^4 + x + 1$ is irreducible. If it were reducible, then it would have a factor of degree 1 or 2. Since $m(0) = m(1) = 1$, it does not have linear factors. So, if it is reducible, the only possibility left is that it is the square of the only irreducible polynomial of degree 2, that is $(x^2 + x + 1)^2 = x^4 + x^2 + 1$. This does not coincide with $m(x)$, hence $m(x)$ is irreducible. Hence K is a field. Since $\dim_{\mathbb{Z}_2} K = \deg(m(x)) = 4$, K has $2^4 = 16$ elements.

By using the extended Euclidean algorithm we get

$x^4 + x + 1$	1	0
$x^3 + x^2$	0	1
$x^2 + x + 1$	1	$x + 1$
x	x	$x^2 + x + 1$
1	$x^2 + x + 1$	$x^3 + x$

Thus $(x^3 + x^2)^{-1} = x^3 + x$. $\square$

Example 5.2.3 Let us continue to investigate $K = \mathbb{Z}_2[x]/(x^4 + x + 1)$ for a while. We know that, as a finite field, K must have a primitive element, in fact $\phi(15) = 8$ of them. The polynomial $x^4 + x + 1$ is very convenient since x is one of the primitive elements of K. Let us compute powers of x and place all elements of K in the following table below.

Note that $x^{15} = 1$, so logs are manipulated mod 15. We have now two different representations of elements of K: as tuples (or polynomials) and as powers. The first representation is best for calculating additions and the second for calculating multiplications and inverses.

Construction of the field K

4-tuple	Polynomial	Power of x	Logarithm
0000	0		
1000	1	1	0
0100	x	x	1
0010	x^2	x^2	2
0001	x^3	x^3	3
1100	$1+x$	x^4	4
0110	$x+x^2$	x^5	5
0011	x^2+x^3	x^6	6
1101	$1+x+x^3$	x^7	7
1010	$1+x^2$	x^8	8
0101	$x+x^3$	x^9	9
1110	$1+x+x^2$	x^{10}	10
0111	$x+x^2+x^3$	x^{11}	11
1111	$1+x+x^2+x^3$	x^{12}	12
1011	$1+x^2+x^3$	x^{13}	13
1001	$1+x^3$	x^{14}	14

The following calculations clarify the use of this table:

1. $(1+x^2)^{-1} = (x^8)^{-1} = x^{-8} = x^{15-8} = x^7 = 1+x+x^3$.
2. $\log(x+x^2+x^3) = 11$ and $\log(1+x+x^2+x^3) = 12$. Thus $\log(x+x^2+x^3)$ $(1+x+x^2+x^3) = (11+12) \bmod 15 = 8$, hence $(x+x^2+x^3)(1+x+x^2+x^3) = 1+x^2$.

Theorem 5.2.1 allows us to construct a field of cardinality p^n for any prime p and any positive integer n. All we need to do is to take $\mathbb{Z}_p$ and an irreducible polynomial $m(x)$ of degree n. Then $\mathbb{Z}_p[x]/(m(x))$ is the desired field. In this book we will not prove that for any p and any positive integer n such a polynomial indeed exists (although it does!). Moreover, for any prime p and positive integer n the field of p^n elements is unique up to an isomorphism. This is why it is denoted $GF(p^n)$ and called the Galois[3] field of cardinality p^n. Again, proving its uniqueness is beyond the scope of this book.

Theorem 5.2.2 *For any prime p and any positive integer n there exists a unique, up to an isomorphism, field $GF(p^n)$ consisting of p^n elements.*

In the advanced encryption standard (AES) algorithm adopted in 2001, the field $GF(2^8)$ is used for calculations. This field is constructed with the use of the irreducible polynomial $m(x) = x^8 + x^4 + x^3 + x + 1$.

---

[3] See Sect. 3.1.3 for a brief historic note about this mathematician.

Exercises

1. Use the extended Euclidean algorithm to find the (multiplicative) inverse of $\beta = 1 + x + x^2 + x^3$ in $F = \mathbb{Z}_2[x]/(x^5 + x^3 + 1)$.
2. Let $F = \mathbb{Z}_3[x]/(x^2 + 2x + 2)$.

   (a) Prove that F is a field.
   (b) List all elements of F.
   (c) Show that $2x + 1$ is a primitive element in F by calculating all powers of $2x + 1$ and constructing the "logarithm table" as in Example 5.2.3.
   (d) Using the "logarithm table" which you created in part (c), calculate

$$2x^7(x + 1)^{-5}(2x + 2) + (x + 2)^5.$$

   (e) How many primitive elements are there in the field F? List them all.

3. Generate a field consisting of 16 elements using GAP. It will give you:

```
gap> F:=GaloisField(2^4);
GF(2^4)
gap> AsList(F);
[ 0*Z(2), Z(2)^0, Z(2^2), Z(2^2)^2, Z(2^4), Z(2^4)^2, Z(2^4)^3, Z(2^4)^4,
  Z(2^4)^6, Z(2^4)^7, Z(2^4)^8, Z(2^4)^9, Z(2^4)^11, Z(2^4)^12, Z(2^4)^13,
  Z(2^4)^14 ]
```

   (a) Explain why $Z(2^4)^5$ and $Z(2^4)^{10}$ are not listed among the elements.
   (b) Using GAP find the polynomial in $\mathbb{Z}_2[x]$ of smallest degree of which $Z(2^4)^7$ is a root.

4°. Let $x^5 + x^4 + x^2 + x + 1 \in \mathbb{Z}_2$.

   (a) Show that this polynomial is irreducible over $\mathbb{Z}_2$.
   (b) How many elements does the field $K = \mathbb{Z}_2[x]/(x^5 + x^4 + x^2 + x + 1)$ have? Which of the fields $GF(p^n)$ is it isomorphic to?
   (c) Find $(x^3 + 1)^{-1}$ in K using extended Euclidean algorithm.

5°. A field F with 9 elements can be constructed from $\mathbb{Z}_3$ as $\mathbb{Z}_3[x]/(x^2 + 1)$.

   (a) Show that $g = 2x + 1$ is a primitive element in F by calculating all powers of $2x + 1$.
   (b) Create the logarithm table for this field.
   (c) How many primitive elements are in the field F? List them all.

5.2.2   Minimal Annihilating Polynomials

Let F and K be two fields such that $F \subseteq K$. We say that F is a *subfield* of K and that K is an *extension* of F if the addition and multiplication in K, being restricted to F, coincide with the operations in F of the same name.

Example 5.2.4 Elements 0 and 1 of $\mathbb{Z}_2$ can be identified with constant polynomials 0 and 1 of $K = \mathbb{Z}_2[x]/(x^4 + x + 1)$. So $\mathbb{Z}_2$ is a subfield of $K = \mathbb{Z}_2[x]/(x^4 + x + 1)$.

Definition 5.2.1 Let $F \subset K$ be an extension of fields and $a \in K$. We say that a polynomial $f(t) \in F[t]$ is an *annihilating polynomial* of a if a is a root of $f(t)$, i.e., $f(a) = 0$. (Please note that the coefficients of $f(t)$ lie in F while a is an element of K.) A polynomial $f(t) \in F[t]$ is called the *minimal annihilating polynomial* of a over F if it is an annihilating polynomial which is monic and of minimal possible degree.

Example 5.2.5 In the extension $\mathbb{R} \subseteq \mathbb{C}$, check that the polynomial $f(t) = t^2 - 2t + 2$ is the minimal annihilating polynomial for $a = 1 + i$ over $\mathbb{R}$.

Solution Indeed, we have $f(a) = (1 + i)^2 - 2(1 + i) + 2 = 2i - 2 - 2i + 2 = 0$ so $f(t)$ is annihilating for a. At the same time there can be no linear annihilating polynomial. Such polynomial would have real coefficients and hence would be of the form $g(t) = t - r$, where $r \in \mathbb{R}$. Substituting a will give $(1 + i) - r = 0$ which is not possible.   □

Exercise 5.2.1 Every complex number has an annihilating polynomial over $\mathbb{R}$ which is at most quadratic.

Example 5.2.6 In the extension $\mathbb{Z}_2 \subseteq \mathbb{Z}_2[x]/(x^4 + x + 1)$ the polynomial $f(t) = t^4 + t + 1$ is the minimal annihilating polynomial for x.

Solution We note first that $f(x) = x^4 + x + 1 \equiv 0 \mod x^4 + x + 1$, hence $f(t)$ is an annihilating polynomial for x. On the other hand, if it were possible to find an annihilating polynomial of degree 3 or smaller, say $g(t) = \alpha t^3 + \beta t^2 + \gamma t + \delta 1$ with at least one coefficient nonzero, then

$$\alpha x^3 + \beta x^2 + \gamma x + \delta 1 = 0,$$

which means that $1, x, x^2, x^3$ are linearly dependent over $\mathbb{Z}_2$. But this was a basis of $\mathbb{Z}_2[x]/(x^4 + x + 1)$, so we have drawn a contradiction.   □

Theorem 5.2.3 *Let $F \subset K$ be an extension of fields such that $\dim_F K = n$ and $a \in K$. Then the minimal annihilating polynomial for a has degree at most n.*

Proof Let us consider the first $n + 1$ powers of a, that is, $1 = a^0, a, a^2, \ldots, a^n$. Since the dimension of K over F is n, these $n + 1$ vectors must be linearly dependent over F. Thus there exist $c_0, c_1, \ldots, c_n \in F$, not all zero, such that

$$c_0 1 + c_1 a + c_2 a^2 + \ldots + c_n a^n = 0.$$

This is the same to say as $f(a) = 0$ for $f(t) = c_0 + c_1 t + \ldots + c_n t^n$ from $F[t]$, so we have found an annihilating polynomial of degree at most n. $\square$

Theorem 5.2.4 *Let $F \subset K$ be an extension of fields and $a \in K$. Then*

(i) *The minimal annihilating polynomial of a is irreducible over F.*
(ii) *Every annihilating polynomial of a is a multiple of the minimal annihilating polynomial of a.*

Proof (i) Suppose that $f(t)$ is the minimal annihilating polynomial of a and that it is reducible, i.e., $f(t) = g(t)h(t)$, where $g(t)$ and $h(t)$ can be considered monic and each of degree strictly less than $\deg(f)$. Then $0 = f(a) = g(a)h(a)$, whence (no zero divisors in K) either $g(a) = 0$ or $h(a) = 0$, which contradicts the minimality of $f(t)$.

(ii) Suppose that $f(t)$ is the minimal annihilating polynomial of a and $g(t)$ is any other annihilating polynomial of a. Let us divide $g(t)$ by $f(t)$ with remainder:

$$g(t) = f(t)q(t) + r(t), \qquad r = 0 \text{ or } \deg(r) < \deg(f).$$

We claim that $r = 0$. Otherwise we substitute a to obtain $g(a) = f(a)q(a) + r(a)$ or $0 = 0 + r(a)$, from which $r(a) = 0$ but the degree of $r(t)$ is strictly smaller than that of $f(t)$, and thus we have arrived at a contradiction. $\square$

For calculating the minimal annihilating polynomial we use the linear dependency relationship algorithm (see description of it in Sect. 10.1). Suppose we need to find the minimal annihilating polynomial of an element $a \in K$ over a subfield F of K. Suppose $n = \dim_F K$. We choose any basis B of K over F. Then every element $x \in K$ can be represented by its coordinate column $[x]_B$ relative to the basis B. For element $a \in K$ we consider the matrix $A = ([1]_B \ [a]_B \ [a^2]_B \ \ldots [a^n]_B)$. Its columns are linearly dependent (as any $n + 1$ vector in n-dimensional vector space). By row reducing A to its reduced echelon form we find the first k such that $\{[1]_B, [a]_B, [a^2]_B, \ldots, [a^k]_B\}$ is linearly dependent. This reduced row echelon form will also give us coefficients $c_0, c_1, c_2, \ldots, c_{k-1}$ such that $c_0[1]_B + c_1[a]_B + c_2[a^2]_B + \ldots + c_{k-1}[a^{k-1}]_B + [a^k]_B = \mathbf{0}$. Then $f(x) = x^k + c_{k-1}x^{k-1} + \ldots + c_1 x + c_0$ is the minimal annihilating polynomial of a over F.

Example 5.2.7 In the extension $\mathbb{Z}_2 \subseteq \mathbb{Z}_2[x]/(x^4 + x + 1)$, find the minimal annihilating polynomial of $a = 1 + x + x^3$.

Solution We calculate the coordinate tuples of the following powers of a:

$$a^0 = (1 + x + x^3)^0 = 1 \qquad\qquad \rightarrow 1000$$
$$a^1 = (1 + x + x^3)^1 = 1 + x + x^3 \rightarrow 1101$$
$$a^2 = (1 + x + x^3)^2 = 1 + x^3 \qquad \rightarrow 1001$$
$$a^3 = (1 + x + x^3)^3 = x^2 + x^3 \qquad \rightarrow 0011$$
$$a^4 = (1 + x + x^3)^4 = 1 + x^2 + x^3 \rightarrow 1011$$

These five are already linearly dependent, so we do not have to compute any further powers. Now we use Linear Dependency Relationship Algorithm to find linear dependency between these tuples. We place them as columns in a matrix and take it to the row reduced echelon form:

$$\begin{pmatrix} 1 & 1 & 1 & 0 & 1 \\ 0 & 1 & 0 & 0 & 0 \\ 0 & 0 & 0 & 1 & 1 \\ 0 & 1 & 1 & 1 & 1 \end{pmatrix} \xrightarrow{rref} \begin{pmatrix} 1 & 0 & 0 & 0 & 1 \\ 0 & 1 & 0 & 0 & 0 \\ 0 & 0 & 1 & 0 & 0 \\ 0 & 0 & 0 & 1 & 1 \end{pmatrix},$$

from which it follows that $1, a, a^2, a^3$ are linearly independent (hence no annihilating polynomials of degree ≤ 3) and that $a^4 = 1 + a^3$, whence the minimal annihilating polynomial of a will be $f(t) = t^4 + t^3 + 1$.

Exercises

1. What is the dimension of the field $F = GF(2^4)$ over its subfield $F_1 = GF(2^2)$?
2. Let $K = \mathbb{Z}_2[x]/(1 + x + x^4)$ introduced in Example 5.2.3. Find the minimal annihilating polynomial over $\mathbb{Z}_2$ for:

   (a) $\alpha = 1 + x + x^2$;
   (b) $\alpha = 1 + x$.

3. Let K be the field $K = \mathbb{Z}_2[x]/(x^4 + x^3 + 1)$. Then K is an extension of $\mathbb{Z}_2$.

   (a) Create a table for K as in Example 5.2.3. Check that x is a primitive element of this field.
   (b) Find the minimal annihilator polynomials for x, x^3 and x^5 over $\mathbb{Z}_2$.
   (c) Calculate $(x^{100} + x + 1)(x^3 + x^2 + x + 1)^{15} + x^3 + x + 1$ in the most efficient way and represent it as a power of x and as a polynomial in x of degree at most 3.

4°. Let $m(x)$ be a minimal annihilating polynomial of $a \in GF(p^n)$ over $GF(p)$. Prove that $m(x)$ divides $x^{p^n-1} - 1$.
5°. Let $m_1(x), \ldots, m_k(x)$ be all of the distinct minimal annihilating polynomials for nonzero elements of $GF(p^n)$ over $GF(p)$. Prove that

$$\prod_{i=1}^{k} m_i(x) = x^{p^n-1} - 1.$$

5.3 Permutation Polynomials and Applications

5.3.1 Permutation Polynomials

Let $F = \mathbb{Z}_n$ where $n \geq 2$ is a positive integer. A polynomial

$$f(x) = \sum_{i=0}^{k} a_i x^i, \qquad a_i \in F$$

defines the function $f : F \to F$ (which we denote with the same letter f) given by

$$f(\alpha) = \sum_{i=0}^{k} a_i \alpha^i.$$

(We have seen this in Sect. 5.1.1 in case F is a field.) As we have seen this function may not be a bijection (it can be identically zero even if F is a field). Those polynomials for which this function is a bijection have numerous applications.

Definition 5.3.1 A polynomial $f(x) \in F[x]$ is said to be a *permutation polynomial* if the mapping $\alpha \mapsto f(\alpha)$ from F to F is a bijection, i.e., onto and one-to-one.

Example 5.3.1 Let us consider the polynomial $f(x) = x^5 + 2x^2 \in \mathbb{Z}_7[x]$. Then we have

$$f(0) = 0, \quad f(1) = 3, \quad f(2) = 5, \quad f(3) = 2, \quad f(4) = 6, \quad f(5) = 4, \quad f(6) = 1.$$

Hence this is a permutation polynomial. The permutation corresponding to this polynomial will be

$$\pi_f = \begin{pmatrix} 1\,2\,3\,4\,5\,6\,7 \\ 1\,4\,6\,3\,7\,5\,2 \end{pmatrix} = (2\ 4\ 3\ 6\ 5\ 7),$$

i.e., this is a 6-cycle.

Example 5.3.2 The polynomial $f(x) = x(2x + 1)$ is a permutation polynomial in $\mathbb{Z}_8[x]$. We can check it with GAP:

```
gap> s:=[0,0,0,0,0,0,0,0];;
gap> for i in [1..8] do
> s[i]:=(2*(i-1)^2+(i-1)) mod 8;
> od;
gap> s;
[ 0, 3, 2, 5, 4, 7, 6, 1 ]
```

For the general case of this statement see Exercise 1.

Proposition 5.3.1 *Let* F *be a finite field. Then for every permutation polynomial* $f(x)$ *of* $F[x]$, *there exists a unique polynomial* $f^{-1}(x)$ *of* $F[x]$ *such that* $f(f^{-1}(x)) = f^{-1}(f(x)) = x$ *for all* $x \in F$; *it is called the* (compositional) *inverse of* $f(x)$.

Proof Since f is a bijection, it has an inverse f^{-1}. By Exercise 5 of Sect. 5.1.2 this inverse must be a polynomial. □

Example 5.3.3 (*Continuation of Example* 5.3.1) The compositional inverse $f^{-1}(x)$ can be calculated by Lagrange's interpolation. GAP shows:

```
gap> InterpolatedPolynomial(GF(7),[0,3,5,2,6,4,1],[0,1,2,3,4,5,6]);
x_1^5+Z(7)^5*x_1^2
gap> Int(Z(7)^5);
5
```

So $f^{-1}(x) = x^5 + 5x^2$.

If F is not a field, interpolation task becomes tricky. In particular, not every function can be represented by a polynomial (see Exercise 2).

Example 5.3.4 Every linear polynomial $f(x) = ax + b$ with $a \in \mathbb{Z}_n^*$ is a permutation polynomial in $\mathbb{Z}_n^*$. We know this since the compositional inverse is easy to calculate: $f^{-1}(x) = a^{-1}(x - b)$.

Example 5.3.5 If $n = pq$, where p and q are distinct primes, then for any positive integer e such that $\gcd(e, \phi(n)) = 1$ the monomial $f(x) = x^e \mod n$ is a permutation polynomial in $\mathbb{Z}_n[x]$. This must be true since we know from (2.3) that $f^{-1}(x) = x^d \mod n$, where $d = e^{-1} \mod \phi(n)$.

5.3.2 Cryptosystem Based on a Permutation Polynomial

In RSA cryptosystem, as we seen in Example 5.3.5, the public-key encryption consists in the evaluation of a permutation polynomial. We can generalise this as follows. Let $f(x) \in \mathbb{Z}_n[x]$ be a permutation polynomial whose compositional inverse is difficult to calculate without knowing the factorisation of n. The encryption will be the evaluation of $f(m)$ at a given value m which represents the plaintext:

$$c := f(m) \mod n.$$

Obviously, for the permutation polynomial $f(x)$ to be used in a practical public-key scheme there must be an efficient algorithm to evaluate this polynomial. Such an algorithm exists if $f(x)$ has small degree, or small number of terms (then Square-and-Multiply algorithm can be used).

The decryption is also implemented as the evaluation of the inverse polynomial f^{-1}, which is a private key, at the value which represents the ciphertext c:

$$m := f^{-1}(c) \mod n.$$

We can have f^{-1} as a private key since it is not feasible to calculate it without knowing the factorisation of $n = pq$ which is a trapdoor. However, knowing this factorisation finding f^{-1} should be easy. In general, however, it is difficult to construct permutation polynomials whose inverses are known or are not too complicated to construct. J. Schwenk and K. Huber (1998) suggested a scheme in which a much larger class of permutation polynomials can be used for which there is no easy algorithm for calculation of $f^{-1}(c)$ even knowing the prime factorisation of n. The idea is to find a unique root of the equation $f(x) - c = 0$ in $\mathbb{Z}_n$. This algorithm is based on the following lemmas.

Lemma 5.3.1 *Let $n = pq$, where p and q are distinct primes and let $f(x) = a_k x^k + \ldots + a_1 x + a_0$ be a permutation polynomial in $\mathbb{Z}_n$. Then*

$$f_p(x) = \overline{a_k} x^k + \ldots + \overline{a_1} x + \overline{a_0},$$

where $\overline{a_i} = a_i \mod p$, is a permutation polynomial in $\mathbb{Z}_p$.

Proof Let $u \in \mathbb{Z}_p$. Let us choose any $v \in \mathbb{Z}_n$ such that $u = v \mod p$. Since $f(x)$ is a permutation polynomial, there exist $z \in \mathbb{Z}_n$ such that $f(z) = v$. Reducing this equation modulo p we get $f_p(\bar{z}) = u$, where $\bar{z} = z \mod p$. Hence $f_p \colon \mathbb{Z}_p \to \mathbb{Z}_p$ is onto. Since $\mathbb{Z}_p$ is finite, f_p is a bijection, hence a permutation polynomial. $\square$

Lemma 5.3.2 *Let $f(x) \in \mathbb{Z}_p[x]$ be a permutation polynomial. Then for any $a \in \mathbb{Z}_p$ we have*

$$\gcd(f(x) - a, x^p - x) = x - b$$

for some $b \in \mathbb{Z}_p$.

Proof As we know $x^p - x$ splits into linear factors $x^p - x = \prod_{a \in \mathbb{Z}_p}(x - a)$ all of which are distinct. At the same time $f(x) - a$ has only one root, say $b \in \mathbb{Z}_p$ and is therefore equal to $f(x) - a = (x - b)g(x)$, where $g(x)$ has no roots in $\mathbb{Z}_p$ and therefore $\gcd(g(x), x^p - x) = 1$. The statement of the lemma follows. $\square$

Now the decryption procedure for $n = pq$ can proceed as follows. We find

$$\gcd(f_p(x) - a, x^p - x) = x - u, \qquad \gcd(f_q(x) - a, x^q - x) = x - v,$$

and then using the Chinese Remainder theorem we find $c \in \mathbb{Z}_n$ such that $u = c \mod p$ and $v = c \mod q$.

Exercises

1. Prove that $f(x) = x(2x + 1)$ is a permutation polynomial in $\mathbb{Z}_{2^n}[x]$.[4]
2. Let $F = \mathbb{Z}_4$. Consider the function

$$f(x) = \begin{cases} 0 & \text{if } x \in \{0, 1\}; \\ 1 & \text{if } x \in \{2, 3\}. \end{cases}$$

Show that it cannot be represented as a polynomial in $\mathbb{Z}_4[x]$.
3°. Show that permutation polynomials in $\mathbb{Z}_n$ form a group relative to the operation of composition.

---

[4]In Rivest, Ronald L., et al. "The RC6 block cipher". in First Advanced Encryption Standard (AES) Conference. 1998 it was used that polynomial $f(x) = x(2x + 1)$ is a permutation polynomial in $\mathbb{Z}_w[x]$, where w is the word size of the machine.

Secret Sharing

6

> *The very word 'secrecy' is repugnant in a free and open society;*
> *and we are as a people inherently and historically opposed to*
> *secret societies, to secret oaths, and to secret proceedings.*
>
> *John F. Kennedy (1917–1963)*
>
> *Secrecy is the first essential in affairs of state.*
>
> *Cardinal Richelieu (1585–1642)*

Certain cryptographic keys, such as missile launch codes, numbered bank accounts and the secret decoding exponent in an RSA public-key cryptosystem, are so important that they present a dilemma. If too many copies are distributed, one may be leaked. If too few, they might all be lost or accidentally destroyed. Secret sharing schemes invented by Shamir (1979) and Blakley (1979) address this problem and allow arbitrarily high levels of confidentiality and reliability to be achieved. A secret sharing scheme "divides" the secret s into "shares"—one for every user—in such a way that s can be easily reconstructible by any authorised subset of users, but an unauthorised subset of users can extract absolutely no information about s. A secret sharing scheme, for example, can secure a secret over multiple servers and remain recoverable despite multiple server failures.

Secret sharing also provides a mechanism to facilitate a cooperation—in both human and artificial societies—when cooperating agents have different status with respect to the activity and certain actions are only allowed to coalitions that satisfy certain criteria, e.g., to sufficiently large coalitions or coalitions with players of sufficient seniority or to coalitions that satisfy a combination of both criteria. The banking system where the employees are arranged into a hierarchy according to their ranks or designations provides many examples. Simmons,[1] for example, describe the situation of a money transfer from one bank to another. If the sum to be transferred

---

[1] Simmons, G. (1990). How to (really) share a secret. In: Proceedings of the 8th annual international cryptology conference on advances in cryptology (pp. 390–448). London, UK: Springer-Verlag.

© Springer Nature Switzerland AG 2020

A. Slinko, *Algebra for Applications*, Springer Undergraduate Mathematics Series,
https://doi.org/10.1007/978-3-030-44074-9_6

is sufficiently large, this transaction must be authorised by three senior tellers *or* two vice-presidents. However, two senior tellers and a vice-president can also authorise the transaction. Tassa[2] provides another banking scenario. The shares of the vault key may be distributed among bank employees, some of whom are tellers and some are department managers. The bank policy could require the presence of, say, three employees in opening of the vault, but at least one of them must be a departmental manager.

6.1   Introduction to Secret Sharing

6.1.1   Access Structure

More formally, we assume that the set of users is $U = \{1, 2, \ldots, n\}$ and D is the dealer who facilitates secret sharing.[3] It is always assumed that the dealer knows the secret.

Definition 6.1.1 Let 2^U be the power set[4] of the set of all users U. The set $\Gamma \subseteq 2^U$ of all authorised coalitions is called the *access structure* of the secret sharing scheme. An *access structure* Γ is any subset of 2^U such that

$$X \in \Gamma \text{ and } X \subseteq Y, \text{ then } Y \in \Gamma. \tag{6.1}$$

The condition in the definition of an access structure is called the *monotone property*, and it reflects the natural requirement that if a smaller coalition knows the secret, then the larger one will know it too. The access structure is public knowledge, and all users know it.

Let $\Gamma \subseteq 2^U$ be an access structure. A coalition $C \subseteq U$ is called *minimal authorised coalition* if it is authorised and any proper subset of C is not authorised. Due to the monotone property (6.1) the access structure is completely defined by the set $\Gamma_{\min}$ of its minimal authorised coalitions.

We assume that every user participates in at least one minimal authorised coalition. If not, such a user never brings a useful information to any coalition of users, is redundant and called a *dummy*.

Example 6.1.1 The *threshold access structure* "k-out-of-n" consists of all subsets of 2^U consisting of k or more users.

According to Time Magazine, May 4, 1992, a typical threshold access structure was realised in USSR. The three top state officials, the President, the Prime Minister

---

[2]Tassa, T. (2007). Hierarchical threshold secret sharing. Journal of Cryptology, 20, 237–264.
[3]The dealer is not necessarily a person, this can be a computer.
[4]The set of all subsets of U.

and the Minister of Defence, each had the so-called nuclear suitcase and any two of them could authorise a launch of a nuclear warhead. No one of them could do it alone. So it was two-out-of-three threshold scheme.

In a two-out-of-three scheme $U = \{1, 2, 3\}$ and $\Gamma_{\min} = \{\{1, 2\}, \{1, 3\}, \{2, 3\}\}$. We see that all users are equally important. If however, $U = \{1, 2, 3\}$ and $\Gamma_{\min} = \{\{1, 2\}, \{1, 3\}\}$, then user 1 is much more important than the two other users. Without user 1 the secret cannot be accessed. But user 1 is not almighty. To access the secret, she needs to join forces with at least one other user.

Here is a couple of real-life examples.

Example 6.1.2 Consider the situation of a money transfer from one bank to another. If the sum to be transferred is sufficiently large, this transaction must be authorised by three senior tellers or two vice-presidents. However, two senior tellers and a vice-president can also authorise the transaction.

Example 6.1.3 The United Nations Security Council consists of five permanent members and ten non-permanent members. The passage of a resolution requires that all five permanent members vote for it and also at least nine members in total.

We will deal with threshold access structures first. A very elegant construction, by Shamir, realising the threshold access structure is based on Lagrange's interpolation polynomial and will be presented in the next section.

Exercises
1. Let $U = \{1, 2, 3, 4\}$ and $\Gamma_{\min} = \{\{1, 2, 3\}, \{3, 4\}\}$. List all authorised coalitions.
2. Write down the minimal authorised coalitions for the access structure in Example 6.1.2. Assume that the vice-presidents are users 1 and 2 and the senior tellers are users $3, 4, 5$.
3. Find the number of minimal authorised coalitions in Example 6.1.3.
4. Let U_1 and U_2 be disjoint sets of users and let Γ_1 and Γ_2 be access structures over U_1 and U_2, respectively. Let $U = U_1 \cup U_2$. Then
   (a) The *sum* of Γ_1 and Γ_2 is $\Gamma_1 + \Gamma_2 = \{X \subseteq U \mid X \cap U_1 \in \Gamma_1 \text{ or } X \cap U_2 \in \Gamma_2\}$. Prove that $\Gamma_1 + \Gamma_2$ is an access structure.
   (b) The *product* of Γ_1 and Γ_2 is $\Gamma_1 \times \Gamma_2 = \{X \subseteq U \mid X \cap U_1 \in \Gamma_1 \text{ and } X \cap U_2 \in \Gamma_2\}$. Prove that $\Gamma_1 \times \Gamma_2$ is an access structure.

6.1.2 Shamir's Threshold Access Scheme

In this section we will look at one particular application of polynomials to cryptography, namely to secret sharing.

Suppose that the secret is a string of zeros and ones. We may assume that it is the binary representation of a positive integer s. We choose a prime p which is sufficiently large. Then the field $\mathbb{Z}_p$ is large and we may assume that $s \in \mathbb{Z}_p$ without danger that it can be easily guessed. Thus our secret will always be an element of a finite field.

Suppose n users wish to share this secret by dividing it into "pieces" in such a way that *any* k people, where k is a fixed positive integer not exceeding n, can learn the secret from their pieces, but no subset of less than k people can do so. Here the word "dividing" must not be understood literally. Shamir proposed the following elegant solution to this problem. The secret can be "divided into pieces" as follows. The dealer:

1. generates k random coefficients $t_0, t_1, \ldots, t_{k-1} \in \mathbb{Z}_p$ and sets the secret s to be t_0;
2. forms the polynomial $p(x) = \sum_{i=0}^{k-1} t_i x^i \in \mathbb{Z}_p[x]$;
3. gives user i the "piece" $p(i)$, for $i = 1, \ldots, n$. Practically it can be an electronic card where a pair of numbers $(i, p(i))$ is stored.

Now, given any k values for $p(x)$, one can use Theorem 5.1.2 to interpolate and to find all coefficients of $p(x)$ including the secret $t_0 = s$. However, due to Corollary 5.1.1, a subset of $k-1$ values for $p(x)$ provides absolutely no information about s, since for any possible s there is a polynomial of degree $k-1$ consistent with the given values and the possible value of s.

Example 6.1.4 The company Dodgy Dealings Inc. has four directors. According to a clause in the company's constitution any three of them are allowed to get access to the company's secret offshore account. The company sets up a Shamir's threshold access secret sharing scheme for facilitating this clause with the secret password being an element of $\mathbb{Z}_7$. According to this scheme the system administrator issued magnetic cards to the directors as required.

Suppose that three directors with the following magnetic cards

$$\boxed{\begin{array}{c}1\\3\end{array}} \quad \boxed{\begin{array}{c}2\\0\end{array}} \quad \boxed{\begin{array}{c}4\\6\end{array}}$$

gathered to make a withdrawal from their offshore account. Show how the secret password can be calculated.

Solution A quadratic polynomial $p(x) = t_0 + t_1 x + t_2 x^2 \in \mathbb{Z}_7[x]$ satisfies

$$p(1) = 3, \quad p(2) = 0, \quad p(4) = 6;$$

given that, we must find c. Using the Lagrange interpolation formula

$$p(x) = 3\frac{(x-2)(x-4)}{(1-2)(1-4)} + 6\frac{(x-1)(x-2)}{(4-1)(4-2)} = 3\frac{(x+5)(x+3)}{3} + 6\frac{(x+6)(x+5)}{6}$$

$$= (x^2 + x + 1) + (x^2 + 4x + 2) = 2x^2 + 5x + 3,$$

hence the secret is $t_0 = 3$.   □

If in Shamir's scheme the enumeration of users is publicly known, then only the value $p(i)$ must be given to the ith user. In this case the secret s and each share $p(i)$ are both an element of the same field and need the same number of binary digits to encode them. As we will see one cannot do any better.

Exercises

1. According to the three-out-of-four Shamir's threshold secret sharing scheme with the secret in $\mathbb{Z}_7$ the administrator issued electronic cards to the users:

1	2	3	4
4	4	x	0

   (a) Show how the secret can be calculated by users 1,2 and 4.
   (b) Find x and determine the card of user 3.

2. Shamir's secret sharing scheme is set up so that the secret is an element of $\mathbb{Z}_{31}$ and the threshold is 3 which means that any three users are authorised. Show how the secret can be reconstructed from the shares

1	5	7
16	7	22

3. The league club Crawlers United has six senior board members. Each year the club holds an anniversary day, and on this day the senior board members have a duty to open the club vault, take out the club's meagre collection of trophies and put them on display. According to a clause in the club's constitution any four of them are allowed to open the vault. The club sets up a Shamir's threshold access secret sharing scheme for facilitating this clause with the secret password being an element of $\mathbb{Z}_{97}$. According to this scheme the administrator issued electronic cards to the senior board members as required.
   Suppose that four senior board members are gathered to open the vault with the following cards:

1	2	4	6
56	40	22	34

   (a) Show how the secret password can be calculated.
   (b) Guess which cards were given to the two remaining senior board members?
   Hint: Use GAP commands `InterpolatedPolynomial` and `Value`.

4°. Suppose you are an army cryptographer. In the army there are one General and five Lieutenant Generals. Your mission is to design a secret sharing scheme allowing one General and one Lieutenant General *or* five Lieutenant Generals to fire a missile. Accomplish your mission (You may give more than one share to an individual.).

5°. There are four persons A, B, C, D in a room, and one of them is a foreign spy. Other three participants share a secret using the Shamir's threshold scheme with secret in $\mathbb{Z}_p$, where $p = 11$ such that any two of them can recover the secret. The foreign spy chooses his share randomly. As a result, these four participants have the following four shares:

$$\boxed{\begin{array}{c} 1 \\ \hline 7 \end{array}} \quad \boxed{\begin{array}{c} 3 \\ \hline 0 \end{array}} \quad \boxed{\begin{array}{c} 5 \\ \hline 10 \end{array}} \quad \boxed{\begin{array}{c} 7 \\ \hline 9 \end{array}}$$

Find out who is the foreign spy and calculate the secret.

6.2   A General Theory of Secret Sharing Schemes

6.2.1   General Properties of Secret Sharing Schemes

Let us see now how we can define a secret sharing scheme formally.

Let $S_0, S_1, \ldots, S_n$ be finite sets where S_0 will be interpreted as a set of all possible secrets and S_i will be interpreted as a set of all possible shares that can be given to user i. Suppose $|S_i| = m_i$. We may think of a very large table, consisting of up to $M = m_0 m_1 \cdots m_n$ rows, where each row contains a tuple

$$(s_0, s_1, \ldots, s_n), \tag{6.2}$$

where s_i comes from S_i (and all rows are distinct). Mathematically, the set of all such $(n + 1)$-tuples is denoted by the Cartesian product $S_0 \times S_1 \times \ldots \times S_n$. Any subset

$$\mathcal{T} \subseteq S_0 \times S_1 \times \ldots \times S_n$$

of this Cartesian product is called a *distribution table*. Thus $\mathcal{T}$ consists of several rows like the one shown in (6.2). If a secret $s_0 \in S_0$ is to be distributed among users, then one $(n + 1)$-tuple

$$(s_0, s_1, \ldots, s_n) \in \mathcal{T}$$

is chosen by the dealer from $\mathcal{T}$ at random uniformly among those tuples whose first coordinate is s_0. Then user i gets the share $s_i \in S_i$.

There is only one but essential component of a secret sharing scheme that we have not introduced yet. We must ensure that every authorised coalition must be able to recover the secret. Thus we need to have, for every authorised coalition $X = \{i_1, i_2, \ldots, i_k\} \in \Gamma$ a *secret recovery function* (algorithm)

$$f_X : S_{i_1} \times S_{i_2} \times \ldots \times S_{i_k} \to S_0$$

with the property that $f_X(s_{i_1}, s_{i_2}, \ldots, s_{i_k}) = s_0$ for every $(s_0, s_1, s_2, \ldots, s_n) \in \mathcal{T}$. In particular, in the distribution table there cannot be tuples $(s, \ldots, s_{i_1}, \ldots, s_{i_2}, \ldots, s_{i_k}, \ldots)$ with $s \neq s_0$.

Example 6.2.1 Let us consider a secret sharing scheme for $n = 3$ users, $\Gamma = \{\{1, 2\}, \{1, 3\}\}$ with $S_i = \mathbb{Z}_3$ for $i = 0, 1, 2, 3$ and the distribution table

$$
\mathcal{T} =
\begin{bmatrix}
D & 1\ 2\ 3 \\
\hline
0 & 0\ 0\ 0 \\
1 & 1\ 1\ 2 \\
0 & 1\ 2\ 1 \\
1 & 2\ 0\ 0 \\
2 & 2\ 2\ 1 \\
0 & 2\ 1\ 2 \\
2 & 1\ 0\ 0 \\
1 & 0\ 2\ 1 \\
2 & 0\ 1\ 2
\end{bmatrix}.
\tag{6.3}
$$

The two secret recovery functions $s_0 = f_{\{1,2\}}(s_1, s_2)$ and $s_0 = f_{\{1,3\}}(s_1, s_2)$ can be given by the tables

$s_1\ s_2$	$f_{\{1,2\}}(s_1, s_2)$
0 0	0
1 0	2
0 1	2
1 1	1
0 2	1
2 0	1
1 2	0
2 1	0
2 2	2

$s_1\ s_3$	$f_{\{1,3\}}(s_1, s_3)$
0 0	0
1 0	2
0 1	1
1 1	0
0 2	2
2 0	1
1 2	1
2 1	2
2 2	0

respectively. Note that the function $f_{\{2,3\}}$ does not exist. Indeed, when $(s_2, s_3) = (0, 0)$ the secret s_0 can take values $0, 1, 2$ so $f_{\{2,3\}}(0, 0)$ is not defined.

Example 6.2.2 (*n-out-of-n scheme*) Let us design a secret sharing scheme with n users such that the only authorised coalition is the grand coalition, that is the set $U = \{1, 2, \ldots, n\}$. We need a sufficiently large finite field F and set $S_0 = F$ so that it is infeasible to try all secrets one by one. We will also have $S_i = F$ for all $i = 1, \ldots, n$.

To share a secret $s \in F$, the dealer generates $n - 1$ random elements $s_1, s_2, \ldots, s_{n-1} \in F$ and calculates $s_n = s - (s_1 + \ldots + s_{n-1})$. Then he gives share s_i to user i. The distribution table $\mathcal{T}$ will consist of all n-tuples $(s_0, s_1, s_2, \ldots, s_n)$ such that $\sum_{i=1}^{n} s_i = s_0$ and the secret recovery function (in this case the only one) will be $f_U(s_1, s_2, \ldots, s_n) = s_1 + s_2 + \ldots + s_n$.

The distribution table is convenient for defining the secret sharing scheme, however, in practical applications it is normally huge so schemes are normally defined differently.

Definition 6.2.1 A secret sharing scheme realising access structure Γ is called *perfect* if for every non-authorised coalition of users $\{j_1, j_2, \ldots, j_m\} \subset U$, for every sequence of shares $s_{j_1}, s_{j_2}, \ldots, s_{j_m}$ with $s_{j_r} \in S_{j_r}$, and for every two possible secrets $s, s' \in S_0$ the distribution table $\mathcal{T}$ contains as many tuples $(s, \ldots, s_{j_1}, s_{j_2}, \ldots, s_{j_m}, \ldots)$ as tuples $(s', \ldots, s_{j_1}, s_{j_2}, \ldots, s_{j_m}, \ldots)$.

In other words, if the scheme is perfect a non-authorised coalition $X = \{j_1, j_2, \ldots, j_m\}$ with shares $s_{j_1}, s_{j_2}, \ldots, s_{j_m}$ will have no reason to believe that the secret s was more likely chosen than any other secret s'. For example, in Example 6.2.1 if users 2 and 3 have shares 2 and 1, respectively, they will observe the following rows of $\mathcal{T}$

D	1 2 3
0	1 2 1
2	2 2 1
1	0 2 1

and will be unable to determine which row was chosen by the dealer. So the scheme in that example is perfect.

The scheme from Example 6.2.2 is obviously perfect. Let us have another look at the perfect secret sharing scheme invented by Shamir and specify the secret recovery functions.

Example 6.2.3 (Shamir 1979) Suppose that we have n users and the access structure is now $\Gamma = \{X \subseteq U \mid |X| \geq k\}$, i.e., a coalition is authorised if it contains at least k users. Let F be a large finite field and we will have $S_i = F$ for $i = 0, 1, \ldots, n$. Let $a_1, a_2, \ldots, a_n$ be distinct fixed publicly known nonzero elements of F (in the earlier example we took $a_i = i$).

Suppose $s \in F$ is the secret to share. The dealer generates randomly $t_0, t_1, \ldots, t_{k-1} \in F$, sets $s = t_0$, and forms the polynomial

$$p(x) = t_0 + t_1 x + \cdots + t_{k-1} x^{k-1}. \tag{6.4}$$

Then she gives the share $s_i = p(a_i)$ to user i. Note that $s = p(0)$.

Suppose now $X = \{i_1, i_2, \ldots, i_k\}$ be a minimal authorised coalition. Then the secret recovery function is

$$f_X(s_{i_1}, s_{i_2}, \ldots, s_{i_k}) = \sum_{r=1}^{k} s_{i_r} \frac{(-a_{i_1}) \ldots \widehat{(-a_{i_r})} \ldots (-a_{i_k})}{(a_{i_r} - a_{i_1}) \ldots \widehat{(a_{i_r} - a_{i_r})} \ldots (a_{i_r} - a_{i_k})},$$

where the hat over the term means its non-existence. This is the value at zero of the Lagrange's interpolation polynomial

$$\sum_{r=1}^{k} p(a_{i_r}) \frac{(x - a_{i_1}) \ldots \widehat{(x - a_{i_r})} \ldots (x - a_{i_k})}{(a_{i_r} - a_{i_1}) \ldots \widehat{(a_{i_r} - a_{i_r})} \ldots (a_{i_r} - a_{i_k})},$$

which is equal to $p(x)$.

We now may use the idea in Example 6.2.2 to construct a perfect secret sharing scheme for an arbitrary access structure Γ. We will illustrate this method in the following

Example 6.2.4 Let $U = \{1, 2, 3, 4\}$ and $\Gamma_{\min} = \{\{1, 2\}, \{2, 3\}, \{3, 4\}\}$. Let $s \in \mathbb{Z}_p$ be a secret. Firstly we consider three coalitions of users $\{1, 2\}$, $\{2, 3\}$ and $\{3, 4\}$ separately and build two-out-of-two schemes on each of these sets of users. Under the first scheme users 1 and 2 will get shares a and $s - a$, under the second scheme users 2 and 3 get shares b and $s - b$ and under the third scheme users 3 and 4 get shares c and $s - c$. Thus altogether users will get the following shares:

$$1 \leftarrow a,$$
$$2 \leftarrow (s - a, b),$$
$$3 \leftarrow (s - b, c),$$
$$4 \leftarrow s - c.$$

Let us show that this scheme is perfect. For this we have to consider every maximal non-authorised coalition and show that it has no clue about the secret. It is easy to see that every coalition of three or more players is authorised. So the maximal non-authorised coalitions will be $\{1, 3\}$, $\{1, 4\}$, $\{2, 4\}$. The coalition $\{1, 3\}$ will know values a, $s - b$ and c. Since a, b, c were chosen randomly and independently, a, $s - b$ and c are also three random independent values which contain no information about s. Similar for $\{1, 4\}$ and $\{2, 4\}$. Note that under this scheme users 2 and 3 will have to hold as their shares as two elements of $\mathbb{Z}_p$ each. Their shares will be twice as long as the secret (in binary representation).

This can be developed into a general method that allows to prove:

Theorem 6.2.1 *For any access structure Γ there exists a perfect secret sharing scheme which realises it.*

Sketch of Proof Let us consider the set $\Gamma_{\min}$ of all minimal authorised coalitions. Suppose they are $W_1, W_2, \ldots, W_q$ and their cardinalities are $m_1, m_2, \ldots, m_q$. We consider then q separate smaller access structures where the ith one will be defined on the set of users W_i and will be an m_i-out-of-m_i access structure. Let s_i be the share received by user i in this reduced access structure. So, in total, user i receives the vector of shares $(s_1, s_2, \ldots, s_q)$. As the access structure is public knowledge, user i will use his share s_i only when an authorised coalition with his participation contains W_i. If a coalition is not authorised, then it does not contain any of the $W_1, W_2, \ldots, W_q$, and it is possible to show that its participants cannot get any information about the secret. □

Under this method if a user belongs to k minimal authorised coalitions, then she will receive k elements of the field to hold as her share.

Suppose $2^{d-1} \leq |S_0| < 2^d$ or $\lceil \log_2 |S_0| \rceil = d$. Then we can encode elements of S_0 (secrets) using binary strings of length d. In this case we say that the *length* of the secret is d. Similarly we can talk about the lengths of the share that user i has received. We say that the *information ratio* of the secret sharing scheme S is

$$i(S) = \max_{i=1}^{n} \frac{\lceil \log_2 |S_i| \rceil}{\lceil \log_2 |S_0| \rceil}.$$

This number is the maximal ratio of the amount of information that must be conveyed to a participating user to the amount of information that is contained in the secret. In the secret sharing literature it is also common to use the term *information rate*, which is the inverse of the information ratio. The information ratio of the scheme constructed in Theorem 6.2.1 is terrible. For example, for the $(\frac{n}{2} + 1)$-out-of-n scheme (assume that n is even) every user belongs to $\binom{n}{n/2}$ authorised coalitions, which by Stirling's formula grows approximately as $2^n/\sqrt{n}$. More precisely, we will have

$$i(S) \sim \sqrt{\frac{2}{\pi}} \cdot \frac{2^n}{\sqrt{n}},$$

i.e., the information ratio of such scheme grows exponentially with n. We know we can do much better: The information ratio of Shamir's scheme is 1. However, for some access structures the information ratio can be large. It is not exactly known how large it can be.

Exercises

1. Consider the secret sharing scheme with the following distribution table.

s_0	s_1	s_2	s_3	s_4	s_5	s_6
0	0	0	1	1	2	2
0	0	0	2	2	1	1
0	1	1	2	2	0	0
0	1	1	0	0	2	2
0	2	2	0	0	1	1
0	2	2	1	1	0	0
1	0	1	1	2	2	0
1	0	2	2	1	1	0
1	1	2	2	0	0	1
1	1	0	0	2	2	1
1	2	0	0	1	1	2
1	2	1	1	0	0	2

(a) What is the domain of secrets? What is the domain of shares?
(b) Show that the coalition of users $\{1, 2\}$ is authorised but $\{1, 3, 5\}$ is not.
(c) Give the table for the secret recovery function for the coalition $\{1, 2\}$.

2°. Design a perfect secret sharing scheme for the access structure

$$\{\{1, 2\}, \{1, 3\}, \{1, 4\}, \{1, 5\}, \{2, 3, 4, 5\}\}$$

on the set of users $U = \{1, 2, 3, 4, 5\}$.

3°. Let Γ be an access structure on the set of users U. We say that a coalition $X \subseteq U$ is *blocking* if its complement X^c is not authorised. The set $\Gamma^* = \{X \subseteq U \mid X^c \notin \Gamma\}$ is the set of all blocking coalitions. Prove that Γ^* is also an access structure, called the access structure *dual* to Γ.

4°. What is Γ^*, if Γ is k-out-of-n threshold access structure?

6.2.2 Linear Secret Sharing Schemes

Let us look at Shamir's scheme from a different perspective. We can observe that the vector of the shares (where we think that the secret is the share of the dealer) can be obtained by the following matrix multiplication as

$$\begin{bmatrix} 1 & 0 & 0 & \dots & 0 \\ 1 & a_1 & a_1^2 & \dots & a_1^{k-1} \\ 1 & a_2 & a_2^2 & \dots & a_2^{k-1} \\ \dots & \dots & \dots & \dots & \dots \\ 1 & a_n & a_n^2 & \dots & a_n^{k-1} \end{bmatrix} \begin{bmatrix} t_0 \\ t_1 \\ \vdots \\ t_{k-1} \end{bmatrix} = \begin{bmatrix} p(0) \\ p(a_1) \\ \vdots \\ p(a_n) \end{bmatrix} = \begin{bmatrix} s_0 \\ s_1 \\ \vdots \\ s_n \end{bmatrix}, \qquad (6.5)$$

where $p(x)$ is the polynomial (6.4). Since all $a_1, a_2, \ldots, a_n$ are assumed to be different and nonzero, any k rows of the matrix in (6.5) are linearly independent since the determinant of the matrix formed by these rows is the well-known Vandermonde determinant (10.4) which is nonzero. This is why any k users can learn all the coefficients $t_0, t_1, t_2, \ldots, t_{k-1}$ of $p(x)$, including its constant term t_0 (which is the secret).

Let us write (6.5) in the matrix form as $H\mathbf{t} = \mathbf{s}$, where

$$
H = \begin{bmatrix} 1 & 0 & 0 & \cdots & 0 \\ 1 & a_1 & a_1^2 & \cdots & a_1^{k-1} \\ 1 & a_2 & a_2^2 & \cdots & a_2^{k-1} \\ \cdots & \cdots & \cdots & \cdots \\ 1 & a_n & a_n^2 & \cdots & a_n^{k-1} \end{bmatrix}, \quad \mathbf{t} = \begin{bmatrix} t_0 \\ t_1 \\ \vdots \\ t_{k-1} \end{bmatrix}, \quad \mathbf{s} = \begin{bmatrix} s_0 \\ s_1 \\ \vdots \\ s_n \end{bmatrix}, \qquad (6.6)
$$

and denote the rows of H as $\mathbf{h}_0, \mathbf{h}_1, \mathbf{h}_2, \ldots, \mathbf{h}_n$. Then the following is true: The span of a group of distinct rows $\{\mathbf{h}_{i_1}, \mathbf{h}_{i_2}, \ldots, \mathbf{h}_{i_r}\}$, none of which is $\mathbf{h}_0$, contains $\mathbf{h}_0$ if and only if $r \geq k$. We may now define the k-out-of-n access structure as follows:

$$
\Gamma_H = \{\{i_1, i_2, \ldots, i_s\} \subseteq U \mid \mathbf{h}_0 \in \mathrm{span}\{\mathbf{h}_{i_1}, \mathbf{h}_{i_2}, \ldots, \mathbf{h}_{i_s}\}\}. \qquad (6.7)
$$

This can be generalised by considering matrices H other than the one in (6.6).

Theorem 6.2.2 (Linear Secret Sharing Scheme) *Let H be an arbitrary $(n + 1) \times k$ matrix with coefficients in a finite field F. Let $\mathbf{h}_0, \mathbf{h}_1, \ldots, \mathbf{h}_n$ be the rows of H. Let us define a secret sharing scheme on the set of users $U = \{1, 2, \ldots, n\}$ as follows. Choose the coefficients of vector $\mathbf{t} = (t_0, t_1, \ldots, t_{k-1})$ randomly, calculate the vector $\mathbf{s} = (s_0, s_1, \ldots, s_n)$ from the equation $H\mathbf{t} = \mathbf{s}$, declare s_0 to be the secret and $s_1, s_2, \ldots, s_n$ the shares of users $1, 2, \ldots, n$, respectively. Then this is a perfect secret sharing scheme realising the access structure Γ_H defined as in (6.7).*

Proof Suppose $\mathbf{h}_0$ is in the span of $\{\mathbf{h}_{i_1}, \mathbf{h}_{i_2}, \ldots, \mathbf{h}_{i_r}\}$. Then

$$
\mathbf{h}_0 = \lambda_1 \mathbf{h}_{i_1} + \lambda_2 \mathbf{h}_{i_2} + \ldots + \lambda_r \mathbf{h}_{i_r},
$$

where the coefficients $\lambda_1, \lambda_2, \ldots, \lambda_r$ can be found by solving this system of linear equations. Multiplying both sides of this equation by $\mathbf{t}$ we obtain

$$
s_0 = \lambda_1 s_{i_1} + \lambda_2 s_{i_2} + \ldots + \lambda_r s_{i_r}, \qquad (6.8)
$$

hence the secret s_0 can be calculated from the shares of users $i_1, i_2, \ldots, i_r$.

Suppose now that $\mathbf{h}_0$ is not in the span of $\{\mathbf{h}_{i_1}, \mathbf{h}_{i_2}, \ldots, \mathbf{h}_{i_r}\}$. Without loss of generality we may assume that $i_1 = 1, \ldots, i_r = r$, i.e., that there are users $1, 2, \ldots, r$ with their shares $s_1, s_2, \ldots, s_r$ and that $\mathbf{h}_0$ is not a linear combination of the $\mathbf{h}_1, \mathbf{h}_2, \ldots, \mathbf{h}_r$. Let H_r be the matrix with rows $\mathbf{h}_1, \mathbf{h}_2, \ldots, \mathbf{h}_r$ and $\overline{H}_r$ be the matrix with rows $\mathbf{h}_0, \mathbf{h}_1, \ldots, \mathbf{h_r}$. By the assumption we have $\mathrm{rank}(\overline{H}_r) = \mathrm{rank}(H_r) + 1$.

Let also $\mathbf{s}_r$ be the column vector with entries $s_1, s_2, \ldots, s_r$ and $\bar{\mathbf{s}}_r$ be the column vector with entries $s_0, s_1, s_2, \ldots, s_r$. Since the system

$$H_r\mathbf{t} = \mathbf{s}_r \qquad (6.9)$$

is consistent we have $\mathrm{rank}(H_r \mid \mathbf{s}_r) = \mathrm{rank}(H_r)$, where $(H_r \mid \mathbf{s}_r)$ is the augmented matrix of the system (6.9). As the matrix $(\overline{H}_r \mid \bar{\mathbf{s}}_r)$ is obtained by adding just one row to $(H_r \mid \mathbf{s}_r)$ its rank is either the same or larger by 1. On the other hand, it is not smaller than the rank of $\overline{H}_r$. Since $\mathrm{rank}(\overline{H}_r) = \mathrm{rank}(H_r) + 1$ it will be true that $\mathrm{rank}(\overline{H}_r \mid \bar{\mathbf{s}}_r) = \mathrm{rank}(\overline{H}_r)$, and this system is consistent for every s_0. Since the dimension and hence the cardinality—remember F is finite—of the solution is determined by the rank of $\overline{H}_r$ only, we will have the same number of solutions to the equation $\overline{H}_r\mathbf{t} = \bar{\mathbf{s}}_r$ no matter what s_0 was. So members of the coalition $\{i_1, i_2, \ldots, i_r\}$ will be unable to identify s_0, hence this coalition is not authorised.   □

The following corollary will be very useful later.

Corollary 6.2.1 *For an authorised coalition* $X = \{i_1, i_2, \ldots, i_r\}$ *the secret recovery function* f_X *is linear in the shares.*

Proof Due to (6.8) we have

$$f_X(s_{i_1}, \ldots, s_{i_r}) = \lambda_1 s_{i_1} + \lambda_2 s_{i_2} + \ldots + \lambda_r s_{i_r},$$

where $\lambda_1, \lambda_2, \ldots, \lambda_r$ do not depend on the shares.   □

Example 6.2.5 Let $U = \{1, 2, 3\}$ and $\Gamma_{\min} = \{\{1, 2\}, \{1, 3\}\}$. We can realise this access structure by a linear scheme. Consider the matrix

$$H = \begin{bmatrix} 1 & 0 \\ 1 & 1 \\ 1 & -1 \\ 2 & -2 \end{bmatrix}.$$

The dealer may choose two random elements t_0, t_1 from a field Z_p for some large prime p and calculate

$$\begin{bmatrix} s_0 \\ s_1 \\ s_2 \\ s_3 \end{bmatrix} = H \begin{bmatrix} t_0 \\ t_1 \end{bmatrix},$$

where s_0 is taken as the secret, and s_1, s_2 and s_2 are given as shares to users 1,2 and 3, respectively. (Note that $s_0 = t_0$.) If users 1 and 2 come together, they can find t_0 and t_1 from the system of linear equations

$$\begin{bmatrix} 1 & 1 \\ 1 & -1 \end{bmatrix} \begin{bmatrix} t_0 \\ t_1 \end{bmatrix} = \begin{bmatrix} s_1 \\ s_2 \end{bmatrix}$$

because the determinant of this system is nonzero. Similarly, 1 and 3 also can do this. But, if 2 and 3 come together, they will face the system

$$\begin{bmatrix} 1 & -1 \\ 2 & -2 \end{bmatrix} \begin{bmatrix} t_0 \\ t_1 \end{bmatrix} = \begin{bmatrix} s_2 \\ s_3 \end{bmatrix},$$

which has exactly p solutions. Their shares therefore provide them with no information about t_0 and hence s_0.

Exercises

1. Determine the minimal authorised coalitions for the access structure realised by the linear secret sharing scheme with the matrix

$$H = \begin{bmatrix} 1 & 0 \\ 1 & 1 \\ 2 & -2 \\ 3 & 3 \\ 4 & -4 \end{bmatrix}$$

   over $\mathbb{Z}_{11}$.

2. Let F be a sufficiently large field $\mathbb{Z}_p$. Find the access structure which is realised by the linear secret sharing scheme with the matrix

$$H = \begin{bmatrix} 1 & 0 & 0 \\ 1 & 1 & 1 \\ 1 & 2 & 4 \\ 1 & 3 & 9 \\ 0 & 0 & 1 \\ 0 & 0 & 2 \\ 0 & 0 & 3 \end{bmatrix}.$$

3. Let F be a sufficiently large field. Find the access structure which is realised by the linear secret sharing scheme with the matrix

$$H = \begin{bmatrix} 1 & 0 & 0 \\ 1 & 1 & 0 \\ 1 & 2 & 0 \\ 1 & 3 & 3^2 \\ 1 & 4 & 4^2 \\ 1 & 5 & 5^2 \end{bmatrix}.$$

4. Let F be a sufficiently large field. Find the access structure which is realised by the linear secret sharing scheme with the matrix

$$H = \begin{bmatrix} 1 & 0 & 0 \\ 1 & a_1 & 0 \\ 1 & a_2 & 0 \\ 1 & a_3 & a_3^2 \\ 1 & a_4 & a_4^2 \\ 1 & a_5 & a_5^2 \end{bmatrix},$$

where $a_1, \ldots, a_5$ are distinct nonzero elements of the field F.

5. A linear secret sharing scheme for the group of users $U = \{1, 2, 3, 4, 5\}$ is defined by the matrix over $\mathbb{Z}_{31}$:

$$H = \begin{bmatrix} h_0 \\ h_1 \\ h_2 \\ h_3 \\ h_4 \\ h_5 \end{bmatrix} = \begin{bmatrix} 1 & 0 & 0 & 0 \\ 1 & 2 & 3 & 0 \\ 1 & 3 & 3 & 0 \\ 11 & 5 & 2 & 0 \\ 0 & 1 & 1 & 2 \\ 0 & 6 & 1 & 1 \end{bmatrix}.$$

These users got shares 2, 27, 20, 10, 16, respectively, which are also elements of $\mathbb{Z}_{31}$. Let $A = \{1, 2, 3\}$ and $B = \{1, 4, 5\}$ be two coalitions.

(a) Show that one of the coalitions is authorised and the other is not.
(b) Show how the authorised coalition can determine the secret.

6. Let H be an $(n + 1) \times k$ matrix over a field F and Γ_H be the access structure defined by the formula (6.7). Let us represent the ith row $\mathbf{h}_i$ of this matrix as $\mathbf{h}_i = (c_i, \mathbf{h}'_i)$, where $c_i \in F$ is the first coordinate of $\mathbf{h}_i$ and $\mathbf{h}'_i$ is a $(k - 1)$-dimensional row vector of the remaining coordinates. Suppose the coalition $\{i_1, i_2, \ldots, i_r\}$ is not authorised in Γ_H. Then

$$\sum_{j=1}^{r} \lambda_j \mathbf{h}'_j = \mathbf{0} \implies \sum_{j=1}^{r} \lambda_j c_j = 0$$

for all $\lambda_1, \lambda_2, \ldots, \lambda_r$.

7. Let U and V be disjoint sets of k and m users, respectively. Let M and N be two matrices realising linear secret sharing schemes with access structures Γ_M and Γ_N. Find the matrix realising the access structures

(a) $\Gamma_M + \Gamma_N$,
(b) $\Gamma_M \times \Gamma_N$

on the set of users $U \cup V$.

8. Prove that the access structure $\Gamma_{\min} = \{\{1, 2\}, \{2, 3\}, \{3, 4\}\}$ on the set of users $U = \{1, 2, 3, 4\}$ cannot be realised by a linear secret sharing scheme.

9. Let $n > 2$. The access structure with the set of minimal authorised coalitions

$$\Gamma_{\min} = \{\{1, 2\}, \{1, 3\}, \ldots, \{1, n\}, \{2, 3, \ldots, n\}\}$$

on the set of users $U = \{1, 2, \ldots, n\}$ cannot be realised by a linear secret sharing scheme.

10°. (GAP question) A linear secret sharing scheme for the group of users $U = \{1, 2, 3, 4, 5, 6\}$ is defined by the matrix over $\mathbb{Z}_{97}$:

$$H = \begin{bmatrix} h_0 \\ h_1 \\ h_2 \\ h_3 \\ h_4 \\ h_5 \\ h_6 \end{bmatrix} = \begin{bmatrix} 1\ 0\ 0\ 0 \\ 1\ 1\ 1\ 1 \\ 1\ 3\ 9\ 27 \\ 1\ 2\ 4\ 8 \\ 3\ 3\ 8\ 1 \\ 0\ 0\ 5\ 6 \\ 0\ 0\ 3\ 4 \end{bmatrix}.$$

These users got shares 18, 11, 52, 81, 79, 16, respectively, which are also elements of $\mathbb{Z}_{97}$.

(a) Show that the coalition $A = \{1, 3, 5, 6\}$ is authorised and helps it to recover the secret.

(b) Show that the coalition $B = \{1, 4, 5, 6\}$ is not authorised.

6.2.3   Ideal and Non-ideal Secret Sharing Schemes

Given a secret sharing scheme with access structure Γ, a user is called a *dummy* if she does not belong to any minimal authorised coalition in $\Gamma_{\min}$. A dummy user can be removed from any authorised coalition without making it non-authorised.

Theorem 6.2.3 *Let S_0 be the set of possible secrets and S_i be the set of possible shares that can be given to user i in a secret sharing scheme S. If this scheme is perfect and has no dummy users, then $|S_i| \geq |S_0|$ for all $i = 1, \ldots, n$ or $i(S) \geq 1$.*

Proof Let i be an arbitrary user. Since no dummies exist, i belongs to one of the minimal authorised coalitions, say $X = \{i_1, i_2, \ldots, i_k\}$, and with no loss of generality we may assume that $i = i_k$. Suppose that there is a tuple $(s_0, s_1, \ldots, s_n) \in T$ in the distribution table where s_0 is the secret shared and $s_{i_1}, s_{i_2}, \ldots, s_{i_{k-1}}$ are the shares given to users $i_1, i_2, \ldots, i_{k-1}$. Since the scheme is perfect the distribution table contains tuples $(s, \ldots, s_{i_1}, \ldots, s_{i_2}, \ldots, s_{i_{k-1}}, \ldots)$ for every $s \in S_0$. However if we add user $i = i_k$ we get the coalition X which is authorised and can recover the secret. Thus, when the shares $s_{i_1}, s_{i_2}, \ldots, s_{i_{k-1}}$ of users $i_1, i_2, \ldots, i_{k-1}$ are fixed, the secret depends on the share of the user i only. Hence for every possible secret s there is a share $t(s)$ which, if given to the user i, leads to recovery s as the secret by coalition X and can be calculated using the secret recovery function f_X of coalition X, that is

$$f_X(s_{i_1}, \ldots, s_{i_{k-1}}, t(s)) = s.$$

The mapping $t \colon S_0 \to S_i$ is one-to-one as if $t(s_1) = t(s_2)$, then

$$s_1 = f_X(s_{i_1}, \ldots, s_{i_{k-1}}, t(s_1)) = f_X(s_{i_1}, \ldots, s_{i_{k-1}}, t(s_2)) = s_2.$$

Thus S_i has at least as many elements as S_0, that is $|S_i| \geq |S_0|$. $\square$

Definition 6.2.2 A secret sharing scheme $\mathcal{S}$ is called *ideal* if it is perfect and $i(\mathcal{S}) = 1$.

Ideal schemes are the most informationally efficient having their information rate equal to 1. By Theorem 6.2.3 this is the best possible rate for a perfect scheme. An equivalent statement would be that $|S_i| = |S_0|$ for all $i = 1, \ldots, n$. Normally in such cases both secret and shares belong to the same finite field. In particular, this is true for Shamir's secret sharing scheme given in Example 6.2.3. Indeed, if the elements $a_1, a_2, \ldots, a_n$ are publicly known, the secret is $p(0)$ and the share of the ith user is $p(a_i)$ for the polynomial p there defined. More generally,

Theorem 6.2.4 *Any linear secret sharing scheme is ideal.*

Proof We need to recap how the shares in this scheme are defined. We have a (normally large) field F and an $(n + 1) \times k$ matrix H over this field. Then we define a k-dimensional vector $\mathbf{t}$ over F at random and calculate the $(n + 1)$-dimensional vector $H\mathbf{t} = \mathbf{s} = (s_0, s_1, \ldots, s_n)^T$. Here s_0 is the secret and s_i is the share of user i. Both are elements of F. $\square$

However there exist very simple access structures for which there are no ideal secret sharing schemes. Theorem 6.2.4 tells us that we have to look for such examples among nonlinear schemes.

Example 6.2.6 For the access structure Γ of Example 6.2.4 with

$$\Gamma_{\min} = \{\{1, 2\}, \{2, 3\}, \{3, 4\}\}$$

there are no ideal secret sharing schemes realising it.

Proof Suppose on the contrary there is an ideal secret sharing scheme S with the distribution table T realising Γ. Then for some positive integer q we have $|S_i| = q$ for $i = 0, 1, 2, 3, 4$. For any subset $I \subseteq \{0, 1, 2, 3, 4\}$ let T_I be the restriction of T to columns indexed by numbers from I and let $\#T_I$ stand for the number of distinct rows in T_I. Let us firstly note that

$$\#T_{\{1,2\}} = \#T_{\{2,3\}} = \#T_{\{3,4\}} = q^2.$$

Let us consider, for example, $T_{\{1,2\}}$. Take any $s_1 \in S_1$. As in the proof of Theorem 6.2.3 we conclude that for any secret s_0 there will be exactly one value $s_2 \in S_2$ such that $(s_0, s_1, s_2, s_3, s_4)$ is a row in T. Hence there will be exactly q distinct rows in $T_{\{1,2\}}$ with s_1 in column 1. As $|S_1| = q$ there are exactly q^2 distinct rows in $T_{\{1,2\}}$.

Let us now fix arbitrary elements $s_0 \in S_0$ and $s_2 \in S_2$. Since both $\{1, 2\}$ and $\{2, 3\}$ are authorised, there will be unique s_1 and s_3 such that $(s_0, s_1, s_2, s_3, s_4)$ is a row in T. In other words s_1 uniquely determines s_3 in any row of the distribution table. This leads to the coalition $\{1, 4\}$ being authorised. Indeed, since the table T is the public knowledge users 1 and 4 can figure out the share given for user 3 and then can figure out the secret since $\{3, 4\}$ is authorised. □

The construction of the previous theorem leads us to a definition of a generalised linear secret sharing scheme which may not be ideal.

Example 6.2.7 A family $\mathcal{L}$ of subspaces $\{L_0, L_1, \ldots, L_n\}$ is said to satisfy property "all-or-nothing" if for every subset $X \subset \{1, 2, \ldots, n\}$ the span of $\{L_i \mid i \in X\}$ either contains L_0 or has zero intersection with it. Any such family defines a certain access structure, namely

$$\Gamma_{\mathcal{L}} = \{X \subseteq U \mid \operatorname{span}\{L_i \mid i \in X\} \supseteq L_0\}.$$

Now the secret and the shares will be finite-dimensional vectors over F. Let $\{L_0, L_1, \ldots, L_n\}$ be subspaces of F^k satisfying the property all-or-nothing. Let H_i be the matrix whose rows form a basis of L_i. Then we generate random vectors $\mathbf{t}_i$ of the same dimension as $\dim L_i$ and calculate the secret and the shares as $\mathbf{s}_i = H_i \mathbf{t}_i$, $i = 0, 1, \ldots, n$. As in the Theorem 6.2.2 it leads to a perfect secret sharing scheme realising $\Gamma_{\mathcal{L}}$, however it may not be ideal as the following example shows.

Example 6.2.8 Let subspaces L_0, L_1, L_2, L_3, L_4 be the row spaces of the following matrices:

$$H_0^T = \begin{bmatrix} 1 & 0 \\ 0 & 1 \\ 0 & 0 \\ 0 & 0 \\ 0 & 0 \\ 0 & 0 \end{bmatrix}, \ H_1^T = \begin{bmatrix} 0 & 0 \\ 0 & 0 \\ 1 & 0 \\ 0 & 1 \\ 0 & 0 \\ 0 & 0 \end{bmatrix}, \ H_2^T = \begin{bmatrix} 1 & 0 & 0 \\ 0 & 1 & 0 \\ 1 & 0 & 0 \\ 0 & 1 & 0 \\ 0 & 0 & 1 \\ 0 & 0 & 0 \end{bmatrix}, \ H_3^T = \begin{bmatrix} 0 & 0 & 1 \\ 0 & 0 & 0 \\ 0 & 0 & 0 \\ 0 & 1 & 0 \\ 0 & 0 & 1 \\ 1 & 0 & 0 \end{bmatrix}, \ H_4^T = \begin{bmatrix} 0 & 0 \\ 0 & 1 \\ 0 & 0 \\ 0 & 0 \\ 1 & 0 \\ 0 & 1 \end{bmatrix}.$$

This family satisfies the property all-or-nothing. The access structure associated with it can be given by the set of minimal authorised coalitions as follows:

$$\Gamma_{min} = \{\{1, 2\}, \{2, 3\}, \{3, 4\}\}.$$

Since the secret is two dimensional and some shares are three dimensional the information rate of such scheme will be $3/2$. As $3/2 < 2$ this is more efficient secret sharing scheme realising Γ than the one in Example 6.2.4. In fact, it can be proved that the scheme for this example is optimal for Γ in the sense that it gives the best possible information rate.

Exercises

1. Let $\mathcal{T}$ be the distribution table of a perfect ideal secret sharing scheme with the set of user, $U = \{1, 2, \ldots, n\}$, the dealer 0 and the cardinality of the domain of secrets q. Prove that

   (i) If a coalition C is authorised and $C' = C \cup \{0\}$, then $\#\mathcal{T}_{C'} = \#\mathcal{T}_C$;
   (ii) If a coalition C is not authorised and $C' = C \cup \{0\}$, then $\#\mathcal{T}_{C'} = q \cdot \#\mathcal{T}_C$.

2. Prove all the missing details in Example 6.2.8.
3. In this exercise we consider the case, when for an access structure Γ of a secret sharing scheme with the distribution table $\mathcal{T}$ all minimal authorised coalitions have size 2. In this case Γ_{min} can be interpreted as edges of a graph $G(\Gamma)$ defined on $U = \{1, 2, \ldots, n\}$. We assume that this graph is connected. Let also the cardinality of the domain of secrets be q.

   (i)   Show that, if $\{i, j\} \in \Gamma_{min}$, then $\#\mathcal{T}_{\{i, j\}} = q^2$.
   (ii) Prove that $\#\mathcal{T}_{U \cup \{0\}} = q^2$.
   (iii) Prove that if $\{i, j\} \notin \Gamma_{min}$, then $\#\mathcal{T}_{\{i, j\}} = q$.
   (iv) Prove that if $\{i, j\}$ and $\{j, k\}$ are both not authorised, then $\{i, k\}$ is not authorised too.
   (v)   Prove the following theorem

Theorem 6.2.5 *Let Γ be an ideal access structure such that all minimal authorised coalitions have size 2 and $G(\Gamma)$ is connected. Then the complementary graph of $G(\Gamma)$ is a disjoint union of cliques.*[5]

4°. Construct an ideal secret sharing scheme for the access structure

$$\{\{p_1, p_2\}, \{p_1, p_3\}, \{p_2, p_3\}, \{p_4, p_5\}\}$$

on a set of participants $P = \{p_1, p_2, p_3, p_4, p_5\}$.

---

[5]A clique is a subgraph where any two vertices are connected.

6.3   Applications of Secret Sharing

In recent years the importance of secret sharing has significantly increased. Invented by Shamir and Blakley in 1979, initially it was just used for storing securely sensitive data like missile launch codes or decryption keys. Nowadays, secret sharing has acquired many applications and becomes one of the major cryptographic primitives. For example, it is used in:

- **Secure Multiparty Computation (MPC)** which allows n parties to compute a function in n inputs (one per participant) without leaking any extra information about the inputs (electronic voting is one such example).
- **Threshold Cryptography** where the decryption key of a cryptosystem is shared between a group of participants, e.g., employees of a large organisation.
- **Perfectly Secure Message Transmission** over a network that is not trustworthy. In this case there are several channels of communication between parties but some of them can be monitored by an adversary.
- **Multi Cloud Storage**. The Snowden leak has changed the way data is backed up forever. Snowden, who created a widely implemented backup system for the NSA, was given full administrator privileges. For example, IBM's secure cloud uses secret sharing to achieve a reliable and private alternative to the use of a single cloud server.

Here we will show how the threshold cryptography works. The idea belonged to Yvo Desmedt (1988). Suppose a large organisation (e.g., Microsoft or Adobe) uses RSA cryptosystem with modulus n and has an encryption exponent e and decryption exponent d. The decryption exponent is used, as we know, not only for decrypting incoming messages but also for signing messages on behalf of this organisation. No wonder this key is too valuable to trust anybody to be in charge of it. Hence it has to be shared.

Suppose now $X = \{i_1, i_2, \ldots, i_k\}$ be a minimal authorised coalition and d is shared between them with the shares $d_{i_1}, \ldots, d_{i_k}$. Then the secret recovery function, as we know by Corollary 6.2.1 is linear in shares:

$$d = \lambda_{i_1} d_{i_1} + \ldots + \lambda_{i_k} d_{i_k}.$$

Suppose c is a cyphertext. Party j can calculate $c^{d_{ij}}$ and send it to a trusted combiner who will calculate

$$c^d = c^{\lambda_{i_1} d_{i_1} + \ldots + \lambda_{i_k} d_{i_k}} = \prod_{j=1}^{k} (c^{d_{ij}})^{\lambda_{ij}}.$$

The message is decrypted but the decryption key was used but not revealed! An outgoing message can be signed exactly in the same way.

Error-Correcting Codes

7

All sorts of computer errors are now turning up.
You'd be surprised to know the number of doctors
who claim they are treating pregnant men.

Isaac Asimov (1920–1992)

This chapter deals with the problem of reliable transmission of digitally encoded information through an unreliable channel. When we transmit information from a satellite, or an automatic station orbiting the moon, or from a probe on Mars, then for many reasons (e.g., sun-bursts), our message can be distorted. Even the best telecommunication systems connecting numerous information centres in various countries have some nonzero error rate. These are examples of transmission in space. When we save a file on a hard disc and then try to read it one month later, we may find that this file has been distorted (due to, for example, microscopic defects of the disc's surface). This is an example of transmission in time. The channels of transmission in both cases are different but they have one important feature in common: they are not 100% reliable. In some cases even a single mistake in the transmission of a message can have serious consequences. We will show how algebra can help to address this important problem.

We think of a message as a string of symbols of a certain alphabet. The most common is the alphabet consisting of two symbols 0 and 1. It is called the binary alphabet and we can interpret these symbols as elements of the finite field $\mathbb{Z}_2$. Some non-binary alphabets are also used, for example, we can use the symbols of any finite field F. But we will initially concentrate on the binary case.

The symbols of the message are transmitted through the channel one by one. Let us see what can happen to them. Since mistakes in the channel do occur, we assume that, when we transmit 0, with probability $p > 1/2$ we receive 0 and with probability

© Springer Nature Switzerland AG 2020   191
A. Slinko, *Algebra for Applications*, Springer Undergraduate Mathematics Series,
https://doi.org/10.1007/978-3-030-44074-9_7

$1 - p$ we receive 1 as a result of a mistake in the channel. Similarly, we assume that transmitting 1 we get 1 with probability p and 0 with probability $1 - p$. Thus we assume that the probability of a mistake does not depend on the transmitted symbol. In this case the channel is called *symmetric*. In our case we are talking about a binary symmetric channel. It can be illustrated as follows:

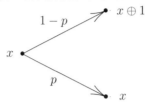

Here the error is modelled by means of addition modulo 2. Let x be the symbol to be transmitted. If transmission is perfect, then x will also be the symbol received, but if a mistake occurs, then the message received will be $x \oplus 1$, where the addition is in the field $\mathbb{Z}_2$. Indeed, $0 \oplus 1 = 1$ and $1 \oplus 1 = 0$. Thus the mistake can be modelled algebraically as the addition of 1 to the transmitted symbol.

In practical situations p is very close to 1, however, even when $p = 0.98$, among any 100 symbols transmitted, on average two will be transmitted with an error. Such channel may not be satisfactory to transfer some sensitive data and an error correction technique must be implemented.

7.1   Binary Error-Correcting Codes

Binary error-correcting codes are used when messages are strings of zeros and ones, i.e., the alphabet is $\mathbb{Z}_2 = \{0, 1\}$.

7.1.1   The Hamming Weight and the Hamming Distance

If we transmit symbols of our message one by one, then there is no way that we can detect an error. That is why we will try to split the message into blocks of symbols of fixed length m. Any block of m symbols

$$a_1 a_2 \dots a_m, \qquad a_i \in \mathbb{Z}_2$$

can be represented by an m-dimensional vector $(a_1, a_2, \dots, a_m) \in \mathbb{Z}_2^m$. Note that according to the long-established tradition in coding theory the messages are written as row vectors—this is different from the convention used in most undergraduate linear algebra courses where elements of $\mathbb{R}^m$ are viewed as column vectors. Since we split all messages into blocks of length m we may consider that all messages have a fixed length m and view them as elements of the m-dimensional vector space $\mathbb{Z}_2^m$ over $\mathbb{Z}_2$. When considering vectors as messages we will often omit commas, e.g., $(1\ 1\ 1\ 0)$ is the vector $(1, 1, 1, 0)$ treated as a message.

As above, mistakes during the transmission can be modelled algebraically. Suppose that a message $\mathbf{a} = (a_1, a_2, \ldots, a_m) \in \mathbb{Z}_2^m$ was transmitted and $\mathbf{b} = (b_1, b_2, \ldots, b_m) \in \mathbb{Z}_2^m$ was received with one mistake in position i. Then

$$\mathbf{b} = \mathbf{a} + \mathbf{e}_i,$$

where $\mathbf{e}_i = (0, \ldots, 0, 1, 0, \ldots, 0)$ with a 1 in the ith position and 0 elsewhere. If positions $i_1, i_2, \ldots, i_k$ were damaged, then

$$\mathbf{b} = \mathbf{a} + \epsilon,$$

where $\epsilon = \mathbf{e}_{i_1} + \cdots + \mathbf{e}_{i_k}$ is a vector with k ones and $m - k$ zeros. In this case ϵ is called the *error vector*.

Definition 7.1.1 The *(Hamming) weight* of a vector $\mathbf{x} \in \mathbb{Z}_2^m$ is the number of nonzero coordinates in $\mathbf{x}$. It is denoted as $wt(\mathbf{x})$.

Proposition 7.1.1 *If a message* $\mathbf{a} = (a_1, a_2, \ldots, a_m) \in \mathbb{Z}_2^m$ *was transmitted and* $\mathbf{b} = (b_1, b_2, \ldots, b_m) \in \mathbb{Z}_2^m$ *was received with k mistakes during the transmission, then* $\mathbf{b} = \mathbf{a} + \epsilon$ *with* $wt(\epsilon) = k$.

Example 7.1.1 If $\mathbf{a} = (0\ 1\ 0\ 1\ 0\ 1)$ and $\mathbf{b} = (0\ 1\ 1\ 0\ 0\ 1)$, then $\epsilon = (0\ 0\ 1\ 1\ 0\ 0)$ with $wt(\epsilon) = 2$.

Another useful concept is that of the *Hamming distance*.

Definition 7.1.2 The Hamming distance between two vectors $\mathbf{x}, \mathbf{y} \in \mathbb{Z}_2^m$ is the number of coordinates in which these two vectors differ. It is denoted as $d(\mathbf{x}, \mathbf{y})$.

Lemma 7.1.1 $d(\mathbf{x}, \mathbf{y}) = wt(\mathbf{x} + \mathbf{y})$.

Proof Let $\mathbf{x} = (x_1, x_2, \ldots, x_m)$ and $\mathbf{y} = (y_1, y_2, \ldots, y_m)$. Then, for all $i = 1, \ldots, m$, we have $x_i + y_i = 0$, if $x_i = y_i$, and $x_i + y_i = 1$, if $x_i \neq y_i$. Hence every i for which $x_i \neq y_i$ increases the weight of $\mathbf{x} + \mathbf{y}$ by one. This proves the lemma. $\square$

We may now reformulate Proposition 7.1.1 as follows:

Proposition 7.1.2 *Suppose that a message* $\mathbf{a} = (a_1, a_2, \ldots, a_m) \in \mathbb{Z}_2^m$ *was transmitted and* $\mathbf{b} = (b_1, b_2, \ldots, b_m) \in \mathbb{Z}_2^m$ *was received. Then the fact that k mistakes occur during the transmission is equivalent to* $d(\mathbf{a}, \mathbf{b}) = k$.

Theorem 7.1.1 (Properties of the Hamming distance) *It is a metric on $\mathbb{Z}_2^m$, which means that*

1. $d(\mathbf{x}, \mathbf{y}) \geq 0$ *and* $d(\mathbf{x}, \mathbf{y}) = 0$ *if and only if* $\mathbf{x} = \mathbf{y}$;
2. $d(\mathbf{x}, \mathbf{y}) = d(\mathbf{y}, \mathbf{x})$ *for all* $\mathbf{x}, \mathbf{y}$;
3. $d(\mathbf{x}, \mathbf{z}) \leq d(\mathbf{x}, \mathbf{y}) + d(\mathbf{y}, \mathbf{z})$ *for all* $\mathbf{x}, \mathbf{y}, \mathbf{z}$;

Proof The first two properties are obvious. Let us prove the third one. Suppose that $x_i \neq z_i$ and the position i contributes 1 to $d(\mathbf{x}, \mathbf{z})$. Then either $x_i = y_i$ and $y_i \neq z_i$ or $x_i \neq y_i$ and $y_i = z_i$. Hence the ith position will also contribute 1 to the sum $d(\mathbf{x}, \mathbf{y}) + d(\mathbf{y}, \mathbf{z})$. Suppose now that $x_i = z_i$ and the position i contributes 0 to $d(\mathbf{x}, \mathbf{z})$. Then either $x_i = y_i = z_i$ and the ith position contributes also 0 to the sum $d(\mathbf{x}, \mathbf{y}) + d(\mathbf{y}, \mathbf{z})$ or $x_i \neq y_i \neq z_i$ and the ith position contributes 2 to the sum $d(\mathbf{x}, \mathbf{y}) + d(\mathbf{y}, \mathbf{z})$. Hence the right-hand side is not smaller than the left-hand side. □

The following sets play a special role in coding theory. For any $\mathbf{x} \in \mathbb{Z}_2^m$ we define $B_k(\mathbf{x}) = \{\mathbf{y} \in \mathbb{Z}_2^m \mid d(\mathbf{x}, \mathbf{y}) \leq k\}$, and we call it the *ball* of radius k with centre $\mathbf{x}$.

Example 7.1.2 Let $\mathbf{a} = (1\ 1\ 1\ 1) \in \mathbb{Z}_2^4$. Then

$$B_1(\mathbf{a}) = \{\mathbf{a}, (1\ 1\ 1\ 0), (1\ 1\ 0\ 1), (1\ 0\ 1\ 1), (0\ 1\ 1\ 1)\}.$$

Theorem 7.1.2 *The cardinality of the ball of radius k with centre $\mathbf{x}$ is*

$$|B_k(\mathbf{x})| = \binom{m}{0} + \binom{m}{1} + \cdots + \binom{m}{k}. \tag{7.1}$$

Proof Let $\mathbf{y} \in B_k(\mathbf{x})$. We may consider the "error vector" $\mathbf{e}$ such that $\mathbf{y} = \mathbf{x} + \mathbf{e}$. Then $\mathbf{y} \in B_k(\mathbf{x})$ if and only if $wt(\mathbf{e}) \leq k$. It is enough to prove that, for each $i = 1, \ldots, k$, there are exactly $\binom{m}{i}$ vectors $\mathbf{e} \in \mathbb{Z}_2^m$ such that $wt(\mathbf{e}) = i$. Indeed, we must choose i positions out of m in the zero vector and change the coordinates there to ones. Hence every vector $\mathbf{e}$ with $wt(\mathbf{e}) = i$ corresponds to an i-element subset of $\{1, 2, \ldots, m\}$. We know that there are exactly

$$\binom{m}{i} = \frac{m!}{i!(m-i)!}$$

such subsets. (see, for example, [1], p.271). Now it is clear that the formula (7.1) counts all "error vectors" of weight at most k, and hence all vectors $\mathbf{y}$ which are at Hamming distance k or less from $\mathbf{x}$.   □

Example 7.1.3 The cardinality of the ball of radius 2 with centre $\mathbf{x}$ is

$$|B_2(\mathbf{x})| = \binom{m}{0} + \binom{m}{1} + \binom{m}{2} = 1 + m + \frac{m(m-1)}{2}.$$

Exercises

1. (a) Consider the following binary vectors:

$$\mathbf{u} = (1\ 1\ 0\ 1\ 1\ 1\ 0), \quad \mathbf{v} = (1\ 0\ 0\ 0\ 1\ 1\ 1).$$

   Determine the Hamming weights of $\mathbf{u}, \mathbf{v}$. Find $d(\mathbf{u}, \mathbf{v})$.

   (b) The vector $\mathbf{x} = (0\ 1\ 1\ 1\ 0\ 1\ 0\ 1\ 1\ 0)$ was sent through a binary channel and $\mathbf{y} = (0\ 1\ 0\ 1\ 0\ 1\ 1\ 1\ 1\ 0)$ was received. How many mistakes have occurred? Write down the error vector.

2. List all vectors of $B_2(\mathbf{x}) \subset \mathbb{Z}_2^4$, where $\mathbf{x} = (1\ 0\ 1\ 0)$.

3. Let $\mathbf{x}$ be a word in $\mathbb{Z}_2^7$. How many elements are there in the ball $B_3(\mathbf{x})$ of radius 3?

4. Explain why the cardinality of $B_k(\mathbf{x})$ does not depend on $\mathbf{x}$.

5°. Let $\mathbf{u}$ and $\mathbf{v}$ be binary vectors of length n and denote by s the number of positions where $\mathbf{u}$ and $\mathbf{v}$ *both* have a 1. Show that $wt(\mathbf{u} + \mathbf{v}) = wt(\mathbf{u}) + wt(\mathbf{v}) - 2s$.

6°. Show that the set W of vectors of even weight in $\mathbb{Z}_2^n$ is a subspace of the vector space $\mathbb{Z}_2^n$.

7.1.2   Encoding and Decoding. Simple Examples

By now we have already understood the convenience of having all messages of equal length, say m. Longer messages can be split into several shorter ones. The idea of error correction is to increase the length m of a transmitted message and to add to each message several auxiliary symbols, so-called check symbols, which will not bear any information but will help to correct errors. Hence we increase the length of every message from m to n, where $m < n$.

Definition 7.1.3 An error-correcting code $\mathcal{C}$ consists of an encoding function $E \colon \mathbb{Z}_2^m \to \mathbb{Z}_2^n$ and a decoding function $D \colon \mathbb{Z}_2^n \to \mathbb{Z}_2^m \cup \{\text{error}\}$ which satisfies $D(E(\mathbf{x})) = \mathbf{x}$ for all $\mathbf{x} \in \mathbb{Z}_2^m$. Such a code is called a *(binary) (m, n)-code.*

We note that the encoding function is necessarily one-to-one. Indeed, if we had $E(\mathbf{x}_1) = E(\mathbf{x}_2)$, then $\mathbf{x}_1 = D(E(\mathbf{x}_1)) = D(E(\mathbf{x}_2)) = \mathbf{x}_2$ i.e., $\mathbf{x}_1 = \mathbf{x}_2$.

Definition 7.1.4 Elements of $E(\mathbb{Z}_2^m)$ are called *codewords* (or *codevectors*).

Example 7.1.4 *(parity check code)* This code increases the length of a message by 1 adding only one check symbol which is the sum modulo 2 of all other symbols. That is

$$E(x_1, x_2, \ldots, x_m) = (x_1, x_2, \ldots, x_{m+1}),$$

where $x_{m+1} = x_1 + \cdots + x_m$. Note that the sum of all coordinates for any of the codevectors is equal to 0:

$$x_1 + \cdots + x_{m+1} = (x_1 + \cdots + x_m) + (x_1 + \cdots + x_m) = 0.$$

Let us see now what happens if one mistake occurs. In this case for the received vector $\mathbf{y} = (y_1, y_2, \ldots, y_{m+1})$ we will get

$$y_1 + \cdots + y_{m+1} = x_1 + \cdots + x_{m+1} + 1 = 0 + 1 = 1.$$

Hence if we organise the decoding as follows:

$$D(y_1, y_2, \ldots, y_{m+1}) = \begin{cases} (y_1, y_2, \ldots, y_m), & \text{if } y_1 + y_2 + \cdots + y_{m+1} = 0 \\ \text{error} & \text{if } y_1 + y_2 + \cdots + y_{m+1} = 1, \end{cases}$$

this code will detect any single error.

Example 7.1.5 (*triple repetition code*) This code increases the length of a message threefold by repeating every symbol three times:

$$E(x_1, x_2, \ldots, x_m) = (x_1, x_2, \ldots, x_m, x_1, x_2, \ldots, x_m, x_1, x_2, \ldots, x_m),$$

Decoding may be organised as follows. To decide on the first symbol the algorithm inspects y_1, y_{m+1}, and y_{2m+1}. If the majority (two or three) of these symbols are 0s, then the decoding algorithm decides that a 0 was transmitted, while if the majority of symbols are 1s, then the algorithm decides that a 1 was sent. This code will correct any single error but will fail to correct some double ones.

The ability of a particular code $\mathcal{C} = (E, D)$ to detect or correct errors depends on the geometric properties of the set of codewords $E(\mathbb{Z}_2^m) \subset \mathbb{Z}_2^n$ and the properties of the decoding function D.

Definition 7.1.5 Suppose that for all $\mathbf{y} \in \mathbb{Z}_2^n$ the vector $\mathbf{x} = D(\mathbf{y})$ is such that the vector $E(\mathbf{x})$ is the closest (in respect to the Hamming distance) codeword to $\mathbf{y}$ (any of them if there are several within the same distance), then we say that the decoding function D satisfies *maximum likelihood decoding*.

Maximum likelihood decoding is based on the assumption (that can be proved) that, under the assumption that mistakes are random and independent, in the symmetric channel with $p > \frac{1}{2}$ the probability of k mistakes during the transmission is less than the probability of j mistakes if and only if $j > k$. Therefore, if our assumption on the distribution of mistakes is true, the maximum likelihood decoding minimises the probability of the decoder making a mistake and will always be assumed.

Theorem 7.1.3 *For a code* C *the following statements are equivalent:*

(a) C *detects all combinations of* k *or fewer errors;*
(b) *For any codeword* $\mathbf{x}$ *the ball* $B_k(\mathbf{x})$ *does not contain codewords different from* $\mathbf{x}$;
(c) *The minimum distance between any two codewords is at least* $k + 1$.

Proof We will prove that $(c) \Rightarrow (b) \Rightarrow (a) \Rightarrow (c)$. Suppose that the minimum distance between any two codewords is at least $k + 1$. Then, for any codeword $\mathbf{x}$, the ball $B_k(\mathbf{x})$ does not contain any other codeword, hence $(c) \Rightarrow (b)$. Further, if a combination of k or fewer errors occurs, by Proposition 7.1.2 the received vector $\mathbf{y}$ will be in $B_k(\mathbf{x})$. As there are no codevectors in $B_k(\mathbf{x})$, other than $\mathbf{x}$, the error will be detected, hence $(b) \Rightarrow (a)$. Finally, for a maximum likelihood decoder to be able to detect all combination of k or fewer errors, for any codeword $\mathbf{x}$ all vectors in $B_k(\mathbf{x})$ must not be codewords. Hence the distance between any two codewords is at least $k + 1$, thus $(a) \Rightarrow (c)$.   □

Theorem 7.1.4 *For a code* C *the following statements are equivalent:*

(a) C *corrects all combinations of* k *or fewer errors,*
(b) *For any two codewords* $\mathbf{x}$ *and* $\mathbf{y}$ *of* C *the balls* $B_k(\mathbf{x})$ *and* $B_k(\mathbf{y})$ *do not intersect.*
(c) *The minimum distance between any two codewords of* C *is at least* $2k + 1$.

Proof We will prove that $(c) \Rightarrow (b) \Rightarrow (a) \Rightarrow (c)$. Suppose that the minimum distance between any two codewords is at least $2k + 1$. Then, for any two codewords $\mathbf{x}$ and $\mathbf{y}$ the balls $B_k(\mathbf{x})$ and $B_k(\mathbf{y})$ do not intersect. Indeed, if they did, then for a certain $\mathbf{z} \in B_k(\mathbf{x}) \cap B_k(\mathbf{y})$

$$d(\mathbf{x}, \mathbf{z}) \le k, \qquad d(\mathbf{y}, \mathbf{z}) \le k.$$

This, by the triangle inequality,

$$d(\mathbf{x}, \mathbf{y}) \le d(\mathbf{x}, \mathbf{z}) + d(\mathbf{z}, \mathbf{y}) = k + k = 2k,$$

which is a contradiction, hence $(c) \Rightarrow (b)$. Further, if no more than k mistakes happen during the transmission of a vector $\mathbf{x}$, the received vector $\mathbf{y}$ will be in the ball $B_k(\mathbf{x})$ and will not be in the ball of radius k for any other codeword. Hence $\mathbf{y}$ is closer to $\mathbf{x}$ than to any other codevector. Since the decoding is a maximum likelihood decoding $\mathbf{y}$ will be decoded to $\mathbf{x}$ and all mistakes will be corrected. Thus $(b) \Rightarrow (a)$.

On the other hand, it is easy to see that if the distance d between two codewords $\mathbf{x}$ and $\mathbf{y}$ does not exceed $2k$, then a certain combinations of k or fewer errors will not be corrected. To show this let us change d coordinates of $\mathbf{x}$, one by one, and convert it into $\mathbf{y}$:

$$\mathbf{x} = \mathbf{x}_0 \to \mathbf{x}_1 \to \cdots \to \mathbf{x}_k \to \cdots \to \mathbf{x}_d = \mathbf{y}.$$

Then $\mathbf{x}_k$ will be no further from $\mathbf{y}$ than from $\mathbf{x}$. Hence if k mistakes take place and the received vector is $\mathbf{x}_k$, then it may be decoded as $\mathbf{y}$ (even if $d = 2k$). This shows that $(a) \Rightarrow (c)$.   □

Exercises

1. Consider the triple repetition $(4, 12)$-code. Find necessary and sufficient condition on the error vector $\mathbf{e} = (e_1, e_2, \ldots, e_{12})$ for the message to be decrypted correctly. Give an example of an error vector $\mathbf{e}$ of Hamming weight 4 which the code corrects.

2. Let $m = m_1 m_2$ be composite. Let us consider a two-dimensional $m_1 \times m_2$ array and write our messages into this array (in any but fixed way). To every message $\mathbf{a} = (a_1, a_2, \ldots, a_m)$ we add $m_1 + m_2$ additional symbols $e_1, e_2, \ldots, e_{m_1}$ and $f_1, f_2, \ldots, f_{m_2}$, where e_i is the sum (modulo 2) of all symbols in row i and f_j be the sum of all symbols in column j. Thus we have an (m, n)-code, where $n = m + m_1 + m_2$. Show that this code can correct all single errors and detect all triple ones.

3°. Let $A(n, d)$ denote the maximum possible size of a binary code of length n and minimum Hamming distance d. Then

$$A(n, d) \leq \frac{2^n}{\sum_{i=0}^{t} \binom{n}{i}},$$

where $t = \lfloor \frac{d-1}{2} \rfloor$. This inequality is known as the *Sphere Packing Bound*.

4°. Prove that $A(n, d) \leq 2A(n-1, d)$.

7.1.3   Minimum Distance, Minimum Weight. Linear Codes

Let $\mathcal{C} = (E, D)$ be an (m, n)-code, where $E \colon \mathbb{Z}_2^m \to \mathbb{Z}_2^n$ is the encoding function, $D \colon \mathbb{Z}_2^n \to \mathbb{Z}_2^m$ is the (maximum likelihood) decoding function and $D \circ E = id$ or $D(E(\mathbf{x})) = \mathbf{x}$ holds for all $\mathbf{x} \in \mathbb{Z}_2^m$. We observed that the set of codewords $E(\mathbb{Z}_2^m)$ is an important object. It is so important that it is often identified with the code itself and also denoted $\mathcal{C}$. We will also do this when it invites no confusion and the encoding function is clear from the context. We saw that it is extremely important to spread $\mathcal{C} = E(\mathbb{Z}_2^m)$ in $\mathbb{Z}_2^n$ uniformly and that the most important characteristic of $\mathcal{C}$ is the *minimum distance* between any two codewords of $\mathcal{C}$

$$d_{\min}(\mathcal{C}) = \min_{\mathbf{a} \neq \mathbf{b} \in \mathcal{C}} d(\mathbf{a}, \mathbf{b}).$$

We may now reformulate Theorems 7.1.3 and 7.1.4 as follows:

Theorem 7.1.5 *A code $\mathcal{C}$ detects all combinations of k or fewer errors if and only if $d_{\min}(\mathcal{C}) \geq k + 1$ and corrects all combinations of k or fewer errors if and only if $d_{\min}(\mathcal{C}) \geq 2k + 1$.*

The following table shows the error-correcting capabilities of codes depending on their minimum distance.

$d_{\min}$	1	2	3	4	5	6	7	8	9
Errors detected	0	1	2	3	4	5	6	7	8
Errors corrected	0	0	1	1	2	2	3	3	4

Now let us consider several examples.

Example 7.1.6 Let $H_1 = \begin{bmatrix} 1 & 1 \\ 1 & -1 \end{bmatrix}$ and let us define inductively: $H_{k+1} = \begin{bmatrix} H_k & H_k \\ H_k & -H_k \end{bmatrix}$. Then H_n is a matrix of order $2^n \times 2^n$. For example,

$$H_2 = \begin{bmatrix} 1 & 1 & 1 & 1 \\ 1 & -1 & 1 & -1 \\ 1 & 1 & -1 & -1 \\ 1 & -1 & -1 & 1 \end{bmatrix}, \qquad H_3 = \begin{bmatrix} 1 & 1 & 1 & 1 & 1 & 1 & 1 & 1 \\ 1 & -1 & 1 & -1 & 1 & -1 & 1 & -1 \\ 1 & 1 & -1 & -1 & 1 & 1 & -1 & -1 \\ 1 & -1 & -1 & 1 & 1 & -1 & -1 & 1 \\ 1 & 1 & 1 & 1 & -1 & -1 & -1 & -1 \\ 1 & -1 & 1 & -1 & -1 & 1 & -1 & 1 \\ 1 & 1 & -1 & -1 & -1 & -1 & 1 & 1 \\ 1 & -1 & -1 & 1 & -1 & 1 & 1 & -1 \end{bmatrix}.$$

It can be proved by induction that any two distinct rows of H_n are orthogonal (see Exercise 2). This, in turn, is equivalent to the matrix equation

$$H_n H_n^T = nI_n, \tag{7.2}$$

where I_n is the identity $n \times n$ matrix.

Definition 7.1.6 An $n \times n$ matrix H with entries from $\{+1, -1\}$ satisfying (7.2) is called an *Hadamard matrix*.

The orthogonality of rows of H_n means that any two rows of H_n coincide in 2^{n-1} positions and also differ in 2^{n-1} positions. Hence if we replace each -1 with a 0, we will have a set of vectors with minimum distance 2^{n-1}. For example, if we do this with the rows of H_3 shown above we will get eight vectors with minimum distance 4. We can use these vectors for the construction of a code. For example,

$$\begin{aligned}
(0\,0\,0) &\to (1\,1\,1\,1\,1\,1\,1\,1), \\
(1\,0\,0) &\to (1\,0\,1\,0\,1\,0\,1\,0), \\
(0\,1\,0) &\to (1\,1\,0\,0\,1\,1\,0\,0), \\
(0\,0\,1) &\to (1\,0\,0\,1\,1\,0\,0\,1), \\
(1\,1\,0) &\to (1\,1\,1\,1\,0\,0\,0\,0), \\
(1\,0\,1) &\to (1\,0\,1\,0\,0\,1\,0\,1), \\
(0\,1\,1) &\to (1\,1\,0\,0\,0\,0\,1\,1), \\
(1\,1\,1) &\to (1\,0\,0\,1\,0\,1\,1\,0).
\end{aligned}$$

We obtain a $(3, 8)$-code with minimum distance 4.

In fact, we can do even better as the following exercise shows.

Exercise 7.1.1 We may consider the matrix $\begin{bmatrix} H_3 \\ -H_3 \end{bmatrix}$ and replace in this matrix each -1 by a 0. Then we will obtain 16 vectors which may be used to construct a $(4, 8)$-code with minimum distance 4.

When, in 1969, the Mariner spacecraft sent pictures to Earth, the matrix H_5 was used to construct 64 codewords of length 32 with minimum distance 16. Each pixel had a darkness given by a 6-bit number. Each of them was changed to one of the 64 codewords and transmitted. This code could correct any combination of 7 errors. Since the signals from Mariner were fairly weak such an error-correcting facility was really needed.

We may also define the *minimum weight* of the code by

$$wt_{\min}(\mathcal{C}) = \min_{\mathbf{0} \neq \mathbf{a} \in \mathcal{C}} wt(\mathbf{a}).$$

This concept will be also quite important, especially for linear codes.

We remind to the reader of the definition of a subspace. Let F be a field and V be a vector space over F. A subset $W \subseteq V$ is a *subspace* if for any two vectors $\mathbf{u}, \mathbf{v} \in W$ and any two scalars $\alpha, \beta \in F$ the linear combination $\alpha\mathbf{u} + \beta\mathbf{v}$ is also an element of W. In this case W becomes a vector space in its own right.

Exercise 7.1.2 Let W be the set of all vectors from $\mathbb{Z}_2^n$ whose sum of all coordinates is equal to zero. Show that W is a subspace of $\mathbb{Z}_2^n$.

Definition 7.1.7 An error-correcting code $\mathcal{C} = (E, D)$ is called *linear* if $E : \mathbb{Z}_2^m \to \mathbb{Z}_2^n$ is a linear transformation from $\mathbb{Z}_2^m$ into $\mathbb{Z}_2^n$. For a binary field, where the only scalars are 0 and 1, this means that

$$E(\mathbf{x} + \mathbf{y}) = E(\mathbf{x}) + E(\mathbf{y})$$

for all $\mathbf{x}, \mathbf{y} \in \mathbb{Z}_2^m$.

Exercise 7.1.3 Prove that the parity check code is linear.

An important property of a linear code is formulated in the following proposition.

Proposition 7.1.3 *For any linear code the set of codewords $\mathcal{C}$ is a subspace of $\mathbb{Z}_2^n$. In particular, the zero vector $\mathbf{0}$ is a codeword.*

Proof We will prove that C is a subspace of $\mathbb{Z}_2^n$ if we show that the sum of any two codewords is again a codeword. (As our coefficients come from $\mathbb{Z}_2$, linear combinations are reduced to sums.) Let $\mathbf{b}, \mathbf{c}$ be two codewords. Then $\mathbf{b} = E(\mathbf{x})$ and $\mathbf{c} = E(\mathbf{y})$ and

$$\mathbf{b} + \mathbf{c} = E(\mathbf{x}) + E(\mathbf{y}) = E(\mathbf{x} + \mathbf{y}) \in C.$$

In particular, $\mathbf{0} = \mathbf{b} + \mathbf{b} \in C.$ □

For a linear code finding the minimum distance is much simplified.

Theorem 7.1.6 *For any linear code C*

$$d_{\min}(C) = wt_{\min}(C).$$

Proof Suppose $d_{\min}(C) = d(\mathbf{a}, \mathbf{b})$. Then as we know from Lemma 7.1.1 $d(\mathbf{a}, \mathbf{b}) = wt(\mathbf{a} + \mathbf{b})$, and since $\mathbf{a} + \mathbf{b} \in C$ we get

$$d_{\min}(C) \geq wt_{\min}(C).$$

On the other hand, if $wt_{\min}(C) = wt(\mathbf{a})$, then, again by Lemma 7.1.1, $wt(\mathbf{a}) = d(\mathbf{0}, \mathbf{a})$, and hence

$$d_{\min}(C) \leq wt_{\min}(C).$$

This completes the proof.   □

Theorem 7.1.6 is very useful. There are $M = 2^m$ codewords in any (m, n)-code. To find the minimum distance we need to perform $M(M - 1)/2$ calculations of distance while to find the minimum weight we need only M such calculations.

Example 7.1.7 For the following $(3, 6)$-code C

$$
\begin{aligned}
\mathbf{0} = (0\,0\,0) &\rightarrow (0\,0\,0\,0\,0\,0) = \mathbf{0} \\
\mathbf{a}_1 = (1\,0\,0) &\rightarrow (1\,0\,0\,1\,0\,0) = \mathbf{c}_1 \\
\mathbf{a}_2 = (0\,1\,0) &\rightarrow (0\,1\,0\,1\,1\,1) = \mathbf{c}_2 \\
\mathbf{a}_3 = (0\,0\,1) &\rightarrow (0\,0\,1\,0\,1\,1) = \mathbf{c}_3 \\
\mathbf{a}_1 + \mathbf{a}_2 = (1\,1\,0) &\rightarrow (1\,1\,0\,0\,1\,1) = \mathbf{c}_1 + \mathbf{c}_2 \\
\mathbf{a}_1 + \mathbf{a}_3 = (1\,0\,1) &\rightarrow (1\,0\,1\,1\,1\,1) = \mathbf{c}_1 + \mathbf{c}_3 \\
\mathbf{a}_2 + \mathbf{a}_3 = (0\,1\,1) &\rightarrow (0\,1\,1\,1\,0\,0) = \mathbf{c}_2 + \mathbf{c}_3 \\
\mathbf{a}_1 + \mathbf{a}_2 + \mathbf{a}_3 = (1\,1\,1) &\rightarrow (1\,1\,1\,0\,0\,0) = \mathbf{c}_1 + \mathbf{c}_2 + \mathbf{c}_3,
\end{aligned}
$$

it is easy to see that it is linear. We see that $C = Span\{\mathbf{c}_1, \mathbf{c}_2, \mathbf{c}_3\}$, and $d_{\min}(C) = wt_{\min}(C) = wt(\mathbf{c}_1) = 2$.

Exercises

1. Prove by induction that in the sequence of matrices $H_1, H_2, \ldots, H_n, \ldots$ all matrices are Hadamard matrices.

2. Let H be a Hadamard matrix. Let us construct a $2n \times n$ matrix $\begin{bmatrix} H \\ -H \end{bmatrix}$ and then replace each -1 by a 0. Prove that in the resulting matrix every two distinct rows have Hamming distance of at least $n/2$ between them.

3. Let $E_i \colon \mathbb{Z}_2^3 \to \mathbb{Z}_2^7$, $i = 1, 2$ be the encoding mappings of the codes C_1 and C_2, respectively, given by

    (a) $E_1(\mathbf{a}) = (a_1, a_2, a_3, a_1 + a_2, a_2 + a_3, a_1 + a_3, 0)$,
    (b) $E_2(\mathbf{a}) = (a_1, a_2, a_3, a_1 + a_2, a_2, a_1 + a_2 + a_3, 1)$.

    Which code is linear and which is not?

4. Show that in a binary linear code, either all codewords have even Hamming weight or exactly half of the codewords have even Hamming weight.

5°. Prove that if $n \times n$ Hadamard matrix exists, then n is 1 or 2 or a multiple of 4.

6°. Prove that in any binary linear code C, either all codewords begin with a 0 or exactly half the codewords begin with a 0.

7.1.4   Matrix Encoding Technique

Let $C = (E, D)$ be a linear (m, n)-code. Let us consider the vectors $\mathbf{e}_1, \mathbf{e}_2, \ldots, \mathbf{e}_m$ of the standard basis of $\mathbb{Z}_2^m$, where $\mathbf{e}_i = (0 \ldots 1 \ldots 0)$ is the vector which has the only one nonzero element 1 in the ith position. Let us consider the vectors

$$E(\mathbf{e}_1) = \mathbf{g}_1, \ \ldots \ , E(\mathbf{e}_m) = \mathbf{g}_m,$$

which encode the simplest possible messages $\mathbf{e}_1, \mathbf{e}_2, \ldots, \mathbf{e}_m$. These vectors are important since in the linear code they fully determine the encoding function. Indeed, for an arbitrary message vector $\mathbf{a} = (a_1, a_2, \ldots, a_m)$ we have

$$E(\mathbf{a}) = E(a_1\mathbf{e}_1 + a_2\mathbf{e}_2 + \cdots + a_m\mathbf{e}_m)$$
$$= a_1 E(\mathbf{e}_1) + \cdots + a_m E(\mathbf{e}_m) = a_1\mathbf{g}_1 + a_2\mathbf{g}_2 + \cdots + a_m\mathbf{g}_m.$$

Hence the subspace of all codewords C is spanned by $\{\mathbf{g}_1, \mathbf{g}_2, \ldots, \mathbf{g}_m\}$. We can now represent the encoding function by means of matrix multiplication

$$E(\mathbf{a}) = a_1\mathbf{g}_1 + a_2\mathbf{g}_2 + \cdots + a_m\mathbf{g}_m = \mathbf{a}G, \tag{7.3}$$

where

$$G = \begin{bmatrix} \mathbf{g}_1 \\ \mathbf{g}_2 \\ \vdots \\ \mathbf{g}_m \end{bmatrix}$$

is the matrix with rows $\mathbf{g}_1, \mathbf{g}_2, \ldots, \mathbf{g}_m$. Equation (7.3) shows that the code is the row space of the matrix G, i.e., $C = \mathrm{Row}(G)$.

Definition 7.1.8 Let $C = (E, D)$ be a linear (m, n)-code. Then the matrix G such that

$$E(\mathbf{a}) = \mathbf{a}G,$$

for all $\mathbf{a} \in \mathbb{Z}_2^m$, is called the *generator matrix* of C.

Example 7.1.8 Suppose the encoding function of an $(2, 4)$-code is

$$E(\mathbf{a}) = (a_1, a_2, a_2, a_1 + a_2).$$

Then

$$E(\mathbf{a}) = (a_1, a_2, a_2, a_1 + a_2) = a_1(1, 0, 0, 1) + a_2(0, 1, 1, 1) = (a_1, a_2) \begin{bmatrix} 1 & 0 & 0 & 1 \\ 0 & 1 & 1 & 1 \end{bmatrix}.$$

Proposition 7.1.4 *Let $C = (E, D)$ be a linear (m, n)-code with generator $m \times n$ matrix G. Then the rows of G are linearly independent. Moreover, $\mathrm{rank}(G) = m$ and $\dim C = m$.*

Proof It is enough to prove linear independence of the rows $\mathbf{g}_1, \mathbf{g}_2, \ldots, \mathbf{g}_m$. The two remaining statements will then follow. Suppose on the contrary that $a_1 \mathbf{g}_1 + a_2 \mathbf{g}_2 + \cdots + a_m \mathbf{g}_m = \mathbf{0}$ with not all a_i's being zero. Then, since E is linear,

$$\begin{aligned} \mathbf{0} &= a_1 \mathbf{g}_1 + a_2 \mathbf{g}_2 + \cdots + a_m \mathbf{g}_m \\ &= a_1 E(\mathbf{e}_1) + a_2 E(\mathbf{e}_2) + \cdots + a_m E(\mathbf{e}_m) = E(a_1 \mathbf{e}_1 + \cdots + a_m \mathbf{e}_m). \end{aligned}$$

Since the standard basis is linearly independent, we have $\mathbf{a} = a_1 \mathbf{e}_1 + \cdots + a_m \mathbf{e}_m \neq \mathbf{0}$. This contradicts the fact E is one-to-one, since we have $E(\mathbf{0}) = \mathbf{0}$ and $E(\mathbf{a}) = \mathbf{0}$. $\square$

Example 7.1.9 (*parity check code revisited*) The parity check $(m, m + 1)$-code is linear. Indeed, if the sum of coordinates for both $\mathbf{x}$ and $\mathbf{y}$ is zero, then the same is true for $\mathbf{x} + \mathbf{y}$. We have

$$\begin{aligned} E(\mathbf{e}_1) &= (1\ 0\ \ldots 0\ 1), \\ E(\mathbf{e}_2) &= (0\ 1\ \ldots 0\ 1), \\ &\quad \cdots \\ E(\mathbf{e}_m) &= (0\ 0\ \ldots 1\ 1). \end{aligned}$$

Hence

$$G = \begin{bmatrix} 1 & 0 & \dots & 0 & 1 \\ 0 & 1 & \dots & 0 & 1 \\ .. & .. & \dots .. & .. \\ 0 & 0 & \dots & 1 & 1 \end{bmatrix} = [I_m \; \mathbf{1}_m],$$

where I_m is the $m \times m$ identity matrix and $\mathbf{1}_m$ is the m-dimensional column of 1s.

Example 7.1.10 (*triple repetition code*) The triple repetition code $(m, 3m)$-code is also linear. We have

$$E(\mathbf{e}_1) = (1 \; 0 \; \dots 0 \; 1 \; 0 \; \dots 0 \; 1 \; 0 \; \dots 0),$$
$$E(\mathbf{e}_2) = (0 \; 1 \; \dots 0 \; 0 \; 1 \; \dots 0 \; 0 \; 1 \; \dots 0),$$
$$\dots$$
$$E(\mathbf{e}_m) = (0 \; 0 \; \dots 1 \; 0 \; 0 \; \dots 1 \; 0 \; 0 \; \dots 1).$$

Hence

$$G = \begin{bmatrix} 1 & 0 & \dots & 0 & 1 & 0 & \dots & 0 & 1 & 0 & \dots & 0 \\ 0 & 1 & \dots & 0 & 0 & 1 & \dots & 0 & 0 & 1 & \dots & 0 \\ .. & .. & \dots .. & .. & .. & .. & \dots .. & .. & .. & .. & \dots \\ 0 & 0 & \dots & 1 & 0 & 0 & \dots & 1 & 0 & 0 & \dots & 1 \end{bmatrix} = [I_m \, I_m \, I_m]$$

Example 7.1.11 Let us define a linear $(3, 5)$-code by its generator matrix

$$G = \begin{bmatrix} 1 & 0 & 0 & 0 & 1 \\ 0 & 1 & 0 & 1 & 0 \\ 0 & 0 & 1 & 1 & 1 \end{bmatrix}.$$

Then the encoding function is

$$E(a_1, a_2, a_3) = (a_1, a_2, a_3) \begin{bmatrix} 1 & 0 & 0 & 0 & 1 \\ 0 & 1 & 0 & 1 & 0 \\ 0 & 0 & 1 & 1 & 1 \end{bmatrix} = (a_1, a_2, a_3, a_2 + a_3, a_1 + a_3).$$

We see that the codeword $E(\mathbf{a})$, which encodes $\mathbf{a}$, consists of the vector $\mathbf{a}$ itself embedded into the first three coordinates and two additional symbols.

Definition 7.1.9 A linear (m, n)-code $\mathcal{C} = (E, D)$ is called *systematic* if, for any $\mathbf{a} \in \mathbb{Z}_2^m$, the first m symbols of the codeword $E(\mathbf{a})$ are the symbols of the word $\mathbf{a}$, i.e.,

$$E(a_1, a_2, \dots, a_m) = (\underbrace{a_1, a_2, \dots, a_m}_{\text{info symbols}}, \underbrace{b_1, b_2, \dots, b_{n-m}}_{\text{check symbols}}).$$

The symbols of $\mathbf{a}$ in $E(\mathbf{a})$ are called the *information symbols* and the remaining symbols are called the *check symbols*. These are the auxiliary symbols which we mentioned earlier.

Proposition 7.1.5 *For a linear (m, n)-code to be systematic, it is necessary and sufficient that its generator matrix has the form $G = (I_m \; A)$, where A is an $m \times (n - m)$ matrix.*

Proof Any systematic code $\mathcal{C} = (E, D)$ must encode $\mathbf{e}_1, \mathbf{e}_2, \ldots, \mathbf{e}_m$ into vectors

$$E(\mathbf{e}_i) = \mathbf{g}_i = (0, \ldots, 1, \ldots, 0, a_{i1}, a_{i2}, \ldots, a_{i\,n-m}),$$

Hence

$$G = \begin{bmatrix} \mathbf{g}_1 \\ \mathbf{g}_2 \\ \vdots \\ \mathbf{g}_m \end{bmatrix} = \begin{bmatrix} 1 \; 0 \ldots 0 \; a_{11} \; \ldots \; a_{1n-m} \\ 0 \; 1 \ldots 0 \; a_{21} \; \ldots \; a_{2n-m} \\ \ldots \ldots \ldots \; \ldots \; \ldots \; \ldots \\ 0 \; 0 \ldots 1 \; a_{m1} \; \ldots \; a_{mn-m} \end{bmatrix} = [I_m \; A].$$

The converse is easy and left as an exercise.   □

Definition 7.1.10 Two (m, n)-codes $\mathcal{C}_1 = (E_1, D_1)$ and $\mathcal{C}_2 = (E_2, D_2)$ are called *equivalent* if, for every $\mathbf{a} \in \mathbb{Z}_2^m$, their respective codewords $E_1(\mathbf{a})$ and $E_2(\mathbf{a})$ differ only in the order of symbols, moreover the permutation that is required to obtain $E_1(\mathbf{a})$ from $E_2(\mathbf{a})$ does not depend on $\mathbf{a}$.

Example 7.1.12 The two codes

$$\begin{array}{ll} (0\,0) \rightarrow (0\,0\,0\,0) & (0\,0) \rightarrow (0\,0\,0\,0) \\ (0\,1) \rightarrow (0\,1\,0\,1) & (0\,1) \rightarrow (0\,1\,0\,1) \\ (1\,0) \rightarrow (1\,0\,0\,1) & (1\,0) \rightarrow (0\,1\,1\,0) \\ (1\,1) \rightarrow (1\,1\,0\,0) & (1\,1) \rightarrow (0\,0\,1\,1) \end{array}$$

are equivalent. The permutation that must be applied to the symbols of the first code to obtain the second is $(1\;3)(2\;4)$.

It is clear that two equivalent codes have the same minimum distance.

Theorem 7.1.7 *Let C be a linear (m, n)-code with minimum distance d. Then there is a systematic linear (m, n)-code with the same minimum distance d.*

Proof Let C be a linear (m, n)-code with generator matrix G. When we perform elementary row operations over the rows of G we do not change $\mathrm{Row}(G)$ and hence the set of codewords (it will change the encoding function however).

We may, therefore, assume that our matrix G is already in its reduced row echelon form. Since G has full rank (its rows are linearly independent), we must have m pivot columns which are the m columns of the identity matrix I_m. Let the positions of these columns be $i_1, i_2, \ldots, i_m$. Then in a codeword $E(\mathbf{a})$ we will find our information symbols $a_1, a_2, \ldots, a_m$ in positions $i_1, i_2, \ldots, i_m$. Moving these columns (and hence the respective coordinates) to the first m positions, we will obtain a systematic code which is equivalent to the given one.   □

Example 7.1.13 Let $\mathcal{C}$ be a (3,6)-code with the generator matrix

$$G = \begin{bmatrix} 1\ 0\ 1\ 0\ 1\ 1 \\ 0\ 1\ 1\ 1\ 1\ 0 \\ 0\ 0\ 0\ 1\ 1\ 1 \end{bmatrix}.$$

Then reducing G to its reduced row echelon form

$$G = \begin{bmatrix} 1\ 0\ 1\ 0\ 1\ 1 \\ 0\ 1\ 1\ 1\ 1\ 0 \\ 0\ 0\ 0\ 1\ 1\ 1 \end{bmatrix} \rightarrow \begin{bmatrix} 1\ 0\ 1\ 0\ 1\ 1 \\ 0\ 1\ 1\ 0\ 0\ 1 \\ 0\ 0\ 0\ 1\ 1\ 1 \end{bmatrix} = G'$$

gives us a generator matrix G' of a new code with the same minimum distance. It is equivalent to the systematic code with the generator matrix

$$G'' = \begin{bmatrix} 1\ 0\ 0\ 1\ 1\ 1 \\ 0\ 1\ 0\ 1\ 0\ 1 \\ 0\ 0\ 1\ 0\ 1\ 1 \end{bmatrix},$$

which is G' with columns 3 and 4 swapped.

Example 7.1.14 The following matrix

$$G = \begin{bmatrix}
1\ 0\ 0\ 0\ 0\ 0\ 0\ 0\ 0\ 0\ 0\ 0\ 1\ 1\ 1\ 0\ 1\ 1\ 1\ 0\ 0\ 0\ 1\ 0 \\
0\ 1\ 0\ 0\ 0\ 0\ 0\ 0\ 0\ 0\ 0\ 0\ 1\ 0\ 1\ 1\ 0\ 1\ 1\ 1\ 0\ 0\ 0\ 1 \\
0\ 0\ 1\ 0\ 0\ 0\ 0\ 0\ 0\ 0\ 0\ 0\ 1\ 1\ 0\ 1\ 1\ 0\ 1\ 1\ 1\ 0\ 0\ 0 \\
0\ 0\ 0\ 1\ 0\ 0\ 0\ 0\ 0\ 0\ 0\ 0\ 1\ 0\ 1\ 0\ 1\ 1\ 0\ 1\ 1\ 1\ 0\ 0 \\
0\ 0\ 0\ 0\ 1\ 0\ 0\ 0\ 0\ 0\ 0\ 0\ 1\ 0\ 0\ 1\ 0\ 1\ 1\ 0\ 1\ 1\ 1\ 0 \\
0\ 0\ 0\ 0\ 0\ 1\ 0\ 0\ 0\ 0\ 0\ 0\ 1\ 0\ 0\ 0\ 1\ 0\ 1\ 1\ 0\ 1\ 1\ 1 \\
0\ 0\ 0\ 0\ 0\ 0\ 1\ 0\ 0\ 0\ 0\ 0\ 1\ 1\ 0\ 0\ 0\ 1\ 0\ 1\ 1\ 0\ 1\ 1 \\
0\ 0\ 0\ 0\ 0\ 0\ 0\ 1\ 0\ 0\ 0\ 0\ 1\ 1\ 1\ 0\ 0\ 0\ 1\ 0\ 1\ 1\ 0\ 1 \\
0\ 0\ 0\ 0\ 0\ 0\ 0\ 0\ 1\ 0\ 0\ 0\ 1\ 1\ 1\ 1\ 0\ 0\ 0\ 1\ 0\ 1\ 1\ 0 \\
0\ 0\ 0\ 0\ 0\ 0\ 0\ 0\ 0\ 1\ 0\ 0\ 1\ 0\ 1\ 1\ 1\ 0\ 0\ 0\ 1\ 0\ 1\ 1 \\
0\ 0\ 0\ 0\ 0\ 0\ 0\ 0\ 0\ 0\ 1\ 0\ 1\ 1\ 0\ 1\ 1\ 1\ 0\ 0\ 0\ 1\ 0\ 1 \\
0\ 0\ 0\ 0\ 0\ 0\ 0\ 0\ 0\ 0\ 0\ 1\ 0\ 1\ 1\ 1\ 1\ 1\ 1\ 1\ 1\ 1\ 1\ 1
\end{bmatrix}.$$

is the generator matrix of the famous Golay code. This is a $(12, 24)$-code and its minimum distance is 8. It was used by the Voyager I and Voyager II space-crafts during 1979–1981 to provide error correction when the Voyagers transmitted to Earth colour pictures of Jupiter and Saturn.

Exercises

1. The encoding function $E: \mathbb{Z}^4 \to \mathbb{Z}^7$ of the linear code is

$$E(\mathbf{a}) = (a_1, a_2, a_3, a_1 + a_2 + a_4, a_2 + a_3, a_1 + a_3 + a_4, a_4).$$

Construct the generator matrix.
2. Check by inspection that the Golay code is systematic.
3. Show that elementary row operations performed on a generator matrix G do not change the set of codewords and, in particular, the minimum distance of the code.
4. Let $\mathcal{C}_1$ be the (3,6) linear code with the following generator matrix over $\mathbb{Z}_2$

$$G = \begin{bmatrix} 1 & 0 & 1 & 0 & 1 & 0 \\ 1 & 1 & 0 & 0 & 1 & 1 \\ 1 & 1 & 1 & 0 & 0 & 0 \end{bmatrix}.$$

(a) Encode (1 1 1) and show that $\mathcal{C}_1$ is NOT systematic.
(b) Find the generator matrix of another $(3, 6)$ linear code $\mathcal{C}_2$, which is systematic and equivalent to $\mathcal{C}_1$.
(c) List all codewords of $\mathcal{C}_2$ and determine its minimum distance.

5°. Consider the linear code $\mathcal{C}: \mathbb{Z}_3 \to \mathbb{Z}_6$ which has the generating matrix

$$\begin{bmatrix} 1 & 0 & 0 & 1 & 1 & 1 \\ 0 & 1 & 0 & 1 & 0 & 1 \\ 0 & 0 & 1 & 0 & 1 & 1 \end{bmatrix}$$

(a) Write down all codewords of $\mathcal{C}$.
(b) Determine the minimal weight of the nonzero codeword and therefore determine the minimum distance between two distinct codewords.
(c) Which of the following are codewords

$$(1\ 1\ 1\ 0\ 0\ 1), \quad (0\ 1\ 0\ 1\ 0\ 0), \quad (1\ 0\ 1\ 1\ 0\ 0), \quad (1\ 1\ 0\ 1\ 1\ 1), \quad (1\ 0\ 0\ 0\ 0\ 1)?$$

(d) Which of the words in (c) can be corrected uniquely using maximum likelihood decoding? Correct these!

7.1.5   Parity Check Matrix

The generator matrix of a code is a great tool for the sender since with its help the encoding can be done by means of matrix multiplication. All she needs is to store the generator matrix which contains all the information about the encoding function. However the generator matrix is not very useful at the receiving end. On the receiving end we need another matrix—the parity check matrix which we will introduce below.

Definition 7.1.11 Let $\mathcal{C}$ be a linear (m, n)-code. An $(n - m) \times n$ matrix H is called a *parity check matrix* of $\mathcal{C}$ if $\mathbf{x} \in \mathcal{C}$ if and only if $H\mathbf{x}^T = 0$.

By definition, the null-space of H is $\text{Null}(H) = \{\mathbf{x} \mid H\mathbf{x}^T = 0\}$. Therefore we can reformulate the above definition as follows: an $(n - m) \times n$ matrix H is a parity check matrix of $\mathcal{C}$ if and only if $\mathcal{C} = \text{Null}(H)$.

Having this matrix at the receiving end we may quickly check if the received vector $\mathbf{y}$ was the codevector by calculating its *syndrome* $S(\mathbf{y}) = H\mathbf{y}^T$. Then $\mathbf{y} \in \mathcal{C}$ if and only if $S(\mathbf{y}) = \mathbf{0}$. If the syndrome is the zero vector, the decoder assumes that no mistakes happen. Later we will learn how the syndrome $S(\mathbf{y})$, if nonzero, can help to correct mistakes that occurred.

But, firstly, we have to learn how to construct such a matrix given the generator matrix G. We will assume that our code is systematic and G has the form $G = (I_m\ A)$, where A is an arbitrary $m \times (n - m)$ matrix. In other words,

$$G = \begin{bmatrix} \mathbf{g}_1 \\ \mathbf{g}_2 \\ \vdots \\ \mathbf{g}_m \end{bmatrix} = \begin{bmatrix} 1 & 0 & \dots & 0 & a_{11} & \dots & a_{1n-m} \\ 0 & 1 & \dots & 0 & a_{21} & \dots & a_{2n-m} \\ \dots & \dots & \dots & \dots & \dots & \dots & \dots \\ 0 & 0 & \dots & 1 & a_{m1} & \dots & a_{mn-m} \end{bmatrix}.$$

Let us assume for a moment that an $(n - m) \times n$ parity check matrix H exists. Since $\mathbf{g}_i \in \mathcal{C}$, for any $i = 1, 2, \dots, m$, we must have $H\mathbf{g}_i^T = \mathbf{0}$ and hence $HG^T = 0$. We also have $GH^T = (HG^T)^T = \mathbf{0}$. This means that all columns of H must be solutions to the system of linear equations $G\mathbf{x}^T = 0$. Since G is already in its reduced row echelon form, we separate variables to obtain

$$x_1 = -a_{11}x_{m+1} - \dots - a_{1n-m}x_n$$
$$x_2 = -a_{21}x_{m+1} - \dots - a_{2n-m}x_n$$
$$\dots$$
$$x_m = -a_{m1}x_{m+1} - \dots - a_{mn-m}x_n$$

(of course in $\mathbb{Z}_2$ we have $-a_{ij} = a_{ij}$ however we would like to leave a possibility of a non-binary alphabet). Setting, as usual, the values of the free variables to be

$$\begin{bmatrix} x_{m+1} \\ x_{m+2} \\ \vdots \\ x_n \end{bmatrix} = \begin{bmatrix} 1 \\ 0 \\ \vdots \\ 0 \end{bmatrix}, \begin{bmatrix} 0 \\ 1 \\ \vdots \\ 0 \end{bmatrix}, \dots, \begin{bmatrix} 0 \\ 0 \\ \vdots \\ 1 \end{bmatrix},$$

we obtain a basis $\{\mathbf{f}_1, \mathbf{f}_2, \ldots, \mathbf{f}_{n-m}\}$ for the solution space of the system $G\mathbf{x}^T = 0$ calculating

$$
\mathbf{f}_1^T =
\begin{bmatrix}
-a_{11} \\
-a_{21} \\
\vdots \\
-a_{m1} \\
1 \\
0 \\
\vdots \\
0
\end{bmatrix},
\quad
\mathbf{f}_2^T =
\begin{bmatrix}
-a_{12} \\
-a_{22} \\
\vdots \\
-a_{m2} \\
0 \\
1 \\
\vdots \\
0
\end{bmatrix},
\quad \ldots, \quad
\mathbf{f}_{n-m}^T =
\begin{bmatrix}
-a_{1n-m} \\
-a_{2n-m} \\
\vdots \\
-a_{mn-m} \\
0 \\
0 \\
\vdots \\
1
\end{bmatrix}.
$$

We will show that the matrix H with rows $\{\mathbf{f}_1, \mathbf{f}_2, \ldots, \mathbf{f}_{n-m}\}$ is a parity check matrix for this code. Indeed, $H\mathbf{g}_i^T = \mathbf{0}$, hence for any codeword $\mathbf{c} \in C$ we have $\mathbf{c} = a_1\mathbf{g}_1 + a_2\mathbf{g}_2 + \cdots + a_m\mathbf{g}_m$ and

$$
H\mathbf{c}^T = H(a_1\mathbf{g}_1 + a_2\mathbf{g}_2 + \cdots + a_m\mathbf{g}_m)^T = a_1 H\mathbf{g}_1^T + \cdots + a_m H\mathbf{g}_m^T = 0.
$$

Therefore $C \subseteq \mathrm{Null}(H)$. On the other hand, since H has rank $n - m$, we get $\dim \mathrm{Null}(H) = n - (n - m) = m = \dim C$. Hence $\mathrm{Null}(H) = C$ and H is indeed a parity check matrix for C. We see that H has the form

$$
H = \begin{bmatrix} -A \\ I_{n-m} \end{bmatrix}^T = (-A^T \mid I_{n-m}).
$$

We have proved:

Theorem 7.1.8 *Let C be a linear (m, n)-code. If $G = (I_m \mid A)$ is a generator matrix of C, then $H = (-A^T \mid I_{n-m})$ is a parity check matrix of C.*

This works in the other direction too: given an $(n - m) \times n$ matrix $H = (A \mid I_{n-m})$, where A is an $(n - m) \times m$ matrix, we can construct a linear (m, n)-code C with the generator matrix $G = (I_m \mid -A^T)$ and it will have H as its parity check matrix.

Example 7.1.15 Suppose that the generator matrix for a binary $(4, 7)$-code is

$$
G = \begin{bmatrix}
1 & 0 & 0 & 0 & 1 & 0 & 1 \\
0 & 1 & 0 & 0 & 0 & 1 & 1 \\
0 & 0 & 1 & 0 & 1 & 1 & 0 \\
0 & 0 & 0 & 1 & 0 & 1 & 0
\end{bmatrix} = (I_4 \mid A).
$$

Then

$$
H = \begin{bmatrix}
1 & 0 & 1 & 0 & 1 & 0 & 0 \\
0 & 1 & 1 & 1 & 0 & 1 & 0 \\
1 & 1 & 0 & 0 & 0 & 0 & 1
\end{bmatrix} = (A^T \mid I_3).
$$

If we encode $\mathbf{a} = (1\ 0\ 1\ 0)$, we get

$$\mathbf{b} = \mathbf{a}G = (1\ 0\ 1\ 0)\begin{bmatrix} 1\ 0\ 0\ 0\ 1\ 0\ 1 \\ 0\ 1\ 0\ 0\ 0\ 1\ 1 \\ 0\ 0\ 1\ 0\ 1\ 1\ 0 \\ 0\ 0\ 0\ 1\ 1\ 1\ 1 \end{bmatrix} = (1\ 0\ 1\ 0\ 0\ 1\ 1).$$

We have $S(\mathbf{b}) = H\mathbf{b}^T = (0\ 0\ 0)^T = \mathbf{0}$ but for $\mathbf{c} = \mathbf{b} + \mathbf{e}_2 = (1\ 1\ 1\ 0\ 0\ 1\ 1)$ we have $S(\mathbf{c}) = H\mathbf{c}^T = (0\ 1\ 1)^T \neq (0\ 0\ 0)^T$. If $\mathbf{c}$ was received, this would show that one or more mistakes happened.

Let $\mathbf{h}_i$ be the ith column of the parity check matrix, that is, $H = (\mathbf{h}_1\ \mathbf{h}_2 \ldots \mathbf{h}_n)$. We know that a vector $\mathbf{b} \in \mathbb{Z}_2^n$ is a codevector if and only if $S(\mathbf{b}) = \mathbf{0}$.

Let $\mathbf{a}$ be a codevector and suppose $\mathbf{b} = \mathbf{a} + \mathbf{e}$. We may treat $\mathbf{b}$ as the codevector $\mathbf{a}$ with an error. Our goal is to determine how the syndrome $S(\mathbf{b})$ of the vector $\mathbf{b} \in \mathbb{Z}_2^n$ depends on the codevector $\mathbf{a}$ and on the error vector $\mathbf{e}$. We will find that it does not depend on $\mathbf{a}$ at all! This will allow us to develop a method of error correction.

Lemma 7.1.2 *Let $\mathbf{a}$ be a codevector and suppose $\mathbf{b} = \mathbf{a} + \mathbf{e}$, where the error vector $\mathbf{e}$ has Hamming weight s and ones in positions $i_1, i_2, \ldots, i_s$, which corresponds to s mistakes in the corresponding positions. Then*

$$S(\mathbf{b}) = \mathbf{h}_{i_1} + \mathbf{h}_{i_2} + \cdots + \mathbf{h}_{i_s}. \tag{7.4}$$

Proof By Proposition 7.1.1 $\mathbf{e} = \mathbf{e}_{i_1} + \mathbf{e}_{i_2} + \cdots + \mathbf{e}_{i_s}$, where $\mathbf{e}_j$ is the jth vector of the standard basis of $\mathbb{Z}_2^n$. Then

$$S(\mathbf{b}) = H\mathbf{b}^T = H(\mathbf{a} + \mathbf{e})^T = \mathbf{0} + H\mathbf{e}^T$$
$$= H(\mathbf{e}_{i_1}^T + \mathbf{e}_{i_2}^T + \cdots + \mathbf{e}_{i_s}^T) = \mathbf{h}_{i_1} + \mathbf{h}_{i_2} + \cdots + \mathbf{h}_{i_s},$$

since $H\mathbf{e}_{i_t}^T = \mathbf{h}_{i_t}$.   □

We see that, indeed, the syndrome of the received vector depends only on the error vector and not on the codevector.

Theorem 7.1.9 *Let $H = (\mathbf{h}_1, \mathbf{h}_2, \ldots, \mathbf{h}_n)$ be an $(n - m) \times n$ matrix with entries from $\mathbb{Z}_2$ such that no two columns of H coincide. Then any binary linear (m, n)-code C with H as its parity check matrix corrects all single errors. If a single error occurs in ith position, then the syndrome of the received vector is equal to the ith column of H, i.e., $\mathbf{h}_i$.*

Proof Suppose that a codevector $\mathbf{a}$ was sent and the vector $\mathbf{b} = \mathbf{a} + \mathbf{e}_i$ was received (which means that a mistake occurred in the ith position). Then due to (7.4)

$$S(\mathbf{b}) = H\mathbf{b}^T = \mathbf{h}_i.$$

We now know where the mistake happened and can correct it.   □

Exercises

1. Let

$$A = \begin{bmatrix} 1\,2\,1\,2\,1 \\ 1\,2\,1\,0\,2 \\ 2\,1\,0\,1\,0 \end{bmatrix}$$

   be a matrix over $\mathbb{Z}_3$.

   (a) Find a basis for the nullspace $\text{Null}(A)$ of this matrix;
   (b) List all vectors of the $\text{Null}(A)$;
   (c) Find among the nonzero vectors of $\text{Null}(A)$ the vector whose weight is minimal.

2. Let us consider a binary code $\mathcal{C}$ given by its parity check matrix

$$H = \begin{bmatrix} 0\,0\,1\,1\,1\,0\,1 \\ 0\,1\,0\,1\,0\,1\,1 \\ 1\,0\,0\,0\,1\,1\,1 \\ 1\,1\,1\,1\,1\,1\,0 \end{bmatrix}.$$

   (a) Compute the generator matrix for $\mathcal{C}$. What is the number of information symbols for this code?
   (b) Will the code $\mathcal{C}$ correct any single mistake?
   (c) Will the code $\mathcal{C}$ correct any two mistakes?
   (d) Will the code $\mathcal{C}$ detect any two mistakes?
   (e) Encode the message vector whose coordinates are all equal to 1;
   (f) Decode $\mathbf{y}_1 = (1\,1\,0\,1\,0\,0\,1)$ and $\mathbf{y}_2 = (1\,1\,0\,1\,1\,0\,0)$;
   (g) Show that a single mistake could not result in receiving the vector $\mathbf{z} = (0\,1\,0\,1\,1\,1\,1)$. Show that two mistakes could result in receiving $\mathbf{z}$.

3°. A linear binary $(3, 6)$-code $\mathcal{C}$ is defined by the following parity check matrix:

$$H = \begin{bmatrix} 1\,0\,0\,0\,1\,1 \\ 0\,1\,0\,1\,0\,1 \\ 0\,0\,1\,1\,1\,0 \end{bmatrix}.$$

   (a) Find the generator matrix of C .
   (b) The parity check matrix H does not allow the presence of the codewords of weight at most 2 (apart from the all zero codeword). Explain why?

7.1.6   The Hamming Codes

Richard Hamming[1] was an American mathematician and computer scientist. He started a new subject within information theory. Hamming codes, Hamming distance and Hamming metric are standard terms used today in coding theory but they are also used in many other areas of mathematics.

We start with the Hamming $(4, 7)$-code. Let us consider the binary 3×7 matrix

$$H = (\mathbf{h}_1 \, \mathbf{h}_2 \, \mathbf{h}_3 \, \mathbf{h}_4 \, \mathbf{h}_5 \, \mathbf{h}_6 \, \mathbf{h}_7) = \begin{bmatrix} 0 & 0 & 0 & 1 & 1 & 1 & 1 \\ 0 & 1 & 1 & 0 & 0 & 1 & 1 \\ 1 & 0 & 1 & 0 & 1 & 0 & 1 \end{bmatrix}, \qquad (7.5)$$

where in the ith column $\mathbf{h}_i$ of H we write the binary representation of i from $i = 1$ to $i = 7$. Theorem 7.1.9 gives us reason to believe that the $(4, 7)$-code with this parity check matrix will be good since by that theorem such a code will correct all single errors. We also note that all nonzero three-dimensional columns are used in the construction of H and every binary 3×8 matrix will have equal columns. This says to us that the code with parity check matrix H must be in some way the optimal $(4, 7)$-code.

Let us find a generator matrix G that will match the parity check matrix H. We know that by row reducing H we do not change the nullspace of H, hence the set of codewords stays the same. We will therefore be trying to obtain a matrix with the identity matrix I_3 in the last three columns in order to apply Theorem 7.1.8. The technique is the same as for finding the reduced row echelon form. We obtain:

$$H = \begin{bmatrix} 0 & 0 & 0 & 1 & 1 & 1 & 1 \\ 0 & 1 & 1 & 0 & 0 & 1 & 1 \\ 1 & 0 & 1 & 0 & 1 & 0 & 1 \end{bmatrix} \longrightarrow \begin{bmatrix} 0 & 0 & 0 & 1 & 1 & 1 & 1 \\ 0 & 1 & 1 & 0 & 0 & 1 & 1 \\ 1 & 0 & 1 & 1 & 0 & 1 & 0 \end{bmatrix} \longrightarrow \begin{bmatrix} 0 & 0 & 0 & 1 & 1 & 1 & 1 \\ 0 & 1 & 1 & 0 & 0 & 1 & 1 \\ 1 & 1 & 0 & 1 & 0 & 0 & 1 \end{bmatrix}$$

$$\longrightarrow \begin{bmatrix} 0 & 1 & 1 & 1 & 1 & 0 & 0 \\ 0 & 1 & 1 & 0 & 0 & 1 & 1 \\ 1 & 1 & 0 & 1 & 0 & 0 & 1 \end{bmatrix} \longrightarrow \begin{bmatrix} 0 & 1 & 1 & 1 & 1 & 0 & 0 \\ 1 & 0 & 1 & 1 & 0 & 1 & 0 \\ 1 & 1 & 0 & 1 & 0 & 0 & 1 \end{bmatrix} = (C \mid I_3).$$

Therefore the generator matrix of this code is

$$G = (I_4 \mid C^T) = \begin{bmatrix} 1 & 0 & 0 & 0 & 0 & 1 & 1 \\ 0 & 1 & 0 & 0 & 1 & 0 & 1 \\ 0 & 0 & 1 & 0 & 1 & 1 & 0 \\ 0 & 0 & 0 & 1 & 1 & 1 & 1 \end{bmatrix}.$$

---

[1] **Richard Wesley Hamming (11 February 1915–7 January 1998)** He participated in the Manhattan Project that produced the first atomic bombs during World War II. There he was responsible for running the IBM computers in Los Alamos laboratory which played a vital role in the project. Later he worked for Bell Labs after which he became increasingly interested in teaching and taught in a number of leading universities in USA. Hamming is best known for his work on error-detecting and error-correcting codes. His fundamental paper on this topic "Error detecting and error correcting codes" appeared in April 1950 in the Bell System Technical Journal.

The vector $\mathbf{a} = (1\ 1\ 1\ 0)$ will be encoded to

$$\mathbf{b} = \mathbf{a}G = (1\ 1\ 1\ 0)\begin{bmatrix} 1\ 0\ 0\ 0\ 0\ 1\ 1 \\ 0\ 1\ 0\ 0\ 1\ 0\ 1 \\ 0\ 0\ 1\ 0\ 1\ 1\ 0 \\ 0\ 0\ 0\ 1\ 1\ 1\ 1 \end{bmatrix} = (1\ 1\ 1\ 0\ 0\ 0\ 0).$$

Suppose that the vector $\mathbf{c} = (1\ 0\ 1\ 0\ 0\ 0\ 0)$ is received. The syndrome of it is

$$S(\mathbf{c}) = H\mathbf{c}^T = \begin{bmatrix} 0\ 0\ 0\ 1\ 1\ 1\ 1 \\ 0\ 1\ 1\ 0\ 0\ 1\ 1 \\ 1\ 0\ 1\ 0\ 1\ 0\ 1 \end{bmatrix}\begin{bmatrix} 1 \\ 0 \\ 1 \\ 0 \\ 0 \\ 0 \\ 0 \end{bmatrix} = \begin{bmatrix} 0 \\ 1 \\ 0 \end{bmatrix} = \mathbf{h}_2$$

Assuming that only one mistake happened, we know that this mistake occurred in the second position. Hence the vector $\mathbf{b} = (1\ 1\ 1\ 0\ 0\ 0\ 0)$ was sent and $\mathbf{a} = (1\ 1\ 1\ 0)$ was the original message.

This code is very interesting. It has $2^4 = 16$ codewords and, since it corrects any single error, it has minimum distance of at least 3. So, if we take a ball $B_1(\mathbf{x})$ of radius one centred at a codeword $\mathbf{x}$, it will not intersect with other similar balls of radius one around other codewords. Due to Theorem 7.1 every such ball will have eight vectors of $\mathbb{Z}_2^7$. In total, these balls will contain $16 \cdot 8 = 128 = 2^7$ vectors, that is all vectors of $\mathbb{Z}_2^7$. The whole space is the union of those unit balls! This means that the Hamming (4,7)-code corrects all single mistakes but not a single double mistake since any double mistake will take you to another ball. Lemma 7.1.2 provides an alternative explanation of why any double mistake will not be corrected. Indeed, the syndrome of a double mistake is the sum of corresponding two columns of H. However, since all three-dimensional vectors are used as columns of H, the sum of any two columns will be a third column. This means that any double mistake will be treated as a single mistake and will not be corrected.

Suppose, for example, that the vector $\mathbf{a}=(1\ 1\ 1\ 0)$ encoded as $\mathbf{b}=(1\ 1\ 1\ 0\ 0\ 0\ 0)$, was sent, and the vector $\mathbf{c} = (0\ 1\ 1\ 0\ 0\ 0\ 1)$ was received with mistakes in the first and the seventh positions. The syndrome of it is

$$S(\mathbf{c}) = H\mathbf{c}^T = \begin{bmatrix} 0\ 0\ 0\ 1\ 1\ 1\ 1 \\ 0\ 1\ 1\ 0\ 0\ 1\ 1 \\ 1\ 0\ 1\ 0\ 1\ 0\ 1 \end{bmatrix}\begin{bmatrix} 0 \\ 1 \\ 1 \\ 0 \\ 0 \\ 0 \\ 1 \end{bmatrix} = \mathbf{h}_1 + \mathbf{h}_7 = \begin{bmatrix} 1 \\ 1 \\ 0 \end{bmatrix} = \mathbf{h}_6.$$

The received vector will be decoded as $(0\ 1\ 1\ 0\ 0\ 1\ 1)$ and then $(0\ 1\ 1\ 0)$. This double mistake will not be corrected as it will mimic as a single mistake in position 6.

The (4, 7) binary Hamming code is the "smallest" code from the infinite family of Hamming codes.

Definition 7.1.12 A binary Hamming $(2^k - k - 1, 2^k - 1)$-code is any code with the parity check matrix H, whose rth column contains the binary representation of the integer r, for $r = 1, 2, \ldots, 2^k - 1$.

Example 7.1.16 The Hamming (11, 15)-code is given by its parity check matrix

$$H = \begin{bmatrix} 0\,0\,0\,0\,0\,0\,0\,1\,1\,1\,1\,1\,1\,1\,1 \\ 0\,0\,0\,1\,1\,1\,1\,0\,0\,0\,0\,1\,1\,1\,1 \\ 0\,1\,1\,0\,0\,1\,1\,0\,0\,1\,1\,0\,0\,1\,1 \\ 1\,0\,1\,0\,1\,0\,1\,0\,1\,0\,1\,0\,1\,0\,1 \end{bmatrix}.$$

Corollary 7.1.1 *A binary Hamming $(2^k - k - 1, 2^k - 1)$-code corrects all single mistakes.*

Exercises

1. We have defined the Hamming (4, 7)-code by means of the parity check matrix H and we computed the generator matrix G, where

$$H = \begin{bmatrix} 1\,0\,1\,0\,1\,0\,1 \\ 0\,1\,1\,0\,0\,1\,1 \\ 0\,0\,0\,1\,1\,1\,1 \end{bmatrix}, \quad G = \begin{bmatrix} 1\,0\,0\,0\,0\,1\,1 \\ 0\,1\,0\,0\,1\,0\,1 \\ 0\,0\,1\,0\,1\,1\,0 \\ 0\,0\,0\,1\,1\,1\,1 \end{bmatrix}.$$

  (a) Encode the vector $\mathbf{u} = (1\,1\,0\,1)$;
  (b) Decode the vector $\mathbf{v} = (1\,0\,0\,1\,1\,0\,0)$;
  (c) Find all strings of length 7 which are decoded to $\mathbf{w} = (1\,1\,0\,1)$.
2. A code that, for some k, corrects all combinations of k mistakes and does not correct any combination of ℓ mistakes for $\ell > k$, is called *perfect*. Prove that all codes of the family of Hamming codes are perfect.
3°. Show that the binary repetition code of length n with encoding function

$$0 \mapsto (0\,0\,\ldots\,0), \quad 1 \mapsto (1\,1\,\ldots\,1)$$

is perfect when n is odd. How many errors does it correct?
4°. Show that the minimum distance of a perfect code must be odd.

7.1.7 Polynomial Codes

There is one particular class of linear codes the construction of which uses some advanced algebra and because of that these codes are very effective. In this section we will consider (m, n)-codes obtained in this way. We will identify our messages (strings of symbols of length m or vectors from $\mathbb{Z}_2^m$) with polynomials of degree at most $m - 1$. More precisely, this identification is given by the formula

$$\mathbf{a} = (a_0, a_1, \ldots, a_{m-1}) \mapsto \mathbf{a}(x) = a_0 + a_1 x + \ldots + a_{m-1} x^{m-1}.$$

Given a message we take its symbols as coefficients of a polynomial. Of course, the message $\mathbf{a}$ can be easily recovered from the polynomial $\mathbf{a}(x)$. Suppose now that we have a polynomial $g(x) = g_0 + g_1 x + \cdots + g_k x^k$, where $k = n - m$. Then we can define an (m, n)-code C as follows. For every $\mathbf{a} = (a_0, a_1, \ldots, a_{m-1}) \in \mathbb{Z}_2^m$ we define

$$E : \mathbf{a} \mapsto \mathbf{a}(x) = a_0 + a_1 x + \ldots + a_{m-1} x^{m-1} \mapsto \mathbf{a}(x) g(x) = b_0 + b_1 x + \ldots + b_{n-1} x^{n-1} \mapsto \mathbf{b},$$

where $\mathbf{b} = (b_0, b_1, \ldots, b_{n-1}) \in \mathbb{Z}_2^n$. Such code is called *a polynomial code* and the polynomial $g(x)$ is called the *generator polynomial* of this code.

Example 7.1.17 Suppose $g(x) = 1 + x^2 + x^3$ and $\mathbf{a} = (1\ 1\ 1\ 0)$. Then

$$\mathbf{a}(x) = 1 + x + x^2, \qquad \mathbf{b}(x) = \mathbf{a}(x) g(x) = 1 + x + x^5$$

and hence $\mathbf{b} = (1\ 1\ 0\ 0\ 0\ 1\ 0)$.

Theorem 7.1.10 *The polynomial code C is linear with the following $m \times n$ generator matrix*

$$G = \begin{bmatrix} g_0 & g_1 & \cdots & g_k & & & \\ & g_0 & g_1 & \cdots & g_k & & \\ & & g_0 & g_1 & \cdots & g_k & \\ & & & \cdots \cdots \cdots \cdots & & \\ & & & & g_0 & g_1 & \cdots & g_k \end{bmatrix}, \qquad (7.6)$$

where all empty places are filled with zeros.

Proof The linearity of the encoding function follows from the distributive law for polynomials. Suppose that $E(\mathbf{a}_1) = \mathbf{b}_1$ and $E(\mathbf{a}_2) = \mathbf{b}_2$ with $\mathbf{a}_1(x)$, $\mathbf{b}_1(x)$, $\mathbf{a}_2(x)$, $\mathbf{b}_2(x)$ being the corresponding polynomials. We need to show that $E(\mathbf{a}_1 + \mathbf{a}_2) = \mathbf{b}_1 + \mathbf{b}_2$. Indeed, we have

$$\mathbf{a}_1 + \mathbf{a}_2 \mapsto \mathbf{a}_1(x) + \mathbf{a}_2(x) \mapsto (\mathbf{a}_1(x) + \mathbf{a}_2(x)) g(x) = \mathbf{a}_1(x) g(x) + \mathbf{a}_2(x) g(x)$$

$$= \mathbf{b}_1(x) + \mathbf{b}_2(x) \mapsto \mathbf{b}_1 + \mathbf{b}_2,$$

as required.

To determine the generator matrix we need to calculate $E(\mathbf{e}_1), \ldots, E(\mathbf{e}_m)$. We have

$$\mathbf{e}_i \mapsto x^{i-1} \mapsto x^{i-1}g(x) = g_0 x^{i-1} + g_1 x^i + \cdots + g_{n-m}x^{n-m+i-1}$$

$$\mapsto (\underbrace{0, \ldots, 0}_{i-1}, g_0, g_1, \ldots, g_{n-m}, 0 \ldots, 0).$$

This must be the ith row of the generator matrix G. This gives us (7.6).   □

Although for a polynomial code the generator matrix (7.6) is easy to obtain, it is sometimes more convenient (and gives more insight) to multiply polynomials and not matrices.

Example 7.1.18 Let $g(x) = 1 + x^2 + x^3$. Using it we can define an $(m, m+3)$-code for all m. Let us choose $m = 4$. Then we obtain a $(4, 7)$-code whose generator matrix will be

$$G = \begin{bmatrix} 1\,0\,1\,1\,0\,0\,0 \\ 0\,1\,0\,1\,1\,0\,0 \\ 0\,0\,1\,0\,1\,1\,0 \\ 0\,0\,0\,1\,0\,1\,1 \end{bmatrix}.$$

To encode $(1\ 1\ 1\ 0)$ we perform the following multiplication of polynomials:

$$(1\,1\,1\,0) \to 1 + x + x^2 \to (1 + x + x^2)(1 + x^2 + x^3) = 1 + x + x^5 \to (1\,1\,0\,0\,0\,1\,0).$$

By row reducing G (when we change the encoding function but not the set of codewords), we get

$$G = \begin{bmatrix} 1\,0\,1\,1\,0\,0\,0 \\ 0\,1\,0\,1\,1\,0\,0 \\ 0\,0\,1\,0\,1\,1\,0 \\ 0\,0\,0\,1\,0\,1\,1 \end{bmatrix} \to \begin{bmatrix} 1\,0\,1\,0\,0\,1\,1 \\ 0\,1\,0\,0\,1\,1\,1 \\ 0\,0\,1\,0\,1\,1\,0 \\ 0\,0\,0\,1\,0\,1\,1 \end{bmatrix} \to \begin{bmatrix} 1\,0\,0\,0\,1\,0\,1 \\ 0\,1\,0\,0\,1\,1\,1 \\ 0\,0\,1\,0\,1\,1\,0 \\ 0\,0\,0\,1\,0\,1\,1 \end{bmatrix}.$$

Since it is now in the form $(I_4 \mid A)$, by Theorem 7.1.8 we may obtain its parity check matrix H as $(A^T \mid I_3)$, that is,

$$H = (A^T \mid I_3) = \begin{bmatrix} 1\,1\,1\,0\,1\,0\,0 \\ 0\,1\,1\,1\,0\,1\,0 \\ 1\,1\,0\,1\,0\,0\,1 \end{bmatrix}.$$

From this we observe that the code which we obtained is equivalent to the Hamming code since $H = (\mathbf{h}_5, \mathbf{h}_7, \mathbf{h}_6, \mathbf{h}_3, \mathbf{h}_4, \mathbf{h}_2, \mathbf{h}_1)$, where $\mathbf{h}_1, \mathbf{h}_2, \ldots, \mathbf{h}_7$ are the columns of the parity check matrix of the Hamming code.

Exercises

Let $g(x) = 1 + x + x^3$. Consider the polynomial $(5, 8)$-code $\mathcal{C}$ with $g(x)$ as generator polynomial. For this code

1. (a) Encode $\mathbf{a} = (1\ 0\ 1\ 0\ 1)$;
   (b) Find the generator matrix G of the code $\mathcal{C}$.
   (c) Find a systematic linear code $\mathcal{C}'$ (in terms of its parity check matrix) which is equivalent to $\mathcal{C}$.

$2°$. Let $g(x)$ be a generator polynomial of a polynomial binary code $\mathbb{C}$ of length n. Prove that all codewords in $\mathbb{C}$ have even weight if and only if $g(x)$ is a multiple of $1 + x$.

7.1.8 Bose–Chaudhuri–Hocquenghem (BCH) Codes

This is one particularly good class of polynomial codes which was discovered independently around 1960 by Bose, Chaudhuri and Hocquenghem. They enable us to correct multiple errors. Since the construction of the generator polynomial for these codes is based on a finite field of certain cardinality, we have to construct one first, say F, and then find its primitive element α.

In Chap. 5 we discussed a method of constructing a field which consists of p^n elements. It is unique up to an isomorphism and denoted by $GF(p^n)$. To construct it we need to take $\mathbb{Z}_p$, find an irreducible polynomial $m(x)$ over $\mathbb{Z}_p$ of degree n and form $F = \mathbb{Z}_p[x]/(m(x))$. There are very good tables of irreducible polynomials over $\mathbb{Z}_p$ of virtually any degree (see, for example, [2]).

BCH codes work equally well for binary and for non-binary alphabets but in this section we put the main emphasis on the binary case. The general case is not much different with only minor changes needed.

As usual we will consider (m, n)-codes, where m denotes the number of information symbols and n the length of codewords. The minimum distance of the code we will denote by d. For BCH codes we, first, have to decide on the length of the codewords n and on the minimum distance d, then m will depend on these two parameters but this dependence is not straightforward.

This restriction on the length is not important in applications because it is not the length of codewords that is practically important (we may divide our messages into segments of any length) but the speed of transmission, which is characterised by the ratio m/n, and the error-correcting capabilities of the code, i.e., the minimum distance d.

We use the extension $\mathbb{Z}_2 \subseteq F$ for the construction, where $F = GF(2^r)$ for some r which is an extension of $\mathbb{Z}_2$. The length of the word n will be taken to the number of elements in the multiplicative group of the field F. As we consider the binary situation, this number can only be $n = 2^r - 1$, where r is an arbitrary positive integer, since the field F of characteristic 2 may have only 2^r elements for some r.

Let α be a primitive element of F. Then it has multiplicative order n and the powers $1 = \alpha^0, \alpha, \alpha^2, \ldots, \alpha^{n-1}$ are all different. To construct $g(x)$ we need to know the minimal annihilating polynomials of $\alpha, \alpha^2, \ldots, \alpha^{d-1}$. Let $m_i(x)$ be the minimal annihilating polynomial of α^i.

Theorem 7.1.11 *The polynomial code of length n with the generator polynomial*

$$g(x) = lcm(m_1(x), m_2(x), \ldots, m_{d-1}(x)) \tag{7.7}$$

has minimum distance at least d. It has $m = n - \deg g$ information symbols.

Proof Since this code is linear, the minimum distance is the same as the minimum weight. Hence it is enough to prove that there are no codewords of weight $d - 1$ or less. Since the code is polynomial, all vectors from $\mathbb{Z}_2^n$ are identified with polynomials of degree smaller than n and the codewords are identified with polynomials which are divisible by $g(x)$. Hence, we have to show that there are no polynomials of degree smaller than n which are multiples of $g(x)$ and have less than d nonzero coefficients. Suppose on the contrary that the polynomial

$$c(x) = c_1 x^{i_1} + c_2 x^{i_2} + \cdots + c_{d-1} x^{i_{d-1}}$$

is a multiple of $g(x)$. Then it will be an annihilating polynomial for $\alpha, \alpha^2, \ldots, \alpha^{d-1}$, i.e.,

$$c(\alpha) = c(\alpha^2) = \ldots = c(\alpha^{d-1}) = 0.$$

This can be rewritten as

$$c_1 \alpha^{i_1} + c_2 \alpha^{i_2} + \cdots + c_{d-1} \alpha^{i_{d-1}} = 0$$
$$c_1 \alpha^{2i_1} + c_2 \alpha^{2i_2} + \cdots + c_{d-1} \alpha^{2i_{d-1}} = 0$$
$$\cdots$$
$$c_1 \alpha^{(d-1)i_1} + c_2 \alpha^{(d-1)i_2} + \cdots + c_{d-1} \alpha^{(d-1)i_{d-1}} = 0.$$

Let us set $\beta_k = \alpha^{i_k}$. We see that the system of homogeneous linear equations

$$\beta_1 x_1 + \beta_2 x_2 + \cdots + \beta_{d-1} x_{d-1} = 0$$
$$\beta_1^2 x_1 + \beta_2^2 x_2 + \cdots + \beta_{d-1}^2 x_{d-1} = 0$$
$$\cdots$$
$$\beta_1^{d-1} x_1 + \beta_2^{d-1} x_2 + \cdots + \beta_{d-1}^{d-1} x_{d-1} = 0$$

has a nontrivial solution $(c_1, c_2, \ldots, c_{d-1})$. This can happen only if the determinant of this system vanishes. This however contradicts to the classical result of the theory of determinants that, for any $k > 1$, the Vandermonde determinant

$$\begin{vmatrix} \beta_1 & \beta_2 & \cdots & \beta_k \\ \beta_1^2 & \beta_2^2 & \cdots & \beta_k^2 \\ \cdots\cdots\cdots\cdots\cdots \\ \beta_1^k & \beta_2^k & \cdots & \beta_k^k \end{vmatrix}$$

is zero if and only if $\beta_s = \beta_t$ for some $s \neq t$ such that $s \leq k$ and $t \leq k$ (see Sect. 10.2 of the Appendix for the proof). Indeed, in our case $k = d - 1$ and $\beta_s = \alpha^{i_s} \neq \alpha^{i_t} = \beta_t$ because $i_s \leq d - 1$ and $i_t \leq d - 1$. This contradiction proves the theorem.   □

The following lemma significantly helps with the calculation of $g(x)$.

Lemma 7.1.3 *Let $\mathbb{Z}_2 \subset F$ be an extension of fields. Let $\alpha \in F$ and let $m(t)$ be the minimal annihilating polynomial of α over $\mathbb{Z}_2$. Then $m(t)$ is also the minimal annihilating polynomial of α^2.*

Proof Let $m(t) = t^k + a_1 t^{k-1} + \cdots + a_k \cdot 1$, where $a_i \in \mathbb{Z}_2$. Then we have

$$m(\alpha) = \alpha^k + a_1 \alpha^{k-1} + \cdots + a_{k-1}\alpha + a_k \cdot 1 = 0. \tag{7.8}$$

We note, first, that $a_i^2 = a_i$ as $0^2 = 0$ and $1^2 = 1$ for $a_i \in \{0, 1\}$. We also note that since $2x = 0$ for all $x \in F$, then $(x + y)^2 = x^2 + y^2$ for all $x, y \in F$ and by induction

$$(x_1 + x_2 + \cdots + x_n)^2 = x_1^2 + x_2^2 + \cdots + x_n^2$$

for all $x_1, x_2, \ldots, x_n \in F$. Now (7.8) implies:

$$\begin{aligned} 0 = (m(\alpha))^2 &= (\alpha^k)^2 + (a_1\alpha^{k-1})^2 + \cdots + (a_k \cdot 1)^2 \\ &= (\alpha^2)^k + a_1(\alpha^2)^{k-1} + \cdots + a_k \cdot 1^2 = m(\alpha^2). \end{aligned}$$

Hence $m(t)$ is also an annihilating polynomial for α^2. Therefore the minimal irreducible polynomial of α^2 must divide $m(t)$. Since $m(t)$ is irreducible, this is possible only if it coincides with $m(t)$.   □

Example 7.1.19 Suppose that we need a code which corrects any two errors and has length 15. Hence $d = 5$, and we need a field containing 16 elements. Such field $K = \mathbb{Z}_2[x]/(x^4 + x + 1)$ was constructed in Example 5.2.3. We saw also that the multiplicative order of x was 15, hence x is a primitive element of F. Let $\alpha = x$.

For correcting any two mistakes we need a code with minimum distance $d = 5$. Theorem 7.1.11 tells us that we need to take the generator polynomial

$$g(t) = \operatorname{lcm}(m_1(t), m_2(t), m_3(t), m_4(t)).$$

Then we know that $m_1(t) = t^4 + t + 1$. By Lemma 7.1.3 we have $m_1(t) = m_2(t) = m_4(t)$. Hence $g(t) = m_1(t)m_3(t)$ and we have to calculate $m_3(t)$ which is the minimal irreducible polynomial for $\beta = x^3$. Using the table in Example 5.2.3, we calculate that $\beta^2 = x^6 = x^2 + x^3$, $\beta^3 = x^9 = x + x^3$, $\beta^4 = x^{12} = 1 + x + x^2 + x^3$. Elements $1, \beta, \beta^2, \beta^3, \beta^4$ must be linearly dependent in the 4-dimensional vector space K and we can find the linear dependency between them using the linear dependency relationship algorithm (see Sect. 10.1). By row reducing the following matrix to its reduced echelon form

$$\begin{bmatrix} 1 & 0 & 0 & 0 & 1 \\ 0 & 0 & 0 & 1 & 1 \\ 0 & 0 & 1 & 0 & 1 \\ 0 & 1 & 1 & 1 & 1 \end{bmatrix} \xrightarrow{rref} \begin{bmatrix} 1 & 0 & 0 & 0 & 1 \\ 0 & 1 & 0 & 0 & 1 \\ 0 & 0 & 1 & 0 & 1 \\ 0 & 0 & 0 & 1 & 1 \end{bmatrix}$$

we find that $m_3(t) = t^4 + t^3 + t^2 + t + 1$. Now we calculate

$$g(t) = (t^4 + t + 1)(t^4 + t^3 + t^2 + t + 1) = t^8 + t^7 + t^6 + t^4 + 1.$$

Now we may say that $m = n - \deg g = 15 - 8 = 7$ and our code C will be a $(7, 15)$-code. It will correct any two errors.

It is easy to construct a parity check matrix for this code. Over K this matrix will be

$$H = \begin{bmatrix} 1 & a & a^2 & a^3 & a^4 & a^5 & a^6 & a^7 & a^8 & a^9 & a^{10} & a^{11} & a^{12} & a^{13} & a^{14} \\ 1 & a^3 & a^6 & a^9 & a^{12} & a^{15} & a^{18} & a^{21} & a^{24} & a^{27} & a^{30} & a^{33} & a^{36} & a^{39} & a^{42} \end{bmatrix} =$$

$$\begin{bmatrix} 1 & a & a^2 & a^3 & a^4 & a^5 & a^6 & a^7 & a^8 & a^9 & a^{10} & a^{11} & a^{12} & a^{13} & a^{14} \\ 1 & a^3 & a^6 & a^9 & a^{12} & 1 & a^3 & a^6 & a^9 & a^{12} & 1 & a^3 & a^6 & a^9 & a^{12} \end{bmatrix} =$$

$$= \begin{bmatrix} 1 & 0 & 0 & 0 & 1 & 0 & 0 & 1 & 1 & 0 & 1 & 0 & 1 & 1 & 1 \\ 0 & 1 & 0 & 0 & 1 & 1 & 0 & 1 & 0 & 1 & 1 & 1 & 1 & 0 & 0 \\ 0 & 0 & 1 & 0 & 0 & 1 & 1 & 0 & 1 & 0 & 1 & 1 & 1 & 1 & 0 \\ 0 & 0 & 0 & 1 & 0 & 0 & 1 & 1 & 0 & 1 & 0 & 1 & 1 & 1 & 1 \\ \hline 1 & 0 & 0 & 0 & 1 & 1 & 0 & 0 & 0 & 1 & 1 & 0 & 0 & 0 & 1 \\ 0 & 0 & 0 & 1 & 1 & 0 & 0 & 0 & 1 & 1 & 0 & 0 & 0 & 1 & 1 \\ 0 & 0 & 1 & 0 & 1 & 0 & 0 & 1 & 0 & 1 & 0 & 0 & 1 & 0 & 1 \\ 0 & 1 & 1 & 1 & 1 & 0 & 1 & 1 & 1 & 1 & 0 & 1 & 1 & 1 & 1 \end{bmatrix}.$$

Let us look at the quality of this code. As it is a $(7, 15)$ code, there are 2^7 codewords in $\mathbb{Z}_2^{15}$. As it is 2-error correcting, each codeword $\mathbf{c}$ must have a ball $B_2(\mathbf{c})$ all for itself. We have

$$|B_2(\mathbf{c})| = \binom{15}{0} + \binom{15}{1} + \binom{15}{2} = 1 + 15 + 105 = 121 < 2^7.$$

The syndromes are 8-dimensional, hence there can be 2^8 of them. So there are at least twice as many syndromes than mistakes to be corrected. This code is far from being perfect. In general the quality of BCH codes grows with the length of the codeword.

A more practical example is a code widely used in European data communication systems. It is a binary $(231, 255)$-code with a guaranteed minimum distance of 7. The field consisting of $2^8 = 256$ elements is used and the encoding polynomial has degree 24.

Exercises

1. Construct a binary (m, n)-code with the length of codewords $n = 15$, which corrects all triple errors, in following steps:

   (a) Using the field $K = \mathbb{Z}_2[x]/(x^4 + x^3 + 1)$, compute the generating polynomial $g(t)$ of a binary BCH code with the length of the codewords $n = 15$ and with a minimum distance 7;
   (b) What is the number m of information symbols?
   (c) Write down the generating matrix G of this BCH code;
   (d) Encode the message which is represented by the string of m ones.

2. In European data communication systems a binary BCH $(231, 255)$-code is used with guaranteed minimum distance 7. Using GAP find the generator polynomial of this code.

3°. Prove that polynomials $m_i(x)$, $i = 1, \ldots, d - 1$, and the generator polynomial $g(x)$ in Theorem 7.1.11 are divisors of $x^{2^n} - 1$.

4°. Let C be a binary BCH code of an odd length n. Prove that exactly one of the following holds:

   (a) Every codeword in C has even weight;
   (b) The word $11\ldots11$ is a codeword.

7.2 Non-binary Error-Correcting Codes

Non-binary codes have many different uses. Any finite field $\mathbb{Z}_p$ can be used as an alphabet of a code if the channel allows us to distinguish p different symbols. Even if it is not, non-binary codes can be used as an intermediate step in construction of good binary codes. Non-binary codes can be also used in construction of fingerprinting codes which we will discuss in the next section.

7.2.1 The Basics of Non-binary Codes

We will again consider (m, n)-codes. The encoding function of such a code will be a mapping (normally linear) $E : F^m \to F^n$ for a certain finite field F which serves as the alphabet. The Hamming weight and the Hamming distance are defined exactly as for binary codes.

Example 7.2.1 Let $\mathbf{u} = (0\ 1\ 2\ 0\ 2\ 1\ 0)$ and $\mathbf{v} = (0\ 2\ 2\ 0\ 2\ 0\ 0)$ be vectors of $\mathbb{Z}_3^7$. Then $wt(\mathbf{u}) = 4$, $wt(\mathbf{v}) = 3$ and

$$d(\mathbf{u}, \mathbf{v}) = wt(\mathbf{u} - \mathbf{v}) = wt(0\ 2\ 0\ 0\ 0\ 1\ 0) = 2.$$

If $\mathbf{u}$ was sent and $\mathbf{v}$ was received, then the error vector is $\mathbf{e} = \mathbf{v} - \mathbf{u} = (0\ 1\ 0\ 0\ 0\ 2\ 0)$.

With non-binary codes we do not have the luxury that $-a = a$ anymore. With ternary codes we have $-a = 2a$ instead! But the following theorem is still true:

Theorem 7.2.1 *A code C detects all combinations of k or fewer errors if and only if $d_{\min}(C) \geq k + 1$ and corrects all combinations of k or fewer errors if and only if $d_{\min}(C) \geq 2k + 1$.*

The error correction capabilities of any code will again be dependent on the minimum distance of the code. And the minimum distance for a linear code will be equal to the minimum weight.

Theorem 7.2.2 *For any linear code C*

$$d_{\min}(C) = wt_{\min}(C).$$

The concepts of generator matrix G and parity check matrix H are the same. A little refinement must be made for finding G from H and the other way around. Namely, if $G = (I_m \mid A)$, then $H = (-A^T \mid I_{n-m})$. Theorem 7.1.9 must be also slightly generalised to allow designing non-binary error-correcting codes capable of correcting all single mistakes.

Theorem 7.2.3 *A linear (non-binary) code with parity check matrix H corrects all single mistakes if and only if no one column of H is a multiple of another column.*

Proof Let $\mathbf{e}$ be an error vector of weight 1. Then $\mathbf{e} = a\mathbf{e}_i$ for some $i = 1, 2, \ldots, m$ and some $0 \neq a \in \mathbb{Z}_p$, i.e.,

$$\mathbf{e} = (0 \ldots 0 \, a \, 0 \ldots 0) = a(0 \ldots 0 \, 1 \, 0 \ldots 0).$$

The syndrome of such a mistake will be

$$H\mathbf{e}^T = a\mathbf{h}_i,$$

where $\mathbf{h}_i$ is the ith column of H. If there is no other column in H that is a multiple of $\mathbf{h}_i$, then we can find both i and a. If there is such a column the identification of the mistake would be impossible.   □

Example 7.2.2 Suppose $p = 3$. We can obtain an analogue of the Hamming code by defining a code by its parity check matrix

$$H = \begin{bmatrix} 0 & 0 & 0 & 0 & 1 & 1 & 1 & 1 & 1 & 1 & 1 & 1 & 1 \\ 0 & 1 & 1 & 1 & 0 & 0 & 0 & 1 & 1 & 1 & 2 & 2 & 2 \\ 1 & 0 & 1 & 2 & 0 & 1 & 2 & 0 & 1 & 2 & 0 & 1 & 2 \end{bmatrix}.$$

The secret behind this matrix is that every nonzero column vector from $\mathbb{Z}_3^3$ is either a column of H or a multiple of such a column. Then this code will be a $(10, 13)$-code that corrects any single mistake. For example, the syndrome

$$H\mathbf{y}^T = \begin{bmatrix} 2 \\ 0 \\ 1 \end{bmatrix} = 2\mathbf{h}_7,$$

for $\mathbf{y} \in \mathbb{Z}_3^{13}$ shows that a mistake happened in the seventh position and it should be corrected by subtracting 2 (or adding 1) to the coordinate y_7.

Exercises

In exercises below, all matrices and codes are ternary, i.e., over $\mathbb{Z}_3$.

1. Suppose the matrix

$$H_1 = \begin{bmatrix} 2 & 2 & 1 & 2 & 1 & 1 \\ 1 & 2 & 1 & 0 & 2 & 1 \\ 2 & 1 & 0 & 1 & 0 & 2 \end{bmatrix}$$

   is taken as a parity check matrix of a ternary error-correcting code $\mathcal{C}_1$. Does this code correct all single errors?

2. Let $\mathcal{C}_2$ be the ternary code with parity check matrix

$$H_2 = \begin{bmatrix} 1 & 2 & 1 & 2 & 1 & 1 \\ 1 & 2 & 1 & 0 & 2 & 2 \\ 2 & 1 & 0 & 1 & 0 & 1 \end{bmatrix}.$$

   (a) Show that it corrects all single errors;
   (b) Find the generator matrix for the code $\mathcal{C}_2$;
   (c) Suppose that the code $\mathcal{C}_2$ was used. Decode the vector $\mathbf{y} = (0\ 2\ 2\ 2\ 2\ 2)$.

3°. Show that a ternary code with the parity check matrix

$$H = \begin{bmatrix} 0 & 0 & 0 & 1 & 0 & 1 \\ 0 & 0 & 2 & 0 & 1 & 0 \\ 0 & 1 & 2 & 2 & 0 & 0 \\ 1 & 0 & 1 & 2 & 0 & 0 \end{bmatrix}$$

   has minimum distance $d = 4$.

4°. The cardinality of the ball of radius t with centre $\mathbf{x}$ in $GF(q)^m$ is

$$|B_k(\mathbf{x})| = \binom{m}{0} + \binom{m}{1}(q - 1) + \cdots + \binom{m}{k}(q - 1)^t.$$

5°. Let $A_q(n, d)$ denote the maximum possible size of a binary code over $GF(q)$ of length n and minimum Hamming distance d. Then

$$A_q(n, d) \leq \frac{q^n}{\sum_{i=0}^{t} \binom{n}{i}(q-1)^i},$$

where $t = \lfloor \frac{d-1}{2} \rfloor$. This inequality is the q-ary version of the *Sphere Packing Bound*.

7.2.2   Reed–Solomon (RS) Codes

No changes at all should be made for polynomial codes and BCH codes. Among non-binary BCH codes Reed–Solomon codes are of special practical importance. They are also widely used for building up other good codes, including good binary codes.

Definition 7.2.1 Let F be a finite field of $q = p^r$ elements and α be any of its primitive elements. Let $d > 1$ be a positive integer. A *Reed–Solomon (or RS) code* over F is a polynomial $(q - d, q - 1)$-code with the generator polynomial

$$g(x) = (x - \alpha)(x - \alpha^2) \ldots (x - \alpha^{d-1}). \tag{7.9}$$

Theorem 7.2.4 *The Reed–Solomon $(q - d, q - 1)$-code with the generator polynomial (7.9) has a minimum distance of at least d.*

Proof We consider the trivial extension of fields $F \subseteq F$. Let $m_i(x)$ be the minimal irreducible polynomial of α^i over F. Then $m_i(x) = x - \alpha^i$ and we see that the RS code is a BCH code. By Theorem 7.1.11 its guaranteed minimum distance is d.   □

Example 7.2.3 Let $F = \mathbb{Z}_2[t]/(t^2 + t + 1)$. Then $F = \{0, 1, \alpha, \beta\}$, where $\alpha = t$ and $\beta = t + 1$. We note that $\beta = \alpha^2$, so α is a primitive element of F. The RS $(2, 3)$-code over F with generator polynomial $g(x) = x + \alpha$ (which is the same as $x - \alpha$) will have minimum distance 2. It will have $4^2 = 16$ codevectors. Let us encode the message $(\alpha \ \beta)$. We have

$$(\alpha \ \beta) \mapsto \alpha + \beta x \mapsto (\alpha + \beta x)(x + \alpha) = \alpha^2 + (\alpha\beta + \alpha)x + \beta x^2 = \beta + \beta x + \beta x^2 = (\beta \ \beta \ \beta).$$

Here is the table of all the codevectors:

$(0\ 0\ 0)$	$(\alpha\ 1\ 0)$	$(0\ \alpha\ 1)$	$(\alpha\ \beta\ 1)$
$(\beta\ \alpha\ 0)$	$(0\ \beta\ \alpha)$	$(\beta\ 1\ \alpha)$	$(1\ 1\ 1)$
$(1\ \beta\ 0)$	$(0\ 1\ \beta)$	$(1\ \alpha\ \beta)$	$(\alpha\ \alpha\ \alpha)$
$(\beta\ 0\ 1)$	$(\alpha\ \beta\ 1)$	$(1\ 0\ \alpha)$	$(\beta\ \beta\ \beta)$

Example 7.2.4 Let $F = \mathbb{Z}_5$. We take $\alpha = 2$ as the primitive element of $\mathbb{Z}_5$. The RS (2,4)-code over F with generator polynomial $g(x) = (x - \alpha)(x - \alpha^2) = (x - 2)(x - 4) = x^2 + 4x + 3$ will have minimum distance 3. It will have $5^2 = 25$ code-vectors:

$$(3\ 4\ 1\ 0), (2\ 1\ 4\ 0), (1\ 3\ 2\ 0), (0\ 3\ 4\ 1), (1\ 1\ 1\ 1), \ldots.$$

The Reed–Solomon codes are among the best known. To substantiate this claim let us prove the following

Theorem 7.2.5 (The Singleton bound) *Let C be a linear (m, n)-code. Then $d_{min}(C) \leq n - m + 1$.*

Proof Let us consider the codeword $E(\mathbf{e}_1) = \mathbf{g}_1$. It has only one nonzero information symbol. It has $n - m$ check symbols which may also be nonzero. In total, $wt(\mathbf{g}_1) \leq n - m + 1$. But

$$d_{min}(C) = wt_{min}(C) \leq wt(\mathbf{g}_1) \leq n - m + 1.$$

The theorem is proved. □

Now we can show that any Reed–Solomon code achieves the Singleton bound.

Theorem 7.2.6 *Let C be a Reed–Solomon (m, n)-code. Then $d_{min}(C) = n - m + 1$.*

Proof Let us consider the Reed–Solomon code C of length n with the generator polynomial

$$g(x) = (x - \alpha)(x - \alpha^2) \ldots (x - \alpha^{d-1}).$$

Let m be the number of information symbols. We know that $d_{min}(C) \geq d$ since d is the guaranteed minimum distance of this code. Since the degree of the generator polynomial is $d - 1$, this will be the number of check symbols of this polynomial code, i.e., $d - 1 = n - m$. Hence $d_{min}(C) \geq d = n - m + 1$. By the previous theorem we obtain $d_{min}(C) = n - m + 1$ and C achieves the Singleton bound. □

As we mentioned good binary codes can be obtained from RS-codes. Let F be a field of 2^r elements, $n = 2^r - 1$. We know that F is an r-dimensional vector space over $\mathbb{Z}_2$ and any element of F can be represented as a binary r-tuple. First we construct a RS (m, n)-code over F and then, in each codeword we replace every element of F with the corresponding binary tuple. We obtain an (rm, rn)-code which is binary. Such codes are very good in correcting bursts of errors (several error occurring at a close proximity) because such multiple errors affect not too many elements of F in codewords of the RS-code and can be therefore corrected. Such codes are used in CD-players because any microscopic defect on a disc results in a burst of errors.

We see that our choice of a code might be a result of the selected model for mistakes: when they are random and independent we use one type of codes, when they are highly dependent (and come in bursts) we use another type of codes.

Example 7.2.5 In Example 7.2.3, using the basis $\{1, \alpha\}$ for F, we may represent the elements of F as follows:

$$0 \to (0\,0), \quad 1 \to (1\,0), \quad \alpha \to (0\,1), \quad \beta = \alpha^2 \to (1\,1).$$

Then we will obtain a binary (4,6)-code with the following codevectors:

$$
\begin{array}{llll}
(0\ 0\ 0\ 0\ 0\ 0) & (0\ 1\ 1\ 0\ 0\ 0) & (0\ 0\ 0\ 1\ 1\ 0) & (0\ 1\ 1\ 1\ 1\ 0) \\
(1\ 1\ 0\ 1\ 0\ 0) & (0\ 0\ 1\ 1\ 0\ 1) & (1\ 1\ 1\ 0\ 0\ 1) & (1\ 0\ 1\ 0\ 1\ 0) \\
(1\ 0\ 1\ 1\ 0\ 1) & (0\ 0\ 1\ 0\ 1\ 1) & (1\ 0\ 0\ 1\ 1\ 1) & (0\ 1\ 0\ 1\ 0\ 1) \\
(1\ 1\ 0\ 0\ 1\ 0) & (0\ 1\ 1\ 1\ 1\ 0) & (1\ 0\ 0\ 0\ 0\ 1) & (1\ 1\ 1\ 1\ 1\ 1).
\end{array}
$$

Example 7.2.6 In the original paper of Irving S. Reed and Gustave Solomon (1960), every codeword of the Reed–Solomon code is a sequence of values of a code polynomial of degree less than k; this is why it is also called the *Reed–Solomon evaluation code*. Let $F = GF(q)$ be a field and $\beta_1, \beta_2, \ldots, \beta_n \in F$ with $n \le q - 1$.

In order to obtain a codeword of the Reed–Solomon code, the message $\mathbf{a}=(a_0, a_1, \ldots, a_{k-1})$ is interpreted as the a polynomial $\mathbf{a}(x)=a_0 + a_1 x + \ldots + a_{k-1}x^{k-1}$ of degree at most $k - 1$ over the field F. Further, this polynomial $\mathbf{a}(x)$ is evaluated at $n \le q - 1$ distinct points:

$$E(\mathbf{a}) := (\mathbf{a}(\beta_1), \ldots, \mathbf{a}(\beta_n)), \qquad (7.10)$$

Usually, we use $n = q - 1$ and $\beta_i = \alpha^i, i = 1, \ldots, q - 1$.

Exercise 7.2.1 Prove that E is a linear function.

A codeword has a zero symbol in the coordinate corresponding to β_i if and only if $\mathbf{a}(\beta_i) = 0$; i.e., if and only if β_i is a root of the equation $\mathbf{a}(x) = 0$. Since $\deg(\mathbf{a}(x)) \le k - 1$, by Proposition 5.1.3 this equation can have at most $k - 1$ roots in $GF(q)$. Therefore a nonzero codeword can have at most $k - 1$ symbols equal to zero, so its weight is at least $n - k + 1$. Since the code is linear, this implies that its minimum distance is at least $d \ge n - k + 1$. But by the Singleton bound, $d \le n - k + 1$; thus $d = n - k + 1$. It can be shown that this code is equivalent to RS-codes that were derived from BCH framework.

We see that the codeword consists of $q - 1$ values of the code polynomial $\mathbf{a}(x)$. However, by Theorem 5.1.2 on polynomial interpolation any k values determine it. However we do not know which values came unchanged. We need to perform several interpolations and see which polynomials appear more often than others. However there are more efficient ways for decoding RS-codes.

Exercises

1. In a series of exercises below we construct a ternary BCH-code of length $n = 8$ with minimum distance 4 using the field $F = \mathbb{Z}_3[x]/(x^2 + 2x + 2)$.

   (a) Show that $\alpha = x$ is a primitive element of F. Build a "table of powers" of α.

   (b) Show that the minimal annihilating polynomials of α, α^2 and α^3 are

   $$m_1(x) = x^2 + 2x + 2, \quad m_2(x) = x^2 + 1 \quad \text{and} \quad m_3(x) = x^2 + 2x + 2,$$

   respectively.

   (c) Determine the generating polynomial $g(x)$ of $\mathcal{C}$.

   (d) How many information symbols does this code have?

   (e) Find the generator matrix G of $\mathcal{C}$.

2°. Find the generator polynomial for a double-error correcting BCH $(17, 26)$-code over $\mathbb{Z}_3$.

3°. Find the generator polynomial for the RS evaluation code.

7.3   Fingerprinting Codes

The rapid growth of the digital economy, facilitated by spread of broadband availability,and rapid increases in computing power and storage capacity, has created a global market for content and rights holders of intellectual property. But it also creates a threat that without adequate means of protection piracy will prevent this market from functioning properly.

Managing intellectual property in electronic environments is not an easy task. On the one hand owners of the content would like to sell it for profit to paying customers but at the same time to protect it from any further illegal distribution. There are many ways to do so. One avenue is opened with the recent development of fingerprinting[2] codes that provide combinatorial and algebraic methods of tracing illegally "pirated" data. The idea is that a codeword might be embedded in the content (software, music, movie) in such a way that any illegally produced copies will reveal the distributor.

For example, such situation emerges in the context of pay TV, where only paying customers should be able to view certain programs. The broadcasted signal is normally encrypted and the decryption keys are sent to the paying subscribers. If an illegal decoder is found, the source of its decryption keys must be identified.

Fingerprinting techniques have been used for quite some time; fingerprints have been embedded in digital video, documents and computer programs. However only recently it became possible to give protection against colluding malicious users. This is what fingerprinting codes are about. This section is largely based on the groundbreaking paper of Boheh and Shaw [5] and also on the paper by Staddon, Stinson and Wei [6].

---

[2]Or watermarking, the war in terminology is currently raging.

7.3.1   The Basics of Fingerprinting

There are numerous ways to embed a codeword identifying the user in the content which is normally represented as a file. A copy of the file sold to the user, can therefore be characterised by a vector $\mathbf{x} = (x_1, x_2, \ldots, x_n) \in \mathbb{Z}_q^n$, specific to this particular copy. This is a fingerprint of this copy. Any subset $C \subset \mathbb{Z}_q^n$ may be used as the set of fingerprints and will be called *a fingerprinting (watermarking) code*.

A malicious coalition of users may try to create a pirate copy of the product by trying to identify the embedded fingerprint and to change it. To achieve this, they might compare their files—for example, using `diff` command—and find positions in which their files differ. These will certainly belong to the code so the coalition may discover some but not all symbols of the fingerprint. They might change the symbols in the identified positions with the goal to produce another legitimate copy of the product that was sold to another user (or has not yet been sold). This way they might "frame" an innocent user.

The owner of the property rights for the content would like to design a scheme that enables the identification of at least one member of the coalition that produced a pirated copy. As the bottom line, the scheme should make it infeasible for a malicious coalition to frame an innocent user by producing their fingerprint. Of course, we have to make an assumption that the malicious coalition is not too large (and here we have clear analogy with error-correcting codes that too are effective if there were not too many mistakes during the transmission).

Let us now proceed to formal definitions.

Definition 7.3.1 Let $X \subseteq \mathbb{Z}_q^n$. For any coordinate i we define the projection

$$P_i(X) = \bigcup_{\mathbf{x} \in X} \{x_i\}.$$

In other words $P_i(X)$ is the set of all ith coordinates of the words from X.

Example 7.3.1 Let $X = \{\mathbf{x}, \mathbf{y}, \mathbf{z}\}$, where

$$\mathbf{x} = (0\ 1\ 2\ 3),$$
$$\mathbf{y} = (0\ 0\ 2\ 2),$$
$$\mathbf{z} = (0\ 1\ 3\ 1).$$

Then $P_1(X) = \{0\}$, $P_2(X) = \{0, 1\}$, $P_3(X) = \{2, 3\}$, $P_4(X) = \{1, 2, 3\}$.

Definition 7.3.2 We also define the *envelope* of X

$$\text{desc}(X) = \{\mathbf{y} \in \mathbb{Z}_q^n \mid y_i \in P_i(X) \text{ for all } i.\}$$

Elements of the envelope are called *descendants* of X and elements from X are called their *parents*. It is clear that $X \subseteq \text{desc}(X)$.

A descendant of a set of vectors $X = \{\mathbf{x}_1, \mathbf{x}_2, \ldots, \mathbf{x}_n\}$ inherits coordinates from vectors in X but may take, say 1-st coordinate from $\mathbf{x}_5$, the second from $\mathbf{x}_2$ and all the rest from $\mathbf{x}_3$. For example, in Example 7.3.1 vector (0 0 3 3) is a descendant of X but vector (0 2 2 2) is not.

Definition 7.3.3 For any positive integer w, we will also define a *restricted envelope* $\text{desc}_w(X)$, which consists of all descendants of subsets of X of cardinality w.

We illustrate the difference between $\text{desc}(X)$ and $\text{desc}_w(X)$ in the following example.

Example 7.3.2 Let $X = \{\mathbf{x}, \mathbf{y}, \mathbf{z}\}$, where

$$\mathbf{x} = (1\ 0\ 0),$$
$$\mathbf{y} = (0\ 1\ 0),$$
$$\mathbf{z} = (0\ 0\ 1).$$

Then any vector in $\mathbb{Z}_2^3$ will be a descendant of X. At the same time (1 1 1) $\notin \text{desc}_2(X)$.

Example 7.3.3 Let $\mathcal{C} \subset \mathbb{Z}_4^4$ be the fingerprinting code consisting of the vectors

$$\mathbf{u} = (0\ 1\ 2\ 3),$$
$$\mathbf{v} = (1\ 2\ 3\ 0),$$
$$\mathbf{w} = (2\ 3\ 0\ 1),$$
$$\mathbf{x} = (3\ 0\ 1\ 2),$$
$$\mathbf{y} = (0\ 0\ 0\ 0),$$
$$\mathbf{z} = (1\ 1\ 1\ 1).$$

The triple $X = \{\mathbf{v}, \mathbf{y}, \mathbf{z}\}$ can produce the descendant $\mathbf{s} = (0\ 2\ 1\ 1)$ but not $\mathbf{t} = (2\ 0\ 1\ 1)$. We see that $\mathbf{s} \in \text{desc}_3(X)$ but $\mathbf{s} \notin \text{desc}_2(X)$. To prove the last statement we note that for $\mathbf{s}$ to be a descendant of a pair of vectors from $\mathcal{C}$, one of them must be either $\mathbf{u}$ or $\mathbf{y}$ (otherwise we cannot get the first coordinate 0). Neither of these two vectors has 2 as their second coordinate. Hence the second vector in this pair must be $\mathbf{v}$. But $P_4(\{\mathbf{u}, \mathbf{v}, \mathbf{y}\})$ does not contain 1. Hence $\mathbf{s} \notin \text{desc}_2(X)$.

Exercises

1. Let $X = \{\mathbf{x}_1, \mathbf{x}_2, \mathbf{x}_3\}$, where

$$\mathbf{x}_1 = (1\ 1\ 1\ 0\ 0\ 0\ 2\ 2\ 2),$$
$$\mathbf{x}_2 = (1\ 1\ 2\ 2\ 0\ 0\ 1\ 1\ 2),$$
$$\mathbf{x}_3 = (1\ 2\ 2\ 0\ 2\ 0\ 1\ 2\ 0).$$

(a) Find the projections $P_i(X)$ for $i = 1, 2, 3$.
(b) Find the number of elements in the envelope $desc(X)$.
(c) Write down a vector $\mathbf{y}$ which belongs to $desc_2(X)$ but for which no parent can be identified.

2. Give an example of a set of vectors X such that $|X| > 1$ and $desc(X) = X$.

3. Suppose $X = \{\mathbf{x}_1, \mathbf{x}_2, \ldots, \mathbf{x}_n\}$ and $P_i(X) = m_i$ for $i = 1, \ldots, n$. Prove that $|desc(X)| = m_1 \cdot \ldots \cdot m_n$.

7.3.2   Frameproof Codes

One goal that immediately comes to our mind is to secure that a coalition of malicious users cannot frame an innocent user. Of course, such protection can be put in place only against resonably small malicious coalitions in a direct analogy with error-correcting codes where the decoder is capable of correcting only a limited number of mistakes.

Definition 7.3.4 A code $\mathcal{C}$ is called w-*frameproof* (w-FP code) if for every subset $X \subset \mathcal{C}$ such that $|X| \leq w$ we have

$$desc(X) \cap \mathcal{C} = X.$$

In other words, a code is w-frameproof if no coalition of size at most w can frame another user, who is not in the coalition, by producing the fingerprint of that user.

Example 7.3.4 The code $\mathcal{C}$ consisting of the n elements of the standard basis of $\mathbb{Z}_q^n$

$$\mathbf{e}_1 = (1\ 0\ 0\ \ldots 0),$$
$$\mathbf{e}_2 = (0\ 1\ 0\ \ldots 0),$$
$$\cdots$$
$$\mathbf{e}_n = (0\ 0\ 0\ \ldots 1)$$

is w-frameproof for any $w = 1, 2, \ldots, n$.

Example 7.3.5 The code in Example 7.3.3 is 3-frameproof. Indeed, the first four users cannot be framed by any coalition to which they do not belong because each of them contains 3 in the position, where all other users have symbols different from 3. It is also easy to see that the two last users cannot be framed by any coalition of three or less users.

The following function will be useful in our proofs. For any two words $\mathbf{u}, \mathbf{v}$ of length n we define $I(\mathbf{u}, \mathbf{v}) = n - d(\mathbf{u}, \mathbf{v})$. In other words, $I(\mathbf{u}, \mathbf{v})$ is the number of coordinates where $\mathbf{u}$ and $\mathbf{v}$ agree.

As in the theory of error-correcting codes, the minimum distance $d_{min}(C)$ between any two distinct codewords is an important parameter.

Theorem 7.3.1 *Let C be a fingerprinting code of length n. Suppose that*

$$d_{min}(C) > n\left(1 - \frac{1}{w}\right),$$

then C is a w-frameproof code.

Proof Suppose that a coalition $X = \{\mathbf{x}_1, \mathbf{x}_2, \ldots, \mathbf{x}_w\}$ can frame an innocent user $\mathbf{y} \in C \setminus X$, that is $\mathbf{y} \in \mathrm{desc}(X)$. Since $\mathbf{y}, \mathbf{x}_i \in C$, for every $i = 1, 2, \ldots, w$ we have $d(\mathbf{y}, \mathbf{x}_i) > n(1 - 1/w)$ and hence we obtain $I(\mathbf{y}, \mathbf{x}_i) = n - d(\mathbf{y}, \mathbf{x}_i) < n - (n - n/w) = n/w$. This means that $\mathbf{y}$ and $\mathbf{x}_i$ coincide in less than n/w positions and, hence less than n/w positions of $\mathbf{y}$ could come from $\mathbf{x}_i$. Since we have exactly w elements in X, it follows now that less than $w \cdot n/w = n$ coordinates in $\mathbf{y}$ can come from vectors of X. Hence at least one coordinate of $\mathbf{y}$, say y_j, does not coincide with the jth coordinates of any of the vectors $\mathbf{x}_1, \mathbf{x}_2, \ldots, \mathbf{x}_w$ and therefore $y_j \notin P_j(X)$. This contradicts the assumption that $\mathbf{y}$ is a descendant of X. $\qquad\square$

Exercises

The code $C \subset \{1, 2, 3\}^6$ consists of six codewords:

$$\mathbf{c}_1 = (1\ 1\ 1\ 1\ 1\ 1), \quad \mathbf{c}_2 = (2\ 2\ 2\ 2\ 2\ 2), \quad \mathbf{c}_3 = (3\ 3\ 3\ 3\ 3\ 3),$$
$$\mathbf{c}_4 = (1\ 2\ 3\ 1\ 2\ 3), \quad \mathbf{c}_5 = (2\ 3\ 1\ 2\ 3\ 1), \quad \mathbf{c}_6 = (3\ 1\ 2\ 3\ 1\ 2).$$

1. Find the minimum distance of this code.
2. Prove that it is 2-frameproof.

7.3.3 Codes with the Identifiable Parent Property

Definition 7.3.5 We say that a code C has the *identifiable parent property* of order w (w-IPP code) if for any $\mathbf{x} \in \mathrm{desc}_w(C)$ the family of subsets

$$\{X \subseteq C \mid |X| \leq w \text{ and } \mathbf{x} \in \mathrm{desc}(X)\} \tag{7.11}$$

has a nonzero intersection.

What this says is that, for any w-IPP code and for any $\mathbf{x} \in \mathrm{desc}_w(\mathcal{C})$ this vector cannot be produced without a participation of a certain user: the one who is in the intersection of the family of subsets (7.11). Therefore this user can be identified. The w-IPP property is stronger than w-frameproofness.

Proposition 7.3.1 *Any code $\mathcal{C}$ with the identifiable parent property of order w is w-frameproof.*

Proof Suppose that the w-IPP property holds but a certain coalition X with no more than w users can frame an innocent user $\mathbf{c} \in \mathcal{C} \setminus X$. Then $\mathbf{c} \in \mathrm{desc}(X)$ and $\mathbf{c} \in \mathrm{desc}(\{\mathbf{c}\})$. Since $\{\mathbf{c}\} \cap X = \emptyset$, this contradicts the w-IPP property.   □

Let us now give a non-trivial example of a w-IPP code.

Example 7.3.6 The following code has the identifiable parent property of order 2 and was constructed with a help of a Reed–Solomon code.

$$\mathbf{c}_1 = (1\ 1\ 1\ 1\ 1),$$
$$\mathbf{c}_2 = (1\ 2\ 2\ 2\ 2),$$
$$\mathbf{c}_3 = (1\ 3\ 3\ 3\ 3),$$
$$\mathbf{c}_4 = (1\ 4\ 4\ 4\ 4),$$
$$\mathbf{c}_5 = (2\ 1\ 2\ 3\ 4),$$
$$\mathbf{c}_6 = (2\ 2\ 1\ 4\ 3),$$
$$\mathbf{c}_7 = (2\ 3\ 1\ 4\ 2),$$
$$\mathbf{c}_8 = (2\ 4\ 3\ 2\ 1),$$
$$\mathbf{c}_9 = (3\ 1\ 4\ 2\ 3),$$
$$\mathbf{c}_{10} = (3\ 2\ 3\ 1\ 4),$$
$$\mathbf{c}_{11} = (3\ 3\ 2\ 4\ 1),$$
$$\mathbf{c}_{12} = (3\ 4\ 1\ 3\ 2),$$
$$\mathbf{c}_{13} = (3\ 4\ 1\ 3\ 2),$$
$$\mathbf{c}_{14} = (4\ 2\ 4\ 3\ 1),$$
$$\mathbf{c}_{15} = (4\ 4\ 2\ 1\ 3).$$

It is really hard to check that this code indeed is 2-IPP but relatively easy to check that $d_{min}(\mathcal{C}) = 4$. As we will see later Theorem 7.3.3 will imply 2-IPP for this code.

Codes with the identifiable parent property normally require a large alphabet. The binary alphabet is the worst one.

Proposition 7.3.2 *There does not exist a binary 2-IPP code $\mathcal{C}$ with $|\mathcal{C}| \geq 3$.*

Proof Suppose $\mathbf{x}, \mathbf{y}, \mathbf{z} \in C$ be three distinct codewords and $X = \{\mathbf{x}, \mathbf{y}, \mathbf{z}\}$. We define a descendant $\mathbf{u}$ in the following way. For each i, we consider the coordinates x_i, y_i, z_i; among them there will be a majority of zeros or a majority of ones. We define u_i to coincide with the majority. Then $\mathbf{u}$ belongs to each of the $\mathrm{desc}(\mathbf{x}, \mathbf{y})$, $\mathrm{desc}(\mathbf{x}, \mathbf{z})$, and $\mathrm{desc}(\mathbf{y}, \mathbf{z})$. However, $\{\mathbf{x}, \mathbf{y}\} \cap \{\mathbf{x}, \mathbf{z}\} \cap \{\mathbf{y}, \mathbf{z}\} = \emptyset$. $\square$

For generalisation of this to alphabets containing q elements see Exercise 1 below.

We see from the Example 7.3.6 that it is not too easy to check that the code in the above example satisfies the identifiable parent property of order 2. But there exists one slightly stronger property that is much easier to check.

Definition 7.3.6 A code C is called w-*traceable* (w-TA code) if for any $\mathbf{y} \in \mathrm{desc}_w(C)$ the inclusion $\mathbf{y} \in \mathrm{desc}(X)$, for some subset $X \subseteq C$ with $|X| = w$, implies the existence of at least one codeword $\mathbf{x} \in X$ such that $d(\mathbf{y}, \mathbf{x}) < d(\mathbf{y}, \mathbf{z})$ for any $\mathbf{z} \in C \setminus X$.

If a code is an w-TA code, we can always trace at least one parent of $\mathbf{y} \in \mathrm{desc}_w(C)$ using a process similar to maximum likelihood decoding for error-correcting codes. Indeed, the following proposition is true.

Proposition 7.3.3 *Suppose that a code C is w-traceable, and $\mathbf{y} \in \mathrm{desc}(X)$ for some subset $X \subseteq C$ with $|X| = w$. Let $\mathbf{x}_1, \mathbf{x}_2, \ldots, \mathbf{x}_k$ be the set of vectors from C such that $d = d(\mathbf{y}, \mathbf{x}_1) = \ldots = d(\mathbf{y}, \mathbf{x}_k)$ and no vector $\mathbf{z} \in C$ satisfies $d(\mathbf{y}, \mathbf{z}) < d$. Then $\{\mathbf{x}_1, \mathbf{x}_2, \ldots, \mathbf{x}_k\} \subseteq X$.*

Proof Suppose $\mathbf{x}_i \notin X$ for some i. Then by the traceability property there must be a vector in $\mathbf{x} \in X$ such that $d(\mathbf{y}, \mathbf{x}) < d(\mathbf{y}, \mathbf{x}_i) = d$, which contradicts to the minimality of d. $\square$

Let us state now one obvious fact.

Lemma 7.3.1 *Let $X = \{\mathbf{x}_1, \mathbf{x}_2, \ldots, \mathbf{x}_w\}$ and $\mathbf{y} \in \mathrm{desc}(X)$. Then there exists $i \in \{1, 2, \ldots, w\}$ such that $I(\mathbf{x}_i, \mathbf{y}) \geq n/w$.*

Proof Suppose on the contrary that $I(\mathbf{x}_i, \mathbf{y}) < n/w$ for all $i \in \{1, 2, \ldots, w\}$. Then $\mathbf{y}$ inherited less than n/w coordinates from each $\mathbf{x}_i$. In total it inherited less than $n \cdot n/w = n$ coordinates from vectors of X and cannot be a descendant of X. $\square$

Theorem 7.3.2 *Any w-TA code C is also an w-IPP code.*

Proof Suppose that the code C is w-traceable. Let $\mathbf{x} \in \mathrm{desc}_w(C)$. Let us consider a family of subsets (7.11). Suppose $\mathbf{y} \in C$ is the closest or one of the closest vectors of C to $\mathbf{x}$, i.e., the distance $d(\mathbf{x}, \mathbf{y})$ is the smallest possible. Because C is w-traceable $\mathbf{y}$ must belong to every subset of the family (7.11), hence its intersection is non-empty and w-IPP property holds. $\square$

Theorem 7.3.3 *Suppose that a code C of length n has a minimum distance*

$$d_{min}(C) > n\left(1 - \frac{1}{w^2}\right).$$

Then C is a w-traceable code and hence has the identifiable parent property of order w.

Proof Let $X \subseteq C$ with $|X| = w$. Suppose $X = \{x_1, x_2, \ldots, x_w\}$. Let us consider any $\mathbf{z} \in C \setminus X$. Then, for any i, $I(\mathbf{z}, \mathbf{x}_i) = n - d(\mathbf{z}, \mathbf{x}_i) < n - (n - n/w^2) = n/w^2$, i.e., the number of coordinates where $\mathbf{z}$ and $\mathbf{x}_i$ agree is less than n/w^2. We now define

$$I(\mathbf{z}, X) = \{j \mid z_j \in P_j(X)\}.$$

We obtain now

$$I(\mathbf{z}, X) \le wI(\mathbf{z}, \mathbf{x}_i) < w \cdot \frac{n}{w^2} = \frac{n}{w}. \tag{7.12}$$

On the other hand, by Lemma 7.3.1, for every $\mathbf{y} \in desc(X)$ we can find a $\mathbf{x}_i$ such that $I(\mathbf{x}_i, \mathbf{y}) \ge n/w$. Thus we obtain $d(\mathbf{x}_i, \mathbf{y}) \le n - n/w = n(1 - 1/w)$ while for any $\mathbf{z} \in C \setminus X$ we will have $I(\mathbf{z}, \mathbf{y}) \le I(\mathbf{z}, X) < n/w$ and hence $d(\mathbf{z}, \mathbf{y}) > n - n/w = n(1 - 1/w)$, proving w-traceability. □

This theorem works only for a reasonably large alphabet.

Exercises
1. Let the size of the alphabet be q. Then there does not exist an w-IPP code C with $|C| > w \ge q$.
2. Using Reed–Solomon code C over $\mathbb{Z}_{17}$ of length 16 with the minimum distance 13, show that there exists a fingerprinting code with the identifiable parent property of order 2 containing 83521 codewords.

Compression

8

Good things, when short, are twice as good.

Baltasar Gracián y Morales (1601–1658)

Compression of files is an important practical question. Memory is always a limiting resource so if our files can be stored in a more economic fashion, this has to be done. Some files, like pictures, contain a lot of redundancy and can be compressed significantly even without loss of quality of pictures. There are numerous ways to do so.

There are three major approaches to measuring the quantity of information in a message of a certain alphabet: probabilistic, combinatorial and algorithmic. Probabilistic view is that information is anything that resolves uncertainty. The more uncertain an event that may or may not take place in the future, the more information is required to resolve the uncertainty. This works well with messages generated by random sources but cannot help answering questions like: "What is the quantity of information in Lev Tolstoy's War and Peace"? or "How much information is needed for the reproduction of a particular form of cockroach?"

Combinatorial approach tries to reduce complex events to some basic ones. Suppose you would like to know if there will be rain tomorrow. You look the weather forecast and get the answer. This is a simple "yes" or "no" situation and it is easy to resolve. Suppose a 1 means "no rain" and a 0 means rain, then one binary digit carries all the information you need. One *bit* is a unit of information expressed as a choice between two possibilities 0 and 1. Asking whether there will be rain tomorrow you ask for one bit of information. Information of more complex events can also be measured in bits. Given a set of possible events we ask how many bits of information is required to individualise each particular event. Suppose n binary digits are sufficient

© Springer Nature Switzerland AG 2020

A. Slinko, *Algebra for Applications*, Springer Undergraduate Mathematics Series,

https://doi.org/10.1007/978-3-030-44074-9_8

to give a distinctive label to every event and you cannot do this with $n - 1$ binary digits. Then we say that every event in the set of events carries n bits of information.

The algorithmic approach is closely related to the concept of (Kolmogorov) complexity. Roughly speaking the longer is the program that we have to write for a computer to output the given message, the less redundancy this massage has and less compressible it is.

Here we give a glimpse of the combinatorial approach to compression describing Huffman's and Fitingof's compression codes. The first needs information about the source, i.e., we need to know how the file was generated. The second type of codes is universal as they can be used when we do not know where the data came from. Boris Fitingof (1966) developed the first such code, and the construction is quite elegant. His paper was inspired by paper of Kolmogorov (1965). However it is fair to consider Fitingof as the founder of the universal encoding.

8.1   Encoding a Known Source

8.1.1   Motivating Example

Example 8.1.1 Suppose that we have a 100,000 character data file that we wish to store in a most economical way. The file contains only six characters a, b, c, d, e, f appearing with the following frequencies:

	a	b	c	d	e	f
frequency in thousands	40	25	12	10	8	5

A binary code encodes each character as a binary string called codeword. We would like to find a binary code that encodes our file using as few bits as possible, i.e., compresses it as much as possible.

Firstly, let us try to use a fixed-length code that encodes every symbol in the same number of binary digits. Since we have only four combinations of binary digits of length 2, namely $00, 01, 10, 11$, we must use combinations of three digits to encode letters, perhaps,

$$a \to 000, \quad b \to 001, \quad c \to 010, \quad d \to 011, \quad e \to 100, \quad f \to 101.$$

Then our file will be stored as a string of binary digits of length 300,000. It will be easy to decode: indeed, the first three digits encode the first symbol, the next three digits encode the second and so on.

Let us now try to use the fact that a and b are the most frequent symbols and e and f are the least frequent. Consider the encoding

$$a \to 0, \quad b \to 10, \quad c \to 110, \quad d \to 1110, \quad e \to 11110, \quad f \to 111110.$$

Then our file will be stored using

$$(40 \cdot 1 + 25 \cdot 2 + 12 \cdot 3 + 10 \cdot 4 + 8 \cdot 5 + 5 \cdot 6) \cdot 1000 = 236{,}000 \text{ bits},$$

which is much more economical. Will we be able to uniquely decode the file? Yes, we will since not a single codeword is the beginning of the other. For example, 0110011110010 can be uniquely decrypted as *acaeab*. The next section explains why.

8.1.2 Prefix Codes

Let X be a finite alphabet. By $W(X)$ we will denote all possible words of finite length in this alphabet. In particular, $W(\mathbb{Z}_2)$ is the set of all words in binary alphabet $\mathbb{Z}_2 = \{0, 1\}$.

Definition 8.1.1 A mapping $\psi \colon X \to W(\mathbb{Z}_2)$ will be called a *binary code* if it is one-to-one. This code is said to be a *prefix code* if for every two symbols $x, y \in X$ neither of the two codewords $\psi(x)$, $\psi(y)$ is the beginning of the other.

The word $\mathbf{w} = x_1 x_2 \ldots x_n$ in alphabet X will be encoded as $\psi(x_1)\psi(x_2)\ldots\psi(x_n)$. A prefix code guarantees that encoded words can be uniquely decoded.

Example 8.1.2 Let $X = \{a, b, c\}$ and $\psi(a) = 1$, $\psi(b) = 01$, $\psi(c) = 00$. This is a prefix code and the message 0001101100 can be uniquely decoded as

$$0001101100 = \psi(c)\psi(b)\psi(a)\psi(b)\psi(a)\psi(c) \longrightarrow cbabac.$$

Every rooted binary tree with n leaves gives us a prefix code for an alphabet X of size n. We assign a 1 to each edge from a parent to its left child and a 0 to each edge from a parent to its right child. Then the set of all terminal vertices can be identified with the set of codewords of a prefix code. Indeed, for any terminal vertex, there is a unique directed path from the root to it. This path gives a string of 0s and 1s which we assign to the terminal vertex. Since we always finish at a terminal vertex, no path is a beginning of the other and therefore no codeword will be a beginning of the other. We have proved

Theorem 8.1.1 *Every rooted binary tree determines a prefix code and every prefix code corresponds to a rooted binary tree.*

Example 8.1.3 For example, the tree below corresponds to the code $\{0, 11, 101, 100\}$:

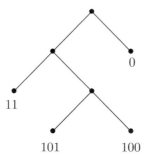

The following theorem tells us how many codewords of a particular length a prefix code may have.

Theorem 8.1.2 (Kraft's Inequality) *Suppose that $|X| = q$. Then a prefix code $\psi \colon X \to W(\mathbb{Z}_2)$ with the lengths of codewords $m_1, m_2, \ldots, m_q$ exists if and only if*

$$\sum_{i=1}^{q} 2^{-m_i} \leq 1. \tag{8.1}$$

Proof We will assume that $m = \max(m_1, \ldots, m_q)$, which means that the longest codeword has length m. Suppose that a prefix code possesses a codeword $\mathbf{u}$ of length i. Then the $2^1 = 2$ words $\mathbf{u}0$ and $\mathbf{u}1$ cannot be codewords. The $2^2 = 4$ words $\mathbf{u}00$, $\mathbf{u}01$, $\mathbf{u}10$ and $\mathbf{u}11$ also cannot be codewords. In general all 2^{k-i} words of length k obtained by extending $\mathbf{u}$ to the right cannot be codewords. If $\mathbf{v}$ is another codeword of length j then it excludes another 2^{k-j} words of length k from being codewords. The codewords $\mathbf{u}$ and $\mathbf{v}$ cannot exclude the same word, otherwise one of them will be the beginning of the other.

Let us denote by S_j the number of codewords of length j. Then, as we just noticed,

$$S_1 \cdot 2^{k-1} + S_2 \cdot 2^{k-2} + \ldots + S_{k-1} \cdot 2$$

words of length k cannot be codewords. This number plus S_k, which is the number of codewords of length k, should be less than or equal to 2^k, which is the total number of words of length k. The existence of a prefix code with the given lengths of codewords implies that the following inequality holds for any $k = 1, \ldots m$:

$$2^k - S_1 \cdot 2^{k-1} - S_2 \cdot 2^{k-2} - \ldots - S_{k-1} \cdot 2 - S_k \geq 0. \tag{8.2}$$

Thus, all these inequalities are necessary conditions for the existence of such a prefix code. But the inequality for $k = m$ is the strongest because it implies all the rest. Indeed, (8.2) implies

$$2^k - S_1 \cdot 2^{k-1} - S_2 \cdot 2^{k-2} - \ldots - S_{k-1} \cdot 2 \geq 0$$

and after dividing by 2 we get

$$2^{k-1} - S_1 \cdot 2^{k-2} - S_2 \cdot 2^{k-3} - \ldots - S_{k-1} \geq 0,$$

i.e., the same inequality for $k - 1$. Thus, indeed, the inequality for $k = m$ implies all other inequalities.

Taking this strongest inequality (8.2) and dividing it by 2^m we get

$$\sum_{j=1}^{m} S_j \cdot 2^{-j} \leq 1. \tag{8.3}$$

This is equivalent to (8.1) as

$$\sum_{j=1}^{m} S_j \cdot 2^{-j} = \sum_{i=1}^{q} 2^{-m_i}.$$

Hence the inequality (8.1) is necessary condition for the existence of a prefix code with lengths of codewords $m_1, m_2, \ldots, m_q$.

Let us show that it is also sufficient. Let S_j be the number of codewords of length j and m be the maximal length of codewords. We will again use (8.1) in its equivalent form (8.3) which implies (8.2) for all $k = 1, \ldots, m$.

Firstly, we take S_1 arbitrary words of length 1. Since (8.2) for $k = 1$ gives $2 - S_1 \geq 0$ we have $S_1 \leq 2$ and we can do this step. Suppose that we have done $k - 1$ steps already and have chosen S_i words of length i for $i = 1, \ldots, k-1$ so that no one word is the beginning of the other. Then the chosen words will prohibit us from choosing

$$S_1 \cdot 2^{k-1} + S_2 \cdot 2^{k-2} + \ldots + S_{k-1} \cdot 2$$

words of length k. But due to (8.2)

$$2^k - \left(S_1 \cdot 2^{k-1} + S_2 \cdot 2^{k-2} + \ldots + S_{k-1} \cdot 2\right) \geq S_k,$$

hence we can find S_k words of length k which are compatible with the words previously chosen. This argument shows that the construction of the code can be completed to the end.   □

Example 8.1.4 Let us consider the following equation:

$$\frac{1}{2^1} + \frac{1}{2^2} + \frac{1}{2^3} + \frac{1}{2^3} = 1. \tag{8.4}$$

If $X = \{a, b, c, d\}$, then according to Theorem 8.1.2 there exists a prefix code $\psi \colon X \to W(\mathbb{Z}_2)$ with the lengths of the codewords 1, 2, 3, 3. Let us choose the codeword $\psi(a) = 0$ of length 1, then we cannot use the words 00 and 01 for the choice of the codeword for b of length 2 and we choose $\psi(b) = 10$. For the choice of codewords for c and d we cannot choose the words $000, 001, 010, 011$ (because of the choice of $\psi(a)$) and the words $100, 101$ (because of the choice of $\psi(b)$), thus we choose the two remaining words of length 3, i.e., $\psi(c) = 110$ and $\psi(d) = 111$.

Example 8.1.5 Suppose now that $X = \{aa, ab, ba, bb\}$. Then $|X| = 4$ and we can use (8.4) again for this situation to define a code $\psi: X \to W(\mathbb{Z}_2)$ as follows:

$$\psi(aa) = 110, \qquad \psi(ab) = 0, \qquad \psi(ba) = 10, \qquad \psi(bb) = 111.$$

The the words *abba* and *baabab* will be encoded as 010 and 1000, respectively. The word 11111001000 can be represented as

$$11111001000 = \psi(bb)\psi(aa)\psi(ab)\psi(ba)\psi(ab)\psi(ab)$$

and therefore it will be decoded to *bbaaabbaabab*.

Exercises

1. Check that the set $\{11, 10, 00, 011, 010\}$ is a set of codewords of a prefix code and construct the corresponding tree.
2. Given the equation

$$\frac{1}{2^2} + \frac{1}{2^2} + \frac{1}{2^3} + \frac{1}{2^3} + \frac{1}{2^4} + \frac{1}{2^4} + \frac{1}{2^4} + \frac{1}{2^4} = 1,$$

   give an example of a prefix code the existence of which can be implied from this equality?
3. Let X be an alphabet consisting of 9 elements. Construct a prefix binary code $\psi: X \to W(\mathbb{Z}_2)$ with the lengths of the codewords: 2, 3, 3, 3, 3, 3, 4, 5, 5 in following steps:

   (a) Use Kraft's inequality to prove that such a code does exist.
   (b) Construct any tree that corresponds to such a code.
   (c) List the codewords (their choice is not unique).

8.1.3 Huffman's Optimal Code

In our motivating example we knew frequencies of symbols. Alternatively, we can assume that the source that generates symbols generates them with known probabilities. Under this assumption we can speak about optimal encoding of these symbols.

Definition 8.1.2 Let $X = \{x_1, x_2, \ldots, x_n\}$ be an alphabet and the source generates x_i with frequency p_i. The average bits per symbol of a prefix code $\psi: X \to W(\mathbb{Z}_2)$

is the sum over all symbols of their frequency times the number of bits in their encodings, i.e.,

$$AB(\psi) = \sum_{i=1}^{n} p_i \cdot |\psi(x_i)|,$$

where $|\psi(x_i)|$ is the length of the codeword encoding symbol x_i.

We would like to find a prefix code that has the lowest possible average bits per symbol.

Definition 8.1.3 For a given alphabet $X = \{x_1, x_2, \ldots, x_n\}$ with probability distribution $p_1, p_2, \ldots, p_n$ the prefix code with the smallest average bits per symbol is called the *optimal code*.

In 1951, David A. Huffman and his MIT information theory classmates were given the choice of a term paper or a final exam. The professor, Robert M. Fano, assigned a term paper on the problem of finding the most efficient binary code. Huffman, unable to prove any codes were the most efficient, was about to give up and start studying for the final when he hit upon the idea of using a frequency-sorted binary tree and quickly proved this method the most efficient. In doing so, the student outdid his professor, who had worked with information theory inventor Claude Shannon to develop a similar code. Huffman avoided the major flaw of the suboptimal Shannon–Fano coding by building the tree from the bottom up instead of from the top down. It starts with n vertices $x_1, x_2, \ldots, x_n$ with no edges.

Step 1 Pick two letters x, y from alphabet X with the smallest frequencies p_x and p_y and create a subtree that has these two letters as leaves adding the root of this subtree denoted z.

Step 2 Set the frequency $p_z = p_x + p_y$. Remove x, y and add z creating new alphabet

$$X' = X \cup \{z\} - \{x, y\}.$$

Note that $|X'| = |X| - 1$.

Repeat this procedure, called merge, with new alphabet X' until an alphabet with only one symbol—the root—is left. The resulting tree corresponds to an optimal code.

The following lemmas will provide a justification of this construction.

Definition 8.1.4 A binary tree is *full* if every node that is not a leaf has two children.

Lemma 8.1.1 *The tree for an optimal code is full.*

Proof If there is a vertex with just one child, then this vertex can be removed and the tree streamlined.   □

Lemma 8.1.2 *Consider the two letters, x and y with the smallest frequencies. Then there is an optimal code in which tree these two letters are sibling leaves at the lowest level.*

Proof Let T be an optimum tree for this code, and let b and c be two sibling leaves at the maximum depth of the tree. (It may be possible that $\{x, y\} \cap \{b, c\} \neq \emptyset$). Such two symbols must exist because T is full. Assume without loss of generality that $p_b \leq p_c$ and $p_x \leq p_y$ (if this is not true, then rename these characters). Since x and y have the two smallest frequencies it follows that $p_x \leq p_b$ (they may be equal) and $p_c \leq p_y$ (may be equal as well). If we swap x with b and y with c we will not increase the average bit number. Indeed, the more frequent b and c will be now encoded by shorter codewords.   □

Example 8.1.6 The optimal tree for Example 8.1.1 is shown on the following picture.

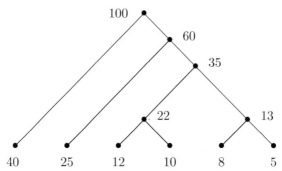

It produces the following Huffman code:

$$a \rightarrow 1, \quad b \rightarrow 01, \quad c \rightarrow 0011, \quad d \rightarrow 0010, \quad e \rightarrow 0001, \quad f \rightarrow 0000.$$

We have

$$(40 \cdot 1 + 25 \cdot 2 + 35 \cdot 4) \cdot 1000 = 230,000 \text{ bits,}$$

which gives us even more efficient encoding than we had before.

Exercises

1. Show that sometimes Huffman's tree may not be unique. Give an example of an alphabet and probabilities of its symbols that are all different but for which there may be several Huffman trees constructed.
2. Construct Huffman's tree for the alphabet $\{a, b, c, d, e, f, g, h\}$ with probabilities

$$\frac{1}{54}, \frac{1}{54}, \frac{2}{54}, \frac{3}{54}, \frac{5}{54}, \frac{8}{54}, \frac{13}{54}, \frac{21}{54},$$

respectively.

3. Consider the code for the alphabet $X = \{x_1, x_2, x_3, x_4, x_5, x_6\}$ given by the table

Source symbol	Probability	Codeword
x_1	0.26	00
x_2	0.24	01
x_3	0.14	100
x_4	0.13	110
x_5	0.12	101
x_6	0.11	111

   (a) Show that this code is an optimal prefix code;
   (b) Show that it is not a Huffman code.

4. Let X be an alphabet consisting of 100 symbols. Find the multiset of codeword lengths of an optimal binary encoding (Huffman code) of the uniform distribution, when every symbol appears with probability

$$p_i = \frac{1}{100}, \qquad (i = 1, 2, ..., 100).$$

That is the source is uniform on 100 symbols. Just find how many codewords of various lengths exist. You do not need to produce the actual code.

5. For every full tree the Kraft's inequality becomes an equality.

6. Prove or disprove: if $|X| = q$ and $m_1, m_2, \ldots, m_q$ be the lengths of the codewords of Huffman's code, then

$$\sum_{i=1}^{q} 2^{-m_i} = 1,$$

i.e., Kraft's inequality for Huffman's code becomes an equality.

8.2 Encoding an Unknown Source

To construct Huffman's code one needs to know the probabilities of symbols, which are never known exactly. It is a drawback of those codes. It came as a surprise that the exact knowledge is not so much needed actually. Boris Fitingof (1966) developed a single code, which is good enough for each Bernoulli source. His paper was inspired by A. N. Kolmogorov (1965). He campaigned enthusiastically for that code and it is fair to consider B. Fitingof the founder of the universal encoding.

8.2.1   Compressing Binary Sequences (Files)

When we need to compress files on a computer, our method in the previous section
will not work. Firstly, our alphabet is already binary, second we have no idea how
the file was generated. However we can get some useful methods if we encode not
symbols of the alphabet which are now 0 and 1 but the words in the alphabet $\mathbb{Z}_2$ of
a fixed length n, which can be identified with vectors in $\mathbb{Z}_2^n$.

Definition 8.2.1 Let n be a positive integer. By a non-uniform (compression) code
we understand a mapping

$$\psi \colon \mathbb{Z}_2^n \to W(\mathbb{Z}_2). \tag{8.5}$$

This means that every word $\mathbf{w}$ from $\mathbb{Z}_2^n$ is encoded into a binary codeword $\psi(\mathbf{w})$.
Note that the length of $\mathbf{w}$ is strictly n while the length of $\psi(\mathbf{w})$ can be arbitrary.
This gives us a chance to compress sequences of bits (files). The code of a sequence
M, which is a word from $W(\mathbb{Z}_2)$, will be obtained as follows. We divide M into
segments of length n and the tail which is of length at most n (but by agreement
it can be viewed also of length n; for example, for English words we may add as
many letters "z" at the end of the message as is needed). Then M is represented as
$M = \mathbf{w}_1 \mathbf{w}_2 \ldots \mathbf{w}_s \ldots$, where $\mathbf{w}_i \in \mathbb{Z}_2^n$ and we define

$$\psi(\mathbf{w}_1)\psi(\mathbf{w}_2) \ldots \psi(\mathbf{w}_s) \ldots \tag{8.6}$$

to be the encoding for M. What we should take care of is that the message (8.6) can
be uniquely decoded and that this decoding is as easy as possible. This is non-trivial
since the words $\psi(\mathbf{w}_1), \ldots \psi(\mathbf{w}_s) \ldots$ may have different lengths and we may not
know, for example, where $\psi(\mathbf{w}_1)$ ends and where $\psi(\mathbf{w}_2)$ starts. We use the same idea
as in the previous section.

Let us recap that a non-uniform code $\psi \colon \mathbb{Z}_2^n \to W(\mathbb{Z}_2)$ is said to be a prefix code
if for every two words $\mathbf{w}_1, \mathbf{w}_2 \in \mathbb{Z}_2^n$ neither of the two codewords $\psi(\mathbf{w}_1), \psi(\mathbf{w}_2)$ is
the beginning of the other.

If our code is a prefix one, then we can decode (8.6) uniquely. Indeed, there will be
only one codeword which is the beginning of (8.6) and that will be $\psi(\mathbf{w}_1)$. Similarly
we decode the rest of the sequence.

We need to look into the structure of $\mathbb{Z}_2^n$.

Let us also recap that the number of ones in a sequence $\mathbf{w} \in \mathbb{Z}_2^n$ is called the
Hamming weight of $\mathbf{w}$, denoted wt($\mathbf{w}$). Let $X = \mathbb{Z}_2^n$ and $X_d = \{\mathbf{x} \in X \mid \text{wt}(\mathbf{x}) = d\}$
be all vectors of Hamming weight d. Then we obtain a partition

$$X = X_0 \cup X_1 \cup \ldots \cup X_n,$$

where $|X_d| = \binom{n}{d}$. This situation needs comprehension from the information theory
point of view.

Exercises

1. Find the Hamming weight of the vectors $\mathbf{u} = 1100101$ and $\mathbf{v} = 0111011$.
2. How many vectors of Hamming weight 3 is there in $X = \mathbb{Z}_2^7$? And vectors of Hamming weight 5?

8.2.2 Information and Information Relative to a Partition

Let Ω be a finite set and $|\Omega|$ be the number of elements in it. Suppose that we want to give an individual label to each element of Ω and each label must be a sequence of zeros and ones. How long must our sequences be so that we have enough labels for all elements of Ω? Since we have exactly 2^n sequences of length n, this number should be taken so that $2^n \geq |\Omega|$. If we aim at sequences of the shortest possible length, we should choose n so that

$$2^n \geq |\Omega| > 2^{n-1}. \tag{8.7}$$

This means that to specify an element of Ω requires n bits of information. The labeling, for example, can be done in the following way. Let $|\Omega| = 2^n$ (or $n = \log_2 |\Omega|$), and $\omega_0, \omega_1, \ldots, \omega_{|\Omega|-1}$ be elements of Ω listed in some order. Then we can think of the correspondence

$$\omega_k \mapsto k \mapsto k_{(2)},$$

where $k_{(2)}$ is the binary representation of k written according to the following convention: if k, in binary, has $\ell < n$ binary digits then $n - \ell$ zeros are added in front of the standard binary representation of k. In other words, the information contained in ω_k is the binary representation of k. Then, under this arrangement, every element of Ω carries exactly n bits of information.

Example 8.2.1 Let $|\Omega| = 16, n = 4$. Then ω_5 can be put in correspondence to 5 and to $5_{(2)} = 0101$. Thus, every element of Ω carries 4 bits of information.

Definition 8.2.2 Information (as defined by Hartley 1928) of an element $\omega \in \Omega$ is by the definition

$$I(\omega) = \log_2 |\Omega|, \tag{8.8}$$

where $|\Omega|$ denote the number of elements in Ω.

Here and further in this section all logarithms will be taken to base 2.

Let $\lceil x \rceil$ be the nearest integer which is greater than or equal to x. Then (8.7) implies $n \geq \log |\Omega| > n - 1$, hence, for an element $\omega \in \Omega$, the integer $\lceil I(\omega) \rceil$ is the minimal number of binary symbols necessary for individualising ω among other elements of Ω.

Let now

$$\Omega = \Omega_1 \cup \Omega_2 \cup \ldots \cup \Omega_n \tag{8.9}$$

be a partition of Ω into n disjoint classes. Let $\pi(\omega)$ denote the class which contains ω.

Definition 8.2.3 Information of an element $\omega \in \Omega$ relative to the given partition is defined as

$$I(\omega) = \log |\pi(\omega)|. \tag{8.10}$$

It can be interpreted as follows. In a partitioned set, to individualise ω we can do it in two stages: firstly, we can individualise the class $\pi(\omega)$ to which ω belongs and second to individualise ω within the class $\pi(\omega)$. Equation (8.10) does the latter. In the extreme case, when there is only one class in the partition, i.e., the set Ω itself, we get the same concept as in Definition 8.2.3.

Example 8.2.2 Let $\Omega = \mathbb{Z}_2^4$ be the four-dimensional vector space over $\mathbb{Z}_2$. Let

$$\Omega_i = \{\mathbf{y} \in \mathbb{Z}_2^4 \mid \mathrm{wt}(\mathbf{y}) = i\},$$

where $\mathrm{wt}(\mathbf{y})$ is the Hamming weight of $\mathbf{y}$. Then $\Omega = \Omega_0 \cup \Omega_1 \cup \Omega_2 \cup \Omega_3 \cup \Omega_4$ is a partition of Ω. Let $\mathbf{u} = 1111$, $\mathbf{v} = 0010$, $\mathbf{w} = 0101$. Then

$$I(\mathbf{u}) = \log |\Omega_4| = \log \binom{4}{4} = \log 1 = 0 \text{ bits,}$$

$$I(\mathbf{v}) = \log |\Omega_1| = \log \binom{4}{1} = \log 4 = 2 \text{ bits,}$$

$$I(\mathbf{w}) = \log |\Omega_2| = \log \binom{4}{2} = \log 6 \approx 2.6 \text{ bits.}$$

Example 8.2.3 Let $\Omega = \mathbb{Z}_2^n$ and

$$\Omega = \Omega_0 \cup \Omega_1 \cup \ldots \cup \Omega_n$$

be the partition for which, as in the previous example, Ω_i consists of vectors of Hamming weight i. Let $\mathbf{z} \in \mathbb{Z}_2^n$ has weight d. Since $|\Omega_d| = \binom{n}{d}$

$$I(\mathbf{z}) = \log |\Omega_d| = \log \binom{n}{d}.$$

If d is small, then

$$I(\mathbf{z}) = \log \binom{n}{d} = \log \frac{n(n-1)\ldots(n-d+1)}{d!} < \log n^d = d \log n,$$

i.e., the information of a vector of a small weight is rather small relative to n. If d is close to n, the information will be small too. It will be maximal for $d = n/2$ in which case, due to the asymptotic formula

$$\binom{n}{n/2} \sim \sqrt{\frac{2}{\pi}} \cdot \frac{2^n}{\sqrt{n}}, \tag{8.11}$$

which can easily be obtained from Stirling's formula (2.2). This implies

$$I(\mathbf{z}) = \log \binom{n}{d} = \log \binom{n}{n/2} \sim n - \frac{1}{2} \log n + \frac{1}{2}(1 - \log \pi) \sim n,$$

which is only slightly smaller than n and asymptotically equal to n.

Proposition 8.2.1 *For the partition (8.9)*

$$\sum_{w \in \Omega} 2^{-(I(w) + \log n)} = 1. \tag{8.12}$$

Proof For any element w of the class Ω_i

$$-(I(w) + \log n) = -\log|\Omega_i| - \log n.$$

Thus

$$\sum_{w \in \Omega} 2^{-(I(w) + \log n)} = \sum_{i=1}^{n} |\Omega_i| 2^{(-\log|\Omega_i| - \log n)} = \sum_{i=1}^{n} \frac{|\Omega_i|}{|\Omega_i| n} = \sum_{i=1}^{n} \frac{1}{n} = 1. \qquad \square$$

We shall see soon what Eq. (8.12) means.

Corollary 8.2.1 *Let $\Omega = \mathbb{Z}_2^n$ and*

$$\Omega = \Omega_0 \cup \Omega_1 \cup \ldots \cup \Omega_n$$

be the partition for which, as in the previous example, Ω_i consists of vectors of Hamming weight i. Then there exists a prefix code $\psi \colon \Omega \to W(\mathbb{Z}_2)$ such that for any $w \in \Omega$ the length of the codeword $\psi(w)$ is $\ell(w) = \lceil I(w) + \log n \rceil$, where $I(w)$ is the information of w relative to the given partition.

Proof By Proposition 8.2.1

$$\sum_{w \in \Omega} 2^{-\lceil I(w) + \log n \rceil} \leq \sum_{w \in \Omega} 2^{-(I(w) + \log n)} = 1,$$

and the result follows from Theorem 8.1.2. $\qquad \square$

This corollary tells us that there is a prefix code that can be organised as follows. To encode the element of Ω we can specify the code of the equivalence class using $\log n$ binary symbols and then specify the label of the element in this class. This is implemented in Fitingof's code. Also existence of the code is not everything. Another important issue is fast decodability. We will see that Fitingof's code is good on this front too.

Exercises

1. How many bits of information does one need to specify one letter of the English alphabet?
2. In a magic trick, there are three participants: the magician, an assistant and a volunteer. The assistant, who claims to have paranormal abilities, is in a sound-proof room. The magician gives the volunteer six blank cards, five white and one blue. The volunteer writes a different integer from 1 to 100 on each card, as the magician is watching. The volunteer keeps the blue card. The magician arranges the five white cards in some order and passes them to the assistant. The assistant then announces the number on the blue card. How does the trick work?

8.2.3   Fitingof's Compression Code. Encoding

We need to compress files when we are short of memory and want to use it effectively. Since computer files are already written as strings of binary digits in this section we will consider the code $\psi\colon \mathbb{Z}_2^n \to W(\mathbb{Z}_2)$ which encodes binary sequences of fixed length n into binary sequences of variable length. The idea of Fitingof's compression is expressed in Example 8.2.3, where it was shown that the information of a vector from $\mathbb{Z}_2^n$ of small (or large) Hamming weight is relatively small compared to n. Therefore if we encode words in such a way that the length of a codeword $\psi(x)$ will be approximately equal to the information of x, then words of small and large Hamming weights will be significantly compressed. This, for example, often works well with photographs.

In this section we will order all binary words of the same length using lexicographic order. This order depends on an order on our binary symbols and we will assume that zero precedes one (denoted $0 \prec 1$).

Definition 8.2.4 Let $\mathbf{y} = y_1 y_2 \ldots y_n$ and $\mathbf{z} = z_1 z_2 \ldots z_n$ be two binary words of the same length. We say that $\mathbf{y}$ is lexicographically earlier than $\mathbf{z}$, and write $\mathbf{y} \prec \mathbf{z}$, if for some $k \geq 0$

$$y_1 = z_1, \ y_2 = z_2, \ \ldots, \ y_k = z_k, \quad \text{but} \quad y_{k+1} \prec z_{k+1},$$

(which means that $y_{k+1} = 0$ and $z_{k+1} = 1$).

This order is called *lexicographic* since it is used in dictionaries to list words. For example, in Oxford Dictionary the word "abash" precedes the word "abate" because the first three letters of these words coincide but the fourth letter "s" of "abash" precedes in the English alphabet the fourth letter "t" of "abate".

For example, all 15 binary words of length 6 and weight 4 will be listed in lexicographic order as shown:

$$001111 \prec 010111 \prec 011011 \prec 011101 \prec 011110 \prec 100111 \prec 101011 \prec 101101$$
$$\prec 101110 \prec 110011 \prec 110101 \prec 110110 \prec 111001 \prec 111010 \prec 111100.$$
$$(8.13)$$

We can refer to these words by just quoting their ordinal numbers. We adopt the agreement that the first word has ordinal number zero. Thus an ordinal number of a word $\mathbf{x}$ is the number of words that are earlier than $\mathbf{x}$. In particular, the ordinal number of 101011 is 6.

Lemma 8.2.1 *Let $X = \mathbb{Z}_2^n$ and $X_d = \{\mathbf{x} \in X \mid wt(\mathbf{x}) = d\}$ be all vectors of weight d. If X_d is ordered lexicographically, then the ordinal number $N(\mathbf{x})$ of $\mathbf{x}$ in X_d can be calculated as*

$$N(\mathbf{x}) = \binom{n-n_d}{1} + \cdots + \binom{n-n_2}{d-1} + \binom{n-n_1}{d}, \qquad (8.14)$$

where 1s in $\mathbf{x}$ occupy the positions $n_1 < n_2 < \ldots < n_d$ (counting from the left).

Proof Firstly, we count all the words of weight d whose n_1-1 leftmost symbols coincide with those of $\mathbf{x}$, i.e., are all zeros, and the position n_1 is also occupied by a zero (this condition secures that all such words are lexicographically earlier than $\mathbf{x}$). Since we have to distribute d ones between $n-n_1$ remaining positions, there will be $\binom{n-n_1}{d}$ such words. Secondly, we have to count all the words whose first n_2-1 symbols coincide with those of $\mathbf{x}$ and which have a zero in the position n_2. There are $\binom{n-n_2}{d-1}$ such words as we have to distribute $d-1$ ones between $n-n_2$ places. Finally, we will have to count all words whose first n_d-1 symbols coincide with those of $\mathbf{x}$ and which have a zero in the position n_d. There will be $\binom{n-n_d}{1}$ such words. All the words that are lexicographically earlier than $\mathbf{x}$ are now counted. As the ordinal number of $\mathbf{x}$ is equal to the number of words which are lexicographically earlier than $\mathbf{x}$, this proves (8.14). □

Example 8.2.4 Consider the word $\mathbf{x} = 101011$ in X_4 where $X = \mathbb{Z}_2^6$. Then $n_1 = 1$, $n_2 = 3, n_3 = 5, n_4 = 6$ and $d = 4$. Then

$$N(\mathbf{x}) = \binom{0}{1} + \binom{1}{2} + \binom{3}{3} + \binom{5}{4} = 0 + 0 + 1 + 5 = 6,$$

which is consistent with (8.13).

We are now ready to describe Fitingof's compression code $\psi: \mathbb{Z}_2^n \to W(\mathbb{Z}_2)$. The idea of this code is to characterise any word $\mathbf{x}$ from $X = \mathbb{Z}_2^n$ by two parameters, namely its Hamming weight d and the ordinal number $N(\mathbf{x})$ of $\mathbf{x}$ in X_d. We partition $X = \mathbb{Z}_2^n$ into $n + 1$ disjoint classes

$$X = X_0 \cup X_1 \cup \ldots \cup X_n, \qquad X_d = \{\mathbf{x} \in X \mid \mathrm{wt}(\mathbf{x}) = d\}.$$

The codeword $\psi(\mathbf{x})$ for $\mathbf{x} \in X_d$ (i.e., for a word $\mathbf{x}$ of weight d) will consists of two parts: $\psi(\mathbf{x}) = \mu(\mathbf{x})\nu(\mathbf{x})$, where $\mu(\mathbf{x})$ is the *prefix* of fixed length $\lceil \log(n + 1) \rceil$, which is the binary code for d, and $\nu(\mathbf{x})$ is the *suffix*, which is the binary code of the ordinal number $N(\mathbf{x})$ of $\mathbf{x}$ in the class X_d consisting of $\lceil \log |X_d| \rceil = \left\lceil \log \binom{n}{d} \right\rceil$ binary symbols. Both parameters together characterise $\mathbf{x}$ uniquely. In total the length of the codeword $\psi(\mathbf{x}) = \mu(\mathbf{x})\nu(\mathbf{x})$ will be

$$\ell(\psi(\mathbf{x})) = \lceil \log(n + 1) \rceil + \left\lceil \log \binom{n}{d} \right\rceil.$$

Asymptotically the length of the word $\psi(\mathbf{x})$ will be

$$\ell(\psi(\mathbf{x})) = I(\mathbf{x}) + o(n), \qquad \frac{o(n)}{n} \to 0,$$

i.e., equal to its information relative to the given partition.

We now state the main theorem of this chapter.

Theorem 8.2.1 (Fitingof's theorem) *There exists a prefix code $\psi\colon \mathbb{Z}_2^n \to W(\mathbb{Z}_2)$ for which the length of the codeword $\psi(\mathbf{x})$ is asymptotically equal to the information of the word $\mathbf{x}$ and for which there exists a decoding procedure of polynomial complexity.*

Proof We have shown already that the length of the codeword $\psi(\mathbf{x})$ is asymptotically equal to the information of the word $\mathbf{x}$. Let us prove that Fitingof's code is a prefix one. Suppose $\psi(\mathbf{x}_1) = \mu(\mathbf{x}_1)\nu(\mathbf{x}_1)$ is a beginning of $\psi(\mathbf{x}_2) = \mu(\mathbf{x}_2)\nu(\mathbf{x}_2)$. We know that the length of $\mu(\mathbf{x}_1)$ is the same as the length of $\mu(\mathbf{x}_2)$, hence $\mu(\mathbf{x}_1) = \mu(\mathbf{x}_2)$ and hence $\mathbf{x}_1$ and $\mathbf{x}_2$ has the same weight. But then the length of $\nu(\mathbf{x}_1)$ is the same as the length of $\nu(\mathbf{x}_2)$ and hence $\psi(\mathbf{x}_1)$ and $\psi(\mathbf{x}_2)$ have the same length. However in such a case one cannot be a beginning of another without being equal.

The proof will be continued in the next section devoted to the decoding algorithm.   □

Let us consider an example.

Example 8.2.5 Consider the Fitingof's compression code $\psi\colon \mathbb{Z}_2^{31} \to W(\mathbb{Z}_2)$. For the vector

$$\mathbf{x} = 0000000100000101000100000000000$$

we will have $\mu(\mathbf{x}) = 00100$ because $\mathrm{wt}(\mathbf{x}) = 4 = 100_{(2)}$ and the prefix must be of length 5 to accommodate all possible weights in the range from 0 to 31. The length of the suffix $\nu(\mathbf{x})$ will be $\lceil \log \binom{31}{4} \rceil = 15$. Further, we will have $n_1 = 8$, $n_2 = 14$, $n_3 = 16$, $n_4 = 20$ and

$$N(\mathbf{x}) = \binom{11}{1} + \binom{15}{2} + \binom{17}{3} + \binom{23}{4} = 9651 = 10010110110011_{(2)},$$

tus the suffix $\nu(\mathbf{x})$ will be 010010110110011. Thus, $\psi(\mathbf{x}) = 00100010010110110011$ has length 20.

Exercises

1. Put the following three words of $\mathbb{Z}_2^7$ in the increasing lexicographic order:

$$\mathbf{a} = 0110111, \quad \mathbf{b} = 0111101, \quad \mathbf{c} = 0111011.$$

2. How many vectors of Hamming weight at least 4 and at most 5 are there in $\mathbb{Z}_2^{10}$?
3. Calculate the ordinal number of the word $\mathbf{w} = 0011011$ in $X_4 \subset \mathbb{Z}_2^7$.
4. Let $\psi \colon \mathbb{Z}_2^{15} \to W(\mathbb{Z}_2)$ be Fitingof's code. Let $\mathbf{x} = 000010100000100$.
   (a) Determine the prefix of $\mathbf{x}$ which shows the Hamming weight of the word?
   (b) How long must be the suffix of the codeword $\psi(\mathbf{x})$?
   (c) Determine the suffix of the codeword $\psi(\mathbf{x})$?
   (d) Encode $\mathbf{x}$, i.e., find $\psi(\mathbf{x})$.
5°. (a) Calculate the size of the class $X_5 \subseteq \mathbb{Z}_2^{15}$ that consists of all vectors of Hamming weight 5;
   (b) Find the ordinal number of the vector

$$\mathbf{x} = 100011000100100$$

in the class X_5.
   (c) What would be the encoding $\psi(\mathbf{x})$ of this vector in the Fitingof's code $\psi \colon \mathbb{Z}_2^{15} \to W(\mathbb{Z}_2)$?

8.2.4 Fitingof's Compression Code. Fast Decoding

To decode a message we have to decode the codewords one by one starting from the first. Suppose the first codeword is $\psi(\mathbf{x})$. First, we separate its prefix $\mu(\mathbf{x})$ (because it is of fixed known length $\lceil \log(n+1) \rceil$) and reconstruct $d = \text{wt}(\mathbf{x})$. Then, knowing d, we calculate the length of the suffix $\nu(\mathbf{x})$, which is $\lceil \log \binom{n}{d} \rceil$. Then looking at the suffix $\nu(\mathbf{x})$ and knowing that it represents the ordinal number $N(\mathbf{x})$ of $\mathbf{x}$ in X_d, we reconstruct $N = N(\mathbf{x})$. Then we are left with the equation

$$\binom{x_d}{1} + \cdots + \binom{x_2}{d-1} + \binom{x_1}{d} = N \qquad (8.15)$$

to solve for $x_d < \ldots < x_2 < x_1$, where $x_i = n - n_i$. This can be done in a fast and elegant way using the properties of Pascal's triangle,[1] part of which is shown below:

$$
\begin{array}{ccccccccccccccc}
 & & & & & & & 1 & & & & & & & \\
 & & & & & & 1 & & 1 & & & & & & \\
 & & & & & 1 & & 2 & & 1 & & & & & \\
 & & & & 1 & & 3 & & 3 & & 1 & & & & \\
 & & & 1 & & 4 & & 6 & & 4 & & 1 & & & \\
 & & 1 & & 5 & & 10 & & 10 & & 5 & & 1 & & \\
 & 1 & & 6 & & 15 & & 20 & & 15 & & 6 & & 1 & \\
1 & & 7 & & 21 & & 35 & & 35 & & 21 & & 7 & & 1
\end{array}
$$

The nth row of this triangle contain the binomial coefficients $\binom{n}{m}$, $m = 0, 1, \ldots n$, where m increases from left to right. These binomial coefficients are defined inductively by the formula

$$
\binom{n}{j} = \binom{n-1}{j} + \binom{n-1}{j-1} \tag{8.16}
$$

and the boundary conditions: $\binom{0}{0} = 1$, and $\binom{0}{m} = 0$ for all $0 \neq m \in \mathbb{Z}$. We also know the explicit formula

$$
\binom{n}{m} = \frac{n!}{m!(n-m)!},
$$

which involves factorials.

The solution of (8.15) will be based on the formula

$$
\binom{n-d}{0} + \binom{n-d+1}{1} + \cdots + \binom{n-1}{d-1} + \binom{n}{d} = \binom{n+1}{d}. \tag{8.17}
$$

We prove it by induction on d for fixed but arbitrary n. If $d = 1$, then (8.17) becomes

$$
\binom{n}{0} + \binom{n}{1} = \binom{n+1}{1} \quad \text{or} \quad 1 + n = n + 1,
$$

which is true. Let us assume that (8.17) is true for $d = k - 1$. Then by induction hypothesis, applied to the first $k - 1$ summands of the left-hand side of (8.17) and using (8.16) we get

$$
\binom{n-k}{0} + \binom{n-k+1}{1} + \cdots + \binom{n-1}{k-1} + \binom{n}{k} =
$$

---

[1]The triangle is called after French mathematician Pascal, although it had been described centuries earlier by Chinese mathematician Yanghui almost 500 years earlier, and the Persian astronomer Omar Khayyám, who is better known for his poetry.

$$\left[\binom{(n-1)-(k-1)}{0} + \binom{(n-1)-(k-1)+1}{1} + \cdots + \binom{n-1}{k-1}\right] + \binom{n}{k} =$$

$$\binom{n}{k-1} + \binom{n}{k} = \binom{n+1}{k},$$

which proves (8.17) for $d = k$. Hence, by induction, (8.17) is proven.

Proposition 8.2.2 *Suppose Eq. (8.15) is satisfied for some* $x_1, \ldots, x_d$ *such that* $x_d < x_{d-1} < \ldots < x_1$. *Then* x_1 *can be found as the largest integer satisfying the inequality*

$$\binom{x_1}{d} \le N.$$

Proof Suppose that $x_1 < m$, where m is the largest integer satisfying $\binom{m}{d} \le N$. Then, since $x_d < x_{d-1} < \cdots < x_1$, we have $x_2 \le x_1 - 1$, $x_3 \le x_1 - 2$ and all the way to $x_d \le x_1 - d + 1$. By (8.17)

$$\binom{x_d}{1} + \cdots + \binom{x_2}{d-1} + \binom{x_1}{d} \le$$

$$\binom{x_1-d+1}{1} + \cdots + \binom{x_1-1}{d-1} + \binom{x_1}{d} = \binom{x_1+1}{d} - 1 < \binom{m}{d} \le N,$$

which is a contradiction. Hence $x_1 = m$. □

Proposition 8.2.2 gives us a fast algorithm for solving Eq. (8.15). Indeed, we find x_1 directly applying Proposition 8.2.2. Then we move the term $\binom{x_1}{d}$ to the right

$$\binom{x_d}{1} + \cdots + \binom{x_2}{d-1} = N - \binom{x_1}{d}$$

and apply Proposition 8.2.2 again to get x_2, etc.

Example 8.2.6 If $d = 4$ and $N = 30$, then for the equation

$$\binom{x_4}{1} + \binom{x_3}{2} + \binom{x_2}{3} + \binom{x_1}{4} = 30$$

we find successively: $\binom{x_1}{4} = 15$ and $x_1 = 6$, $\binom{x_2}{3} = 10$ and $x_2 = 5$, $\binom{x_3}{2} = 3$ and $x_3 = 3$, $\binom{x_4}{1} = 2$ and $x_4 = 2$.

If we needed, for example, to find the word **x** which has ordinal number 30 in $X_4 \subset \mathbb{Z}_2^7$, then according to the Eq. (8.14)

$$\binom{7-n_4}{1} + \binom{7-n_3}{2} + \binom{7-n_2}{3} + \binom{7-n_1}{4} = 30,$$

then we would get $7-n_1 = 6, 7-n_2 = 5, 7-n_3 = 3, 7-n_4 = 2$ or $n_1 = 1, n_2 = 2,$ $n_3 = 4, n_4 = 5$, whence $x = 1101100$.

Exercises

1. Let $\psi\colon \mathbb{Z}_2^{15} \to W(\mathbb{Z}_2)$ be Fitingof's compression code. Decode $\psi(\mathbf{y}) = 00100011110$, i.e., find $\mathbf{y}$.
2. Solve the equation

$$\binom{x_4}{1} + \binom{x_3}{2} + \binom{x_2}{3} + \binom{x_1}{4} = 43,$$

where $x_4 < x_3 < x_2 < x_1$.

3. Let $\psi\colon \mathbb{Z}_2^{15} \to W(\mathbb{Z}_2)$ be Fitingof's code.

   (a) How long is the prefix which shows the Hamming weight of the word?
   (b) Given $\mathbf{x} = 000100001001000$, how long must be the suffix of the codeword $\psi(\mathbf{x})$?
   (c) Encode $\mathbf{x}$, i.e., find $\psi(\mathbf{x})$.
   (d) Decode $\psi(y) = 00100011110$, i.e., find y. Make use of Pascal's triangle.

8.3   Information and Uncertainty

A more traditional measure of information contained in a binary word $\mathbf{x}$ of length n is based on the assumption that this word was generated by a random source. This is of course not always a realistic assumption so our approach here is more general. We will show, however, that in the case of a random source the two approaches are asymptotically equivalent, i.e., when n gets large. Let us consider a random source which sends signal "1" with probability p and signal "0" with probability $1 - p$. Then the measure of uncertainty about what the next signal will be is given by the *binary entropy function*

$$H(p) = -p \log p - (1-p) \log(1-p)$$

(logarithms are to the base 2 and it is assumed that $0 \cdot \log 0 = 0$). The uncertainty is minimal when $p = 0$ or $p = 1$, in which case we essentially do not have any uncertainty and the entropy of such source is zero. If $p = 1/2$, then the uncertainty is maximal and the entropy of such source is equal to 1. We say that one symbol sent from such random source contains $H(p)$ bits of information. Thus we have 1 bit of information from a symbol from a random source only in the case of probability $1/2$. A word of length n contains $nH(p)$ bits of information.

Given a binary word $\mathbf{x}$ of length n consisting of m_1 ones and m_2 zeros we define

$$H(\mathbf{x}) = -\frac{m_1}{n} \log \frac{m_1}{n} - \frac{m_2}{n} \log \frac{m_2}{n}.$$

Of course, if this word was generated from a random source with probability p, then $m_1/n \to p$, when n gets large, and $H(\mathbf{x}) \to H(p)$. The following theorem then shows that the two approaches are equivalent.

Theorem 8.3.1 *For every binary word* $\mathbf{x}$ *of length* n *it is true that*

$$I(\mathbf{x}) = n\left(H(\mathbf{x}) + o(1)\right),$$

where as usual $o(1) \to 0$, *when* $n \to \infty$. *Moreover,* $o(1) \sim \frac{\log n}{n}$.

Proof Let us assume that the Hamming weight of $\mathbf{x}$ is m_1 and let $m_2 = n - m_1$. We will need Stirling's formula (2.2) again. We use it to calculate

$$I(\mathbf{x}) = \left\lceil \log \binom{n}{m_1} \right\rceil \sim \log \binom{n}{m_1} = \log \frac{n!}{(m_1)!(m_2)!} \sim \log \frac{\sqrt{2\pi n}\, n^n e^{-n}}{(\sqrt{2\pi m_1}\, m_1^{m_1} e^{-m_1})(\sqrt{2\pi m_2}\, m_2^{m_2} e^{-m_2})}$$

$$= \frac{1}{2}\log \frac{n}{2\pi m_1 m_2} + \log \frac{1}{\left(\frac{m_1}{n}\right)^{m_1}\left(\frac{m_2}{n}\right)^{m_2}} = \frac{1}{2}\log \frac{n}{2\pi m_1 m_2} - m_1 \log \frac{m_1}{n} - m_2 \log \frac{m_2}{n}.$$

Therefore

$$\frac{I(\mathbf{x})}{n} = \frac{1}{2n}\log \frac{n}{2\pi m_1 m_2} - \frac{m_1}{n}\log \frac{m_1}{n} - \frac{m_2}{n}\log \frac{m_2}{n} = o(1) + H(\mathbf{x}),$$

where $o(1)$ is of order $\log n/n$. The theorem is proved. $\qquad\square$

Appendix A: GAP

9

9.1 Computing with GAP

9.1.1 Starting with GAP

GAP is a system for computational algebra. GAP has been and is developed by the international cooperation of many people, including user contributions. This package is free, and you can install it onto your computer using the instructions from the website www.gap-system.org. A reference manual and tutorial can be found there. There is plenty of information about GAP available online too.

9.1.2 The GAP Interface

Once you have started GAP, you can start working straight away. If you type a simple command (i.e., "quit") followed by a semicolon, GAP will evaluate your command immediately. If you press enter without entering a semicolon, GAP will simply give you a new line to continue entering more input. This is useful if you want to write a more complicated command, perhaps a simple program. If you wanted your simple command to be evaluated, then simply enter a semicolon on the new line and press enter again. Double semicolon executes the command but suppresses the output. Since GAP ignores whitespace, this will work just the same as if you had entered the semicolon in the first place. A semicolon will not always cause GAP to evaluate straight away; GAP is able to work out whether you have finished a complete set of instructions or are part of the way through entering a program.

Another way to interact with GAP, which is particularly useful for things you want to do more than once, is to prepare a collection of commands and programs in a text file. Then you can type the command Read ("MyGAPprog.txt"), and GAP will evaluate all of the instructions in your text file. If your file is not in the same place

© Springer Nature Switzerland AG 2020

A. Slinko, *Algebra for Applications*, Springer Undergraduate Mathematics Series,
https://doi.org/10.1007/978-3-030-44074-9_9

that GAP was launched from, you will have to provide its relative path (for example, "../../GAPprogs/Example1.txt").

9.1.3 Programming in GAP: Variables, Lists, Sets and Loops

You can declare a variable in GAP using the ":=" operator. For example, if you want a variable n to equal 2000, you would enter $n := 2000$, or if you want n to be the product of p and q, you would enter $n := p * q$;. You can also declare lists using the ":=" operator, for example, `zeros := [0,0,0];`. The command `list:=[m..n];` defines the list of integers $m, m + 1, m + 2, \ldots, n$. A list may have several identical numbers in it. Lists have a length given by command `Length(listName);`, and their entries can be referenced individually by typing `listName[index];` (indices start from 1!). In GAP a list of primes ≤ 1000 is stored. It is called "Primes". This is very useful.

```
gap> Primes;
[ 2, 3, 5, 7, 11, 13, 17, 19, 23, 29, 31, 37, 41, 43, 47, 53, 59, 61, 67, 71, 73,
79, 83, 89, 97, 101, 103, 107, 109, 113, 127, 131, 137, 139, 149, 151, 157, 163, 167,
173, 179, 181, 191, 193, 197, 199, 211, 223, 227, 229, 233, 239, 241, 251, 257, 263,
269, 271, 277, 281, 283, 293, 307, 311, 313, 317, 331, 337, 347, 349, 353, 359, 367,
373, 379, 383, 389, 397, 401, 409, 419, 421, 431, 433, 439, 443, 449, 457, 461, 463,
467, 479, 487, 491, 499, 503, 509, 521, 523, 541, 547, 557, 563, 569, 571, 577, 587,
593, 599, 601, 607, 613, 617, 619, 631, 641,643, 647, 653, 659, 661, 673, 677, 683,
691, 701, 709, 719, 727, 733, 739, 743, 751, 757, 761, 769, 773, 787, 797, 809, 811,
821, 823, 827, 829, 839, 853, 857, 859, 863, 877, 881, 883, 887, 907, 911, 919, 929,
937, 941, 947, 953, 967, 971, 977, 983, 991, 997 ]
```

The command

```
gap> Length(Primes);
168
```

gives us the number of primes in this list. We can find the prime in 100th position and the position of 953 in this list as follows:

```
gap> Primes[100];
541
gap> Position(Primes,953);
162
```

Sets cannot contain multiple occurrences of elements, and the order of elements does not matter. Basically GAP views sets as ordered lists without repetitions. The command `Set(list);` converts a list into a set removing duplication.

```
gap> list:=[2,5,8,3,5];
[ 2, 5, 8, 3, 5 ]
gap> Add(list,2);
gap> list;
[ 2, 5, 8, 3, 5, 2 ]
gap> set:=Set(list);
[ 2, 3, 5, 8 ]
```

```
gap> RemoveSet(set,2);
gap> set;
[ 3, 5, 8 ]
```

For loops and while loops exist in GAP. Both have the same format:

```
for (while) [condition] do [statements] od;
```

For example, the following for loop squares all of the entries in the list "boringList" and places them in the same position in the list "squaredList":

```
gap> boringList:=[2..13];
[ 2 .. 13 ]
gap> squaredList:=[1..Length(boringList)];
[ 1 .. 12 ]
gap> for i in [1..Length(boringList)] do
> squaredList[i]:=boringList[i]^2;
> od;
gap> squaredList;
[ 4, 9, 16, 25, 36, 49, 64, 81, 100, 121, 144, 169 ]
```

Here is the example of using a while loop. We want to square the first five numbers of the boringList.

```
gap> boringList:=[2..13];;
gap> i:=1;;
gap> while i<6 do
> boringList[i]:=boringList[i]^2;
> i:=i+1;
> od;
gap> boringList;
[ 4, 9, 16, 25, 36, 7, 8, 9, 10, 11, 12, 13 ]
```

List may contain other lists. Analyse the following program that lists all pairs of twin primes not exceeding 1000. It also illustrates the use of "if-then" command.

```
if [condition] then [statements] fi;
```

Here it is:

```
gap> twinpairs:=[];
[  ]
gap> numbers:=[1..Length(Primes)-1];
[ 1 .. 167 ]
gap> for i in numbers do
> if Primes[i]=Primes[i+1]-2 then
> Add(twinpairs,[Primes[i],Primes[i+1]]);
> fi;
> od;
gap> twinpairs;
[ [ 3, 5 ], [ 5, 7 ], [ 11, 13 ], [ 17, 19 ], [ 29, 31 ], [ 41, 43 ],
  [ 59, 61 ], [ 71, 73 ], [ 101, 103 ], [ 107, 109 ], [ 137, 139 ],
  [ 149, 151 ], [ 179, 181 ], [ 191, 193 ], [ 197, 199 ], [ 227, 229 ],
  [ 239, 241 ], [ 269, 271 ], [ 281, 283 ], [ 311, 313 ], [ 347, 349 ],
  [ 419, 421 ], [ 431, 433 ], [ 461, 463 ], [ 521, 523 ], [ 569, 571 ],
  [ 599, 601 ], [ 617, 619 ], [ 641, 643 ], [ 659, 661 ], [ 809, 811 ],
  [ 821, 823 ], [ 827, 829 ], [ 857, 859 ], [ 881, 883 ] ]
```

9.2   Number Theory

Most of the number-theoretic commands in GAP are self-explanatory. Some of them we have already encountered in the previous section.

9.2.1   Basic Number-Theoretic Algorithms

One of the most important commands is the command `FactorsInt(n);`. It outputs the prime factorisation of *n* or, more precisely, the primes that enter this prime facorisation with their multiplicity. The command `PrintFactorsInt(n);` gives a nicer view of this prime factorisation but you cannot use the output as a list which you can do with the output of `FactorsInt`. The command `DivisorsInt(n);` can be used to find all of the divisors of *n*. The command `PrimeDivisorsInt(n);` finds the set of unique prime divisors of *n*. For example,

```
gap> FactorsInt(571428568);
[ 2, 2, 2, 71428571 ]
gap> PrintFactorsInt(571428568);
2^3*71428571
gap> DivisorsInt(571428568);
[ 1, 2, 4, 8, 71428571, 142857142, 285714284, 571428568 ]
gap> PrimeDivisors(571428568);
[ 2, 71428571 ]
```

The command `IsPrime(n);` answers the question if *n* is prime. The command `NextPrimeInt(n);` gives the smallest prime number that is strictly greater than *n*. The action of the command `PrevPrimeInt(n);` is similar. For example,

```
gap> IsPrime(571428568);
false
gap> NextPrimeInt(571428568);
571428569
gap> PrevPrimeInt(571428568);
571428527
```

The list of primes "Primes" contains only the 168 primes that are smaller than 1000. Using the commands that we have just introduced we can, for example, create a list of the first 5000 primes:

```
gap> biggerPrimes := [];
[  ]
gap> counter := 1;
1
gap> currentPrime := 2;
2
gap> while counter < 5000 do;
> biggerPrimes[counter] := currentPrime;
> counter := counter + 1;
> currentPrime := NextPrimeInt(currentPrime);
> od;
```

The remainder and quotient of n divided by m are given by the commands `RemInt(n,m);` and `QuoInt(n,m);`, respectively. For example,

```
gap> RemInt(9786354,383);
321
gap> QuoInt(9786354,383);
25551
```

The following command does the same thing as `RemInt(n,m)`:

```
gap> 9786354 mod 383;
321
```

The greatest common divisor of a and b is given by `GcdInt(a,b);`. For example,

```
gap> GcdInt(123456789,987654321);
9
```

To find m, n such that $ma + nb = \gcd(a, b)$, use the GAP command `Gcdex(a,b);`. For example,

```
Gcdex(108,801);
```

returns

```
rec( gcd := 9, coeff1 := -37, coeff2 := 5,   coeff3 := 89, coeff4 := -12 )
```

where m =coeff1, n =coeff2 (m_1 =coeff3 and n_1 =coeff4 will also work). Another example,

```
gap> Gcdex(123456789,987654321);
rec( gcd := 9, coeff1 := -8, coeff2 := 1, coeff3 := 109739369,
  coeff4 := -13717421 )
```

To find the least common multiple of m and n, use the GAP command `LcmInt(m, n);`. For example,

```
gap> LcmInt(123456789,987654321);
13548070123626141
```

The Euler's totient function $\phi(n)$ is given by the command `Phi(n);`. For example,

```
gap> Phi(2^15-1); Phi(2^17-1);
27000
131070
```

The Chinese remainder theorem states the existence of the minimal solution $N \geq 0$ of $N = a_1 \mod n_1$, $N = a_2 \mod n_2$, ..., $N = a_k \mod n_k$. The command for finding this solution is ChineseRem($[n_1, n_2, ..., n_k]$, $[a_1, a_2, ..., a_k]$);. For example

```
gap> ChineseRem([5,7],[1,2]);
16
```

GAP does not provide automatic conversion between bases. One way of doing base conversion is to use the *p*-adic numbers package, feel free to investigate this on your own. Another way is to write simple programs. For example, 120789 can be converted to binary as follows:

```
gap> n := 120789;
120789
gap> base := 2;
2
gap> rems := [];
[  ]
gap> pos := 1;
1
gap> while n > 0 do;
> rems[pos] := RemInt(n,base);
> n := QuoInt(n,base);
> pos := pos + 1;
> od;
gap> n;
0
gap> rems;
[ 1, 0, 1, 0, 1, 0, 1, 1, 1, 1, 1, 0, 1, 0, 1, 1, 1 ]
```

That is, 120789 is 11101011111010101 in binary. If you are not sure why the list rems are read in the reverse order, you need to study the Base Conversion algorithm in Chap. 1. As for converting from another base into decimal, you should now be able to do it themselves. Write a simple program to convert 100011100001111100000 from binary to decimal.

The commands RootInt(n,k); and LogInt(n,b); can be used to determine, respectively, the integer part of the kth (positive real) root of n and the logarithm of n to the base b, that is, $\lfloor \sqrt[k]{n} \rfloor$ and $\lfloor \log_b(n) \rfloor$. These should be used instead of computing roots and logarithms as GAP does not support real numbers.

Despite not supporting real numbers GAP can display a complicated fraction as a floating-point real number, e.g.,

```
gap> Float(254638754321/387498765398);
0.657134
```

9.2.2   Arithmetic Modulo m

The multiplicative order of a modulo m is given by OrderMod(a,m);. For example,

```
gap> OrderMod(10,77);
6
```

The command `SmallestRootInt(n);` determines the smallest root of the integer n, which is the integer r of smallest absolute value for which a positive integer k exists such that $n = r^k$. For example, $13^5 = 371293$ and this command gives

```
gap> SmallestRootInt(371293);
13
```

The command `PowerMod(r,e,m);` returns the e-th power of r modulo m. For example,

```
gap> PowerMod(987654321,123456789,987654321123456823);
171767037218848697
```

Calculating this number as

```
987654321^123456789 mod 987654321123456823;
```

will be a mistake. The latter may take centuries (guess why). The command `QuotientMod(r,s,m)` returns the quotient rs^{-1} of the elements r and s modulo m. In particular, using the command `QuotientMod(r,s,m)` is preferable to using s^{-1} mod m. For example,

```
gap> QuotientMod(1,123456789,987654321123456823);
743084182864240163
gap> 123456789^-1 mod 987654321123456823;
743084182864240163
```

For larger moduli, the first command works faster.

The primitive root modulo m is given by `PrimitiveRootMod(m)`, and the discrete log of a to the base b modulo m is given by `LogMod(a,b,m)`. For example,

```
gap> PrimitiveRootMod(23);
5
gap> LogMod(11,5,97);
86
```

The command `RootMod(m,p)` will be especially useful dealing with elliptic curves; it determines whether or not m is a quadratic residue in $\mathbb{Z}_p$ and, if it is, outputs k such that $m = k^2$ mod p.

```
gap> q:=[0,0,0,0,0,0,0,0,0,0,0,0];
[ 0, 0, 0, 0, 0, 0, 0, 0, 0, 0, 0, 0 ]
gap> for i in [1..12] do
> q[i]:=RootMod(i,13);
> od;
gap> q;
[ 1, fail, 9, 11, fail, fail, fail, fail, 3, 7, fail, 8 ]
```

9.2.3   Digitising Messages

In the crypto section we needed to convert messages into numbers. Two small programs LettertoNumber and NumbertoLetter do the trick.[1] They are not part of GAP so you have to execute them before converting.

LtoN (acronym of "Letter to Number") takes any capital letter, which must be put between apostrophes, e.g., "A", and returns the corresponding number in the range [0..25]. Any other argument would return -1 and print out an error message.

```
LtoN:=function(itamar)
   local amith;
   if itamar < 'A' or 'Z' < itamar then
       Print("Out of range\n");
       return -1;
   else
       amith:=INT_CHAR(itamar)-65;
       return amith;
   fi;
end;;
```

NtoL (acronym of "Number to Letter") takes any number, positive or negative, and finds the corresponding letter. The argument must be an integer.

```
NtoL:=function(itamar)
   local amith;
   amith:=CHAR_INT(itamar mod 26+65);
   return amith;
end;;
```

Now we can digitise **ABRACADABRA** to numbers and back:

```
gap> Read("LettertoNumber");
gap> Read("NumbertoLetter");
gap> letters:="ABRACADABRA";
"ABRACADABRA"
gap> numbers:=[1..Length(letters)];
[ 1 .. 11 ]
gap> for i in [1..Length(letters)] do
> numbers[i]:=LtoN(letters[i]);
> od;
gap> numbers;
[ 0, 1, 17, 0, 2, 0, 3, 0, 1, 17, 0 ]
gap> letters2:="ZZZZZZZZZZZ";
"ZZZZZZZZZZZ"
gap> for i in [1..Length(numbers)] do
> letters2[i]:=NtoL(numbers[i]);
> od;
gap> letters2;
"ABRACADABRA"
```

In certain applications, for example in RSA, it is convenient to have all messages of the same length. In such situations the following two programs can be used instead.

---

[1]I owe former student Amith Itamar for writing these.

LtoN1 takes any capital letter, which must be put between apostrophes, e.g., "A", and returns the corresponding number in the range [11..36]. Any other argument would return −1, and print out an error message.

```
LtoN1:=function(itamar)
    local amith;
    if itamar < 'A' or 'Z' < itamar then
        Print("Out of range\n");
        return -1;
    else
        amith:=INT_CHAR(itamar)-65+11;
        return amith;
    fi;
end;;
```

NtoL1 takes any two-digit number, positive or negative, and finds the corresponding letter. The argument must be an integer.

```
NtoL1:=function(itamar)
    local amith;
    amith:=CHAR_INT(itamar-11 mod 26+65);
    return amith;
end;;
```

The following program CNtoL1 written by Joel Laity is very convenient for decryption of messages in RSA. It converts a number with any number of digits into a message. For example,

```
gap> n:=1112131415161718192021222324252627282930313233343536;
1112131415161718192021222324252627282930313233343536
gap> CNtoL1(n);
"A B C D E F G H I J K L M N O P Q R S T U V W X Y Z"
```

Here is the code for it:

```
# CNtoL1 converts a number with an even number of digits to a sequence of characters.
# The last two digits will be converted to a character, two at a time, using the
# function NtoL1 until the entire number is exhausted. The output is a string of the
# characters with spaces in between.

CNtoL1:=function(joel)
    local n, string, temp, i;

    if IsInt(joel) then
        string:=[];
        while joel > 0 do
            n:=joel mod 100;
            joel:= (joel-n)/100;
            Add(string,NtoL1(n));
            Add(string,' ');
        od;
        #reverses the order of the list
        for i in [1..QuoInt(Length(string),2)] do
            temp:=string[i];
            string[i]:=string[Length(string)+1-i];
            string[Length(string)+1-i]:=temp;
```

```
        od;
        #removes extra space
        string:=string{[2..Length(string)]};
        return string;
    else Print("Input must be an integer!");
    fi;
end;;
```

9.3   Matrix Algebra

GAP treats row vectors as a special case of lists.

```
gap> v:=[1,2,3];
[ 1, 2, 3 ]
gap> IsRowVector(v);
true
```

You can calculate linear combinations of vectors as usual.

```
gap> 2*[1,1,1] + [1,2,3];
[ 3, 4, 5 ]
```

A matrix is a list of rows. For example, the matrix

$$A = \begin{bmatrix} 1\ 2\ 3 \\ 4\ 5\ 6 \\ 7\ 8\ 9 \end{bmatrix}$$

will be presented as

```
gap> A:=[[1,2,3],[4,5,6],[7,8,9]];
[ [ 1, 2, 3 ], [ 4, 5, 6 ], [ 7, 8, 9 ] ]
gap> IsMatrix(A);
true
```

We can calculate the vector-matrix product

```
gap> u:=[1,1,1];
[ 1, 1, 1 ]
gap> u*A;
[ 12, 15, 18 ]
```

One has to note that if we multiply the matrix A by a row vector u (which would not be normally defined), it will actually calculate Au^T, e.g.,

```
gap> A*u;
[ 6, 15, 24 ]
```

We can calculate the determinant of A as

```
gap> Determinant(A);
0
```

and calculate the inverse in two ways:

```
gap> B:=[[1,1,1],[0,2,1],[0,0,13]];
[ [ 1, 1, 1 ], [ 0, 2, 1 ], [ 0, 0, 13 ] ]
gap> B^-1;
[ [ 1, -1/2, -1/26 ], [ 0, 1/2, -1/26 ], [ 0, 0, 1/13 ] ]
gap> Inverse(B);
[ [ 1, -1/2, -1/26 ], [ 0, 1/2, -1/26 ], [ 0, 0, 1/13 ] ]
```

Matrices with entries in $\mathbb{Z}_{26}$ can be added, multiplied and inverted adding mod 26 at the end of the line, e.g.,

```
gap> C:=[[1,1,1],[0,3,1],[0,0,5]];
[ [ 1, 1, 1 ], [ 0, 3, 1 ], [ 0, 0, 5 ] ]
gap> C^-1 mod 26;
[ [ 1, 17, 12 ], [ 0, 9, 19 ], [ 0, 0, 21 ] ]
```

9.4   Algebra

9.4.1   Permutations

Permutations in GAP are represented as products of disjoint cycles. For example, the permutation

$$\pi = \begin{pmatrix} 1\,2\,3\,4 \\ 2\,1\,4\,3 \end{pmatrix}$$

will be represented as $(1, 2)(3, 4)$. The identity permutation is represented as ( ). For example:

```
gap> pi:=(1,2)(3,4);
(1,2)(3,4)
gap> pi^2;
()
```

Permutations can also be defined by its last row using the command PermList. For example, permutation π can be defined as

```
gap> pi:=PermList([2,1,4,3]);
(1,2)(3,4)
```

Given a permutation written as a product of disjoint cycles, we may recover its last row using the command ListPerm:

```
gap> tau:=(1,3,4)(2,5,6,7);
(1,3,4)(2,5,6,7)
gap> ListPerm(tau);
[ 3, 5, 4, 1, 6, 7, 2 ]
```

The image $\pi(i)$ is calculated as i^π. For example:

```
gap> c:=PermList([2,3,4,1]);
(1,2,3,4)
gap> 2^c;
3
```

The order of a permutation can be calculated with GAP too.

```
gap> Order(PermList([2,4,5,1,3]));
6
```

The symmetric group can be defined by the command `G:=Symmetric Group(n);`. Then you can ask to generate a random permutation of degree n. Say,

```
gap> G:=SymmetricGroup(9);
Sym( [ 1 .. 9 ] )
gap> Random(G);
(1,9,2,8,4,6,3,7)
```

9.4.2   Elliptic Curves

For working in elliptic curves, first of all we have to read the two files elliptic.gd and elliptic.gi, given in appendix:

```
gap> Read("elliptic.gd");
gap> Read("elliptic.gi");
```

Then an elliptic curve $Y^2 = X^3 + aX + b$ modulo $\mathbb{Z}_p$ can be defined by the command:

```
gap> G:=EllipticCurveGroup(a,b,p);
```

If we try to input parameters for which the discriminant of the cubic $d = -(4a^3 + 27b^2)$ is zero, it will show a mistake. If the discriminant is nonzero, it will generate the group G. To list it, we may use the command `AsList(G);`

```
gap> G:=EllipticCurveGroup(3,2,5);
EllipticCurveGroup(3,2,5)
gap> AsList(G);
[ ( 1, 1 ), ( 1, 4 ), ( 2, 1 ), ( 2, 4 ), infinity ]
```

Now let us consider a larger group.

```
gap> H:=EllipticCurveGroup(17,19,97);
EllipticCurveGroup(17,19,97)
gap> ptsList := AsList(H);
[ ( 2, 35 ), ( 2, 62 ), ( 3, 0 ), ( 4, 32 ), ( 4, 65 ), ( 5, 36 ), ( 5, 61 ),
  ( 7, 44 ), ( 7, 53 ), ( 8, 45 ), ( 8, 52 ), ( 10, 5 ), ( 10, 92 ),
  ( 12, 37 ), ( 12, 60 ), ( 13, 20 ), ( 13, 77 ), ( 14, 24 ), ( 14, 73 ),
```

```
( 16, 33 ), ( 16, 64 ), ( 23, 8 ), ( 23, 89 ), ( 24, 34 ), ( 24, 63 ),
( 25, 8 ), ( 25, 89 ), ( 31, 48 ), ( 31, 49 ), ( 35, 18 ), ( 35, 79 ),
( 36, 40 ), ( 36, 57 ), ( 37, 45 ), ( 37, 52 ), ( 38, 21 ), ( 38, 76 ),
( 40, 0 ), ( 41, 31 ), ( 41, 66 ), ( 44, 3 ), ( 44, 94 ), ( 45, 27 ),
( 45, 70 ), ( 46, 19 ), ( 46, 78 ), ( 47, 47 ), ( 47, 50 ), ( 49, 8 ),
( 49, 89 ), ( 51, 29 ), ( 51, 68 ), ( 52, 45 ), ( 52, 52 ), ( 54, 0 ),
( 55, 2 ), ( 55, 95 ), ( 56, 12 ), ( 56, 85 ), ( 60, 27 ), ( 60, 70 ),
( 63, 2 ), ( 63, 95 ), ( 65, 47 ), ( 65, 50 ), ( 66, 16 ), ( 66, 81 ),
( 68, 39 ), ( 68, 58 ), ( 69, 17 ), ( 69, 80 ), ( 70, 21 ), ( 70, 76 ),
( 71, 25 ), ( 71, 72 ), ( 76, 2 ), ( 76, 95 ), ( 79, 34 ), ( 79, 63 ),
( 81, 4 ), ( 81, 93 ), ( 82, 47 ), ( 82, 50 ), ( 83, 23 ), ( 83, 74 ),
( 85, 30 ), ( 85, 67 ), ( 86, 21 ), ( 86, 76 ), ( 89, 27 ), ( 89, 70 ),
( 91, 34 ), ( 91, 63 ), ( 92, 10 ), ( 92, 87 ), ( 93, 9 ), ( 93, 88 ),
( 96, 1 ), ( 96, 96 ), infinity ]
gap> Size(H);
100
```

In GAP the group of an elliptic curve is represented multiplicatively so we have to multiply points instead of adding them and calculate P^{-1} instead of $-P$:

```
gap> point1:=ptsList[2];
( 2, 62 )
gap> point2:=ptsList[21];
( 16, 64 )
gap> point1 * point2;
( 81, 93 )
gap> point1^-1;
( 2, 35 )
gap> g := Random(G);
( 92, 87 )
gap> g^5;
( 69, 80 )
```

You can determine orders of all elements of the group simultaneously using the command

```
gap> List(ptsList,Order);
[ 50, 50, 2, 25, 25, 25, 25, 25, 25, 50, 50, 50, 50, 50, 50, 10, 10, 50, 50,
  50, 50, 50, 50, 10, 10, 50, 50, 50, 50, 50, 50, 5, 5, 50, 50, 25, 25, 2,
  50, 50, 50, 50, 50, 50, 25, 25, 50, 50, 50, 50, 25, 25, 5, 5, 2, 50, 50,
  25, 25, 50, 50, 50, 50, 25, 25, 50, 50, 50, 50, 10, 10, 50, 50, 50, 50, 25,
  25, 10, 10, 50, 50, 50, 50, 50, 10, 10, 50, 50, 25, 25, 10, 10, 50, 50,
  50, 50, 50, 50, 1 ]
```

The group of an elliptic curve for a ten-digit prime is already too big for GAP; it will not be able to keep the whole group in the memory. For example, the two commands

```
gap> p:=123456791;;
gap> G:=EllipticCurveGroup(123,17,p);
```

will show a mistake. If one wants to calculate in larger groups, a special technique must be applied.

We can find out if the group is cyclic or not.

```
gap> n:=NextPrimeInt(12345);
12347
```

```
gap> G:=EllipticCurveGroup(123,17,n);
EllipticCurveGroup(123,17,12347)
gap> Size(G);
12371
gap> Random(G);
( 11802, 5830 )
gap> P:=Random(G);
gap> Order(P);
12371
gap> IsCyclic(G);
true
```

There is no known polynomial time algorithm which finds a point on the given curve although the following randomised algorithm gives us a point with probability close to 1/2. This algorithm chooses x at random and tries to find a matching y such that (x, y) is on the curve. For example,

```
gap> p:=NextPrimeInt(99921);
99923
gap> G:=EllipticCurveGroup(123,17,p);
EllipticCurveGroup(123,17,99923)
gap> Size(G);
100260
gap> IsCyclic(G);
true
gap> x:=12345;
12345
gap> fx:=(x^3+123*x+17) mod p;
51321
gap> y:=RootMod(fx,p);
fail
gap> x:=1521;
1521
gap> fx:=(x^3+123*x+17) mod p;
42493
gap> y:=RootMod(fx,p);
72372
```

and we obtain a point $(1521, 72372)$ which belongs to G.

It is not so easy to input a point of a given elliptic curve. Suppose we want to input a point $M = (2425, 89535)$ of the curve

$$Y^2 = X^3 + 12345 \quad \text{mod } 95701.$$

We must generate the curve but we also have to explain GAP that M is the point of the curve we have defined. For this we present GAP with the already known point of the target curve (i.e., we can generate a point P on this curve at random) and say that we will input a point of the same curve. See how this can be done in the example below:

```
gap> G:=EllipticCurveGroup(0,12345,95701);;
gap> P:=Random(G);
( 91478, 65942 )
gap> M:=EllipticCurvePoint(FamilyObj(P),[2425,89535]);
( 2425, 89535 )
```

9.4.3 Finite Fields

GAP knows about all the finite fields. To create the finite field Z_p, type GF (p) ; For example,

```
gap>  F:=GF(5);;
gap>  List:=Elements(F);
[ 0*Z(5), Z(5)^0, Z(5), Z(5)^2, Z(5)^3 ]
```

The first element is 0 (GAP makes it clear that this is the zero of $\mathbb{Z}_5$ and not, say, of $\mathbb{Z}_3$). The remaining elements are powers of a primitive element of $\mathbb{Z}_5$, and, in particular, the second element is 1. Type Int (Z (5)) ; to determine the value of $Z(5)$ (as an integer mod 5).

```
gap> Int(Z(5));
2
gap> value:=[0,0,0,0,0];;
gap> for i in [1..5] do
> value[i]:=Int(List[i]);
> od;
gap> value;
[ 0, 1, 2, 4, 3 ]
```

Let us consider also $\mathbb{Z}_7$:

```
gap> F:=GF(7);
GF(7)
gap> Elements(F);
[ 0*Z(7), Z(7)^0, Z(7), Z(7)^2, Z(7)^3, Z(7)^4, Z(7)^5 ]
gap> # Here 0*Z(7)=0, Z(7)^0=1, Z(7)=3, Z(7)^2=2, Z(7)^3=6, Z(7)^4=4, Z(7)^5=5.
gap> # Z(7) is not 2 since 2 is not a primitive element.
```

Addition is carried out using $+$ and multiplication using $*$.

In GAP the generator of $Z(p)$ is chosen as the smallest primitive root mod p, as is obtained from the PrimitiveRootMod function. Here is how to verify this for $p = 3$ and $p = 5$:

```
gap> PrimitiveRootMod(7);
3
gap> p:=123456791;;
gap> PrimitiveRootMod(p);
17
```

To generate a finite field $GF(p^k)$, where p is a prime and $k > 0$ is an integer, we type GF(p$^\wedge$k);. For example,

```
gap> GF4:=GF(4);
GF(2^2)
gap> F:=Elements(GF4);
[ 0*Z(2), Z(2)^0, Z(2^2), Z(2^2)^2 ]
```

Since F^* is a cyclic group, GAP uses a generator of this cyclic group, denoted $Z(p^k)$, to list all elements (except zero) as its powers.

```
gap> GF4:=GF(4);
GF(2^2)
gap> gf4:=Elements(GF4);
[ 0*Z(2), Z(2)^0, Z(2^2), Z(2^2)^2 ]
gap> # Note that GAP lists elements of Z_2 first.
gap> GF8:=GF(8);
GF(2^3)
gap> gf8:=Elements(GF8);
[ 0*Z(2), Z(2)^0, Z(2^3), Z(2^3)^2, Z(2^3)^3, Z(2^3)^4, Z(2^3)^5, Z(2^3)^6 ]
```

Note that $GF(8)$ contains $GF(2)$ but not $GF(4)$. It is a general fact that $GF(p^m)$ contains $GF(p^k)$ as a subfield if and only if $k|m$.

```
gap> GF9:=GF(9);
GF(3^2)
gap> gf9:=Elements(GF9);
[ 0*Z(3), Z(3)^0, Z(3), Z(3^2), Z(3^2)^2, Z(3^2)^3, Z(3^2)^5, Z(3^2)^6,
  Z(3^2)^7 ]
```

Note that GAP lists elements of Z_3 first. Next, let us try adding, subtracting and multiplying field elements in GAP. For example in $GF(9)$:

```
[ 0*Z(3), Z(3)^0, Z(3), Z(3^2), Z(3^2)^2, Z(3^2)^3, Z(3^2)^5, Z(3^2)^6, Z(3^2)^7 ]
gap> gf9[5]+gf9[6]; gf9[5]-gf9[7];
Z(3)
Z(3^2)^3
gap> gf9[5]^2;
Z(3)
```

In a finite field the discrete logarithm of an element z with respect to a root r is the smallest nonnegative integer i such that $r^i = z$. The command $\texttt{LogFFE}(z, r)$ returns this value. (Note that r must be a primitive element of the field for this command to work.) An error is signalled if z is zero, or if z is not a power of r.

```
gap> LogFFE( Z(409)^116, Z(409) ); LogFFE( Z(409)^116, Z(409)^2 );
116; 58
```

9.4.4   Polynomials

It is not too hard to explain to GAP that we now want x to be a polynomial. We can define the polynomial ring $F[x]$ first. For example, we define the polynomial ring in one variable x over $\mathbb{Z}_2$ as follows:

```
gap> R:=PolynomialRing(GF2,["x"]);
PolynomialRing(..., [ x ])
gap> x:=IndeterminatesOfPolynomialRing(R)[1];
x
```

Now GAP will understand the following commands in which we define a polynomial $1 + x + x^3 \in \mathbb{Z}_2$ and substitute the primitive element of $GF(8)$ in it. All calculations therefore will be conducted in the field $GF(8)$:

```
gap> p:=Z(2)+x+x^3;
x^3+x+Z(2)^0
gap> Value(p,Z(2^3));
0*Z(2)
```

This tells us that the generator $\mathbb{Z}(2^3)$ of $GF(8)$ is a root of the polynomial $p(x) = x^3 + x + 1$ over $\mathbb{Z}_2$.

We can factorise polynomials as follows:

```
gap> Factors(x^16+x+1);
[ x^8+x^6+x^5+x^3+Z(2)^0, x^8+x^6+x^5+x^4+x^3+x+Z(2)^0 ]
```

Euclidean and Extended Euclidean algorithms can be performed as follows:

```
gap> g:=x^3+1;
x^3+Z(2)^0
gap> h:=x^4+x^2+1;
x^4+x^2+Z(2)^0
gap> Gcd(g,h);
x^2+x+Z(2)^0
gap> GcdRepresentation(g,h);
[ x, Z(2)^0 ]
gap> x*g+Z(2)^0*h;
x^2+x+Z(2)^0
```

Alternatively we can define the polynomial x as follows. This time we define x as a polynomial from $\mathbb{Q}[x]$:

```
gap> x := Indeterminate(Rationals);
x_1
gap> Factors(x^12-1);
[ x_1-1, x_1+1, x_1^2-x_1+1, x_1^2+1, x_1^2+x_1+1, x_1^4-x_1^2+1 ]
```

When you type x GAP understands what you want to say but still gives the answer in terms of x_1. Another useful command:

```
gap> QuotientRemainder( (x+1)*(x+2)+5, x+1 );
[ x_1+2, 5 ]
```

`InterpolatedPolynomial(R,x,y)` returns, for given lists x and y of elements in a ring R of the same lengths, say, n, the unique polynomial of degree less than n which has value $y[i]$ at $x[i]$, for all $i = 1, 2, \ldots, n$. Note that the elements in x must be distinct. For example,

```
gap> InterpolatedPolynomial( Integers, [ 1, 2, 3 ], [ 5, 7, 0 ] );
-9/2*x^2+31/2*x-6
```

Using GAP we can calculate minimal annihilating polynomials. For example,

```
gap> F:=GF(2^6);
GF(2^6)
gap> elts:=Elements(F);
[ 0*Z(2), Z(2)^0, Z(2^2), Z(2^2)^2, Z(2^3), Z(2^3)^2, Z(2^3)^3, Z(2^3)^4,
  Z(2^3)^5, Z(2^3)^6, Z(2^6), Z(2^6)^2, Z(2^6)^3, Z(2^6)^4, Z(2^6)^5,
  Z(2^6)^6, Z(2^6)^7, Z(2^6)^8, Z(2^6)^10, Z(2^6)^11, Z(2^6)^12, Z(2^6)^13,
  Z(2^6)^14, Z(2^6)^15, Z(2^6)^16, Z(2^6)^17, Z(2^6)^19, Z(2^6)^20,
  Z(2^6)^22, Z(2^6)^23, Z(2^6)^24, Z(2^6)^25, Z(2^6)^26, Z(2^6)^28,
  Z(2^6)^29, Z(2^6)^30, Z(2^6)^31, Z(2^6)^32, Z(2^6)^33, Z(2^6)^34,
  Z(2^6)^35, Z(2^6)^37, Z(2^6)^38, Z(2^6)^39, Z(2^6)^40, Z(2^6)^41,
  Z(2^6)^43, Z(2^6)^44, Z(2^6)^46, Z(2^6)^47, Z(2^6)^48, Z(2^6)^49,
  Z(2^6)^50, Z(2^6)^51, Z(2^6)^52, Z(2^6)^53, Z(2^6)^55, Z(2^6)^56,
  Z(2^6)^57, Z(2^6)^58, Z(2^6)^59, Z(2^6)^60, Z(2^6)^61, Z(2^6)^62 ]
gap> a:=elts[11];
Z(2^6)
gap> MinimalPolynomial(GF(2),a);
x_1^6+x_1^4+x_1^3+x_1+Z(2)^0
gap> MinimalPolynomial(GF(2^3),a);
x_1^2+Z(2^3)*x_1+Z(2^3)
```

So the minimal annihilating polynomial over $\mathbb{Z}_2$ is of degree 6:

$$m(t) = t^6 + t^4 + t^3 + t + 1$$

while the same element has minimal annihilating polynomial

$$m_1(t) = t^2 + \alpha t + \alpha,$$

where $\alpha = Z(2^3)$ is a primitive element of $GF(2^3)$ over $\mathbb{Z}_2$.

Appendix B: Miscellania

10.1 Linear Dependency Relationship Algorithm

This algorithm is based on the following observation.

Lemma 10.1.1 *Let* $A = [\, \mathbf{a}_1, \mathbf{a}_2, \ldots, \mathbf{a}_n \,]$ *and* $B = [\, \mathbf{b}_1, \mathbf{b}_2, \ldots, \mathbf{b}_n \,]$ *be two* $m \times n$ *matrices given by their columns* $\mathbf{a}_1, \mathbf{a}_2, \ldots, \mathbf{a}_n$ *and* $\mathbf{b}_1, \mathbf{b}_2, \ldots, \mathbf{b}_n$. *Suppose that* A *is row reducible to* B. *Then*

$$x_1\mathbf{a}_1 + x_2\mathbf{a}_2 + \cdots + x_n\mathbf{a}_n = \mathbf{0} \quad \text{if and only if} \quad x_1\mathbf{b}_1 + x_2\mathbf{b}_2 + \cdots + x_n\mathbf{b}_n = \mathbf{0}.$$
$$(10.1)$$

In particular, a system of columns $\{\mathbf{a}_{i_1}, \mathbf{a}_{i_2}, \ldots, \mathbf{a}_{i_k}\}$ *is linearly independent if and only if the system* $\{\mathbf{b}_{i_1}, \mathbf{b}_{i_2}, \ldots, \mathbf{b}_{i_k}\}$ *is linearly independent.*

Proof Let $\mathbf{x} = (x_1, x_2, \ldots, x_n)^T$. Then

$$A\mathbf{x} = x_1\mathbf{a}_1 + x_2\mathbf{a}_2 + \cdots + x_n\mathbf{a}_n \quad \text{and} \quad B\mathbf{x} = x_1\mathbf{b}_1 + x_2\mathbf{b}_2 + \cdots + x_n\mathbf{b}_n.$$

Since elementary row operations do not change the solution set of systems of linear equations, we know that

$$A\mathbf{x} = \mathbf{0} \quad \text{if and only if} \quad B\mathbf{x} = \mathbf{0}.$$

Hence (10.1) is true. □

The algorithm is used when we are given a set of vectors $\mathbf{v}_1, \mathbf{v}_2, \ldots, \mathbf{v}_n \in \mathbb{R}^n$ and we need to identify a basis of span$\{\mathbf{v}_1, \mathbf{v}_2, \ldots, \mathbf{v}_n\}$ and express all other vectors as linear combinations of that basis. We form a matrix $[\mathbf{v}_1 \cdots \mathbf{v}_n]$ whose columns are the given vectors and reduce it to the reduced row echelon form where all relationships are transparent.

© Springer Nature Switzerland AG 2020

A. Slinko, *Algebra for Applications*, Springer Undergraduate Mathematics Series,
https://doi.org/10.1007/978-3-030-44074-9_10

Example 10.1.1 The matrix $A = [\mathbf{a}_1, \mathbf{a}_2, \ldots, \mathbf{a}_5]$ with columns $\mathbf{a}_1, \mathbf{a}_2, \ldots, \mathbf{a}_5$ is brought to its reduced row echelon form $R = [\mathbf{r}_1, \mathbf{r}_2, \ldots, \mathbf{r}_5]$ with columns $\mathbf{r}_1, \mathbf{r}_2, \ldots, \mathbf{r}_5$ as follows:

$$A = \begin{bmatrix} 1 & -1 & 0 & 1 & -4 \\ 0 & 2 & 2 & 2 & 0 \\ 2 & 1 & 3 & 1 & 4 \\ 3 & 2 & 5 & 4 & 0 \end{bmatrix} \xrightarrow{\text{rref}} \begin{bmatrix} 1 & 0 & 1 & 0 & 2 \\ 0 & 1 & 1 & 0 & 3 \\ 0 & 0 & 0 & 1 & -3 \\ 0 & 0 & 0 & 0 & 0 \end{bmatrix}.$$

The relationships between columns of R are much more transparent than that of A. For example, we see that $\{\mathbf{r}_1, \mathbf{r}_2, \mathbf{r}_4\}$ is linearly independent (as a part of the standard basis of $\mathbb{R}^4$) and that $\mathbf{r}_1 + \mathbf{r}_2 - \mathbf{r}_3 = \mathbf{0}$ and $\mathbf{r}_5 = 2\mathbf{r}_1 + 3\mathbf{r}_2 - 3\mathbf{r}_4$. Hence we can conclude that $\{\mathbf{a}_1, \mathbf{a}_2, \mathbf{a}_4\}$ is linearly independent, hence a basis of $\text{span}\{\mathbf{a}_1, \mathbf{a}_2, \ldots, \mathbf{a}_5\}$ and that $\mathbf{a}_3 = \mathbf{a}_1 + \mathbf{a}_2$ and $\mathbf{a}_5 = 2\mathbf{a}_1 + 3\mathbf{a}_2 - 3\mathbf{a}_4$.

10.2 The Vandermonde Determinant

The Vandermonde determinant

$$V_n(x_1, x_2, \ldots, x_n) = \begin{vmatrix} 1 & 1 & \cdots & 1 \\ x_1 & x_2 & \cdots & x_n \\ x_1^2 & x_2^2 & \cdots & x_n^2 \\ \vdots & \vdots & \ddots & \vdots \\ x_1^{n-1} & x_2^{n-1} & \cdots & x_n^{n-1} \end{vmatrix} \tag{10.2}$$

plays a significant role in algebra and applications. It can be defined over any field, has a beautiful structure and can be calculated directly for any order.

More precisely, the following theorem is true.

Theorem 10.2.1 *Let F be a field and $a_1, a_2, \ldots, a_n$ be elements of this field. Then the value of the Vandermonde determinant of order $n \geq 2$ is*

$$V_n(a_1, a_2, \ldots, a_n) = \prod_{1 \leq i < j \leq n} (a_i - a_j). \tag{10.3}$$

Proof Since $V_2 = a_2 - a_1$ we get a basis for induction. Suppose the theorem is true for order $n - 1$. Consider the determinant

$$f(x) = \begin{vmatrix} 1 & 1 & \cdots & 1 \\ x & a_2 & \cdots & a_n \\ x^2 & a_2^2 & \cdots & a_n^2 \\ \vdots & \vdots & \ddots & \vdots \\ x^{n-1} & a_2^{n-1} & \cdots & a_n^{n-1} \end{vmatrix}$$

If we expand it using cofactors of the first column we will see that it has degree $n - 1$. Also it is easy to see that $f(a_2) = \ldots = f(a_n) = 0$ since if we replace x with any of the a_i for $i > 1$ we will have a determinant with two equal columns. Hence

$$f(x) = C(x - a_2) \ldots (x - a_n).$$

From the expansion of $f(x)$ by cofactors of the first column we see that $C = V_{n-1}(a_2, \ldots, a_n)$. Hence we have

$$f(x) = V_{n-1}(a_2, \ldots, a_n)(x - a_2) \ldots (x - a_n).$$

Substituting a_1 for x and using the induction hypothesis, we get

$$V_n(a_1, a_2, \ldots, a_n) = f(a_1) = (a_1 - a_2) \ldots (a_1 - a_n) \prod_{2 \le i < j \le n} (a_i - a_j) = \prod_{1 \le i < j \le n} (a_i - a_j). \quad \square$$

Corollary 10.2.1 *If $a_1, a_2, \ldots, a_n$ are distinct elements of the field, the Vandermonde determinant $V_n(a_1, a_2, \ldots, a_n)$ is nonzero.*

The determinant

$$V_n'(x_1, x_2, \ldots, x_n) = \begin{vmatrix} x_1 & x_2 & \cdots & x_n \\ x_1^2 & x_2^2 & \cdots & x_n^2 \\ \vdots & \vdots & \ddots & \vdots \\ x_1^{n-1} & x_2^{n-1} & \cdots & x_n^{n-1} \\ x_1^n & x_2^n & \cdots & x_n^n \end{vmatrix} \quad (10.4)$$

is also sometimes called the Vandermonde determinant. It is closely related to the original Vandermonde determinant as the following theorem states.

Theorem 10.2.2 *Let $a_1, a_2, \ldots, a_n$ be elements of the field F. Then*

$$V_n'(a_1, a_2, \ldots, a_n) = \left(\prod_{i=1}^n a_i \right) V_n(a_1, a_2, \ldots, a_n). \quad (10.5)$$

Proof Immediately follows from Theorem 10.2.1 (exercise). $\square$

10.3   Stirling's Formula

Sometimes one encounters product notation instead of summation notation. By far the most common product is factorial: $n! = 1 \cdot 2 \cdot 3 \cdot \ldots \cdot (n - 1) \cdot n = \prod_{i=1}^n i$. Our goal in this section is to find a good, closed-form estimate of $n!$. The best way to

Fig. 10.1 Illustration of the
proof of Stirling formula

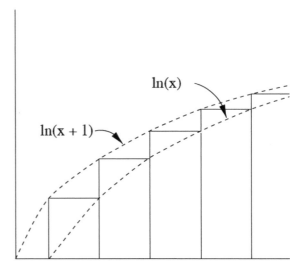

handle a product is to convert it into a sum by taking the logarithm. In the case of
factorial, this gives:

$$\ln n! = \ln(1 \cdot 2 \cdot 3 \cdot \ldots \cdot (n-1) \cdot n) = \ln 1 + \ln 2 + \ln 3 + \ldots + \ln(n-1) + \ln n = \sum_{i=1}^{n} \ln i.$$

We will use the Integral Method to bound the terms of this sum with $\ln x$ and $\ln(x + 1)$
as shown in Fig. 10.1.

This gives bounds on $\ln n!$ as follows:

$$\int_{1}^{n} \ln x \, dx \le \sum_{i=1}^{n} \ln i \le \int_{1}^{n} \ln(x + 1) \, dx$$

or

$$n \ln(\frac{n}{e}) + 1 \le \sum_{i=1}^{n} \ln i \le (n + 1) \ln(\frac{n + 1}{e}) + 1$$

or

$$\left(\frac{n}{e}\right)^{n} e \le n! \le \left(\frac{n + 1}{e}\right)^{n+1} e.$$

We now have a closed form estimate of $n!$, namely, $n! \approx \left(\frac{n}{e}\right)^{n}$. (We are ignoring
stray e and n factors.) A more accurate estimate of $n!$ can be obtained with a more
careful calculation:

Theorem 10.3.1 (Stirling's formula)

$$n! \sim \sqrt{2\pi n} \left(\frac{n}{e}\right)^n, \tag{10.6}$$

which means that

$$\frac{n!}{\sqrt{2\pi n} \left(\frac{n}{e}\right)^n} \to 1,$$

when $n \to \infty$.

Solutions to Exercises

11.1 Solutions to Exercises of Chap. 1

Solutions to Exercises of Sect. 1.1.1

1. (a) The whole set of integers itself does not contain a smallest element.
   (b) The set $\{1/2, 1/3, \ldots, 1/n, \ldots\}$ does not contain a smallest element.
2. Here we just need the Principle of Mathematical Induction. For $n = 1$, the integer $4^n + 15n - 1 = 18$ is divisible by 9. This is a basis for the induction. Suppose that $4^n + 15n - 1$ is divisible by 9 for some $n > 1$. Let us consider $4^{n+1} + 15(n + 1) - 1$ and represent it as $4 \cdot 4^n + 15n + 14 = 4(4^n + 15n - 1) - 45n + 18$. This is now obviously divisible by 9 since both $4(4^n + 15n - 1)$ and $45n + 18$ do (The former by induction hypothesis). Thus $4^{n+1} + 15(n + 1) - 1$ is divisible by 9, and the induction step has been proven.
   For $n = 0$ we have $11^2 + 12^1 = 133$ which is, of course, divisible by 133. This gives us a basis for the induction.
3. We need the Principle of Mathematical Induction again. Suppose now $133 \mid 11^{n+2} + 12^{2n+1}$ (induction hypothesis) and let us consider

$$11^{(n+1)+2} + 12^{2(n+1)+1} = 11^{n+3} + 12^{2n+3}.$$

We rearrange this as follows:

$$11^{n+3} + 12^{2n+3} = 11 \cdot 11^{n+2} + 144 \cdot 12^{2n+1} = 144(11^{n+2} + 12^{2n+1}) - 133 \cdot 11^{n+2}.$$

The right-hand side is divisible by 133. Indeed, the first summand is divisible by 133 by induction hypothesis, and the second is simply a multiple of 133. Thus $11^{n+3} + 12^{2n+3}$ is divisible by 133 which completes the induction step and the proof.

© Springer Nature Switzerland AG 2020

A. Slinko, *Algebra for Applications*, Springer Undergraduate Mathematics Series,

https://doi.org/10.1007/978-3-030-44074-9_11

4. We have $F_0 = 3$ and $F_1 = 5$. We see that $F_0 = F_1 - 2$, and this is a basis for our induction. The induction step

$$F_0 F_1 \ldots F_n = F_{n+1} - 2 \quad \text{implies} \quad F_0 F_1 \ldots F_{n+1} = F_{n+2} - 2$$

will be proved if we could show that $(F_{n+1} - 2)F_{n+1} = F_{n+2} - 2$. We have $(F_{n+1} - 2)F_{n+1} = (2^{2^{n+1}} - 1)(2^{2^{n+1}} + 1) = 2^{2^{n+2}} - 1 = F_{n+2} - 2$.

5. For $k = 1$ we have $3^k = 3$ which is a divisor of $2^3 + 1 = 9$. This gives us a basis for the induction.
   Suppose now $3^k \mid 2^{3^k} + 1$ (induction hypothesis). Then there exists an integer m such that $m \cdot 3^k = 2^{3^k} + 1$ and let us consider

$$2^{3^{k+1}} = (2^{3^k})^3 = (m \cdot 3^k - 1)^3 = m^3 \cdot 3^{3k} - m^2 \cdot 3^{2k+1} + m \cdot 3^{k+1} - 1 = t \cdot 3^{k+1} - 1,$$

where $t = m^3 \cdot 3^{2k-1} - m^2 \cdot 3^k + m$ is an integer. Thus $3^{k+1} \mid 2^{3^{k+1}} + 1$, which proves the induction step.

6. Let M be the minimal number that cannot be represented as required. Then M is between two powers from the list, say $2^k < M < 2^{k+1}$. Since M is minimal, the number $M - 2^k$ can be represented as

$$M - 2^k = 2^{i_1} + \ldots + 2^{i_s},$$

where $i_1 < \ldots < i_s$. Since $M - 2^k < 2^k$, it is clear that $2^k > 2^{i_s}$. Therefore

$$M = 2^{i_1} + \ldots + 2^{i_s} + 2^k$$

is a representation of M as a sum of distinct powers of 2 contrary to what was assumed. This contradiction proves the statement.

7. Let M be the minimal positive integer which can be represented as a sum of distinct powers of 2 in two different ways:

$$M = 2^{i_1} + \ldots + 2^{i_s} = 2^{j_1} + \cdots + 2^{j_t}.$$

Suppose that $i_1 < \ldots < i_s$ and $j_1 < \ldots < j_t$. Then either $i_1 = 0$ or $j_1 = 0$. If not, we can divide both sides by 2 and get two different representations for $M/2$ which contradicts the minimality of M. If $i_1 = j_1 = 0$, then $2^{i_1} = 2^{j_1} = 1$, and subtracting 1 on both sides we would get two different representations for $M - 1$ which again contradicts the minimality of M.
Hence

$$1 + 2^{i_2} + \ldots + 2^{i_s} = 2^{j_1} + \ldots + 2^{j_t}$$

and $j_1 > 0$. But then the left side is odd and the right is even. This contradiction shows that such minimal counterexample M does not exist and all integers can be uniquely represented.

8. Consider a minimal counterexample, i.e., any configuration of discs which cannot be painted as required and which consists of the least possible number of discs. Consider the centres of all discs and consider the convex hull of them. This hull is a convex polygon, and each angle of it is less than 180°. If a disc with the centre O is touched by two other discs with centres P and Q

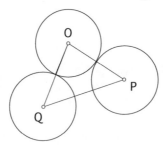

then, as $PQ \geq PO$ and $PQ \geq QO$, then $\angle POQ \geq \angle PQO$ and $\angle POQ \geq \angle QPO$, whence $\angle POQ \geq 60°$. Thus every disc with centre at a vertex of the convex hull cannot be touched by more than three other discs. Remove any of the discs whose centre is at the vertex of the convex hull. Then the rest of the discs can be already painted because the counterexample was minimal. But then the removed disc can be painted as well since it was touched by at most three other discs, and we can choose the fourth colour to paint it. This contradiction proves the statement.

Solutions to Exercises of Sect. 1.1.2

1. The 2007th prime will not be stored in `Primes` so we have to use the command `NextPrimeInt` to find it:

```
gap> p:=1;;
gap> n:=2007;;
gap> for i in [1..n] do
> p:=NextPrimeInt(p);
> od;
gap> p;
17449
```

We see that $p_{2007} = 17449$.
2. The following GAP program

```
gap> k:=1;;
gap> N:=Primes[1];;
gap> while IsPrime(N+1)=true do
> k:=k+1;
```

```
> N:=N*Primes[k];
> od;
gap> k;
6
gap> N:=N+1;
30031
gap> FactorsInt(n);
[ 59, 509 ]
```

It shows that the smallest k for which $N_k = p_1 p_2 \ldots p_k + 1$ is composite is $k = 6$. Then $N_6 = 30031 = 59 \cdot 509$. Both 59 and 509 are greater than $p_6 = 13$.

3. We have $a^3 - 27 = a^3 - 3^3 = (a - 3)(a^2 + 3a + 9)$ and therefore $a - 3$ divides $a^3 - 27$. Hence $a - 3$ divides $a^3 - 17$ if and only if it divides the difference $(a^3 - 27) - (a^3 - 17) = 10$. This happens if and only if

$$a - 3 \in \{\pm 1, \pm 2, \pm 5, \pm 10\},$$

which is equivalent to $a \in \{-7, -2, 1, 2, 4, 5, 8, 13\}$.

4. We will prove the second statement. Let $p > 2$ be a prime. Let us divide it by 6 with remainder: $p = 6k + r$, where $r = 0, 1, 2, 3, 4, 5$. When r takes values $0, 2, 3, 4$, the right-hand side is divisible by 2 or 3, hence in this case p cannot be a prime. Only two possibilities are left: $p = 6k + 1$ and $p = 6k + 5$. Examples of primes of these two sorts are 7 and 11.

5. Let $p = 3k + 1$ be a prime. Then $p > 2$, and hence, it is odd. But then $3k = p - 1$ is even and $3k = 2m$. Due to uniqueness of prime factorisation, k must be divisible by 2, i.e., $k = 2k'$. Therefore $p = 3k + 1 = 6k' + 1$.

6. Here is the program:

```
Primes1:=[];;
Primes3:=[];;
numbers:=[1..168];;
for i in numbers do
if RemInt(Primes[i],4)=1 then
Add(Primes1,Primes[i]);
fi;
if RemInt(Primes[i],4)= 3 then
Add(Primes3,Primes[i]);
fi;
od;
Length(Primes1);
Length(Primes3);
Primes1[32];
Primes3[53];
Position(Primes1,601);
Position(Primes3,607);
```

Run this program yourself to find out the numerical answers.

7. (a) Here is the program and the calculation:

```
gap> NicePrimes:=[];
[  ]
gap> for i in [1..Length(Primes)] do
> if RemInt(Primes[i],6)=5 then
> Add(NicePrimes,Primes[i]);
> fi;
> od;
gap> NicePrimes;
[ 5,  11,  17,  23,  29,  41,  47,  53,  59,  71,  83,  89,  101,  107,
  113,  131,  137,  149,  167,  173,  179,  191,  197,  227,  233,  239,
  251,  257,  263,  269,  281,  293,  311,  317,  347,  353,  359,  383,
  389,  401,  419,  431,  443,  449,  461,  467,  479,   491,  503,  509,
  521,  557,  563,  569,  587,  593,  599,  617,  641,  647,  653,  659,
  677,  683,  701,  719,  743,  761,  773,  797,  809,  821,  827,  839,
  857,  863,  881,   887,  911,  929,  941,  947,  953,  971,  977,  983 ]
```

(b) We know (see Exercise 4) that all primes $p > 3$ fall into two categories: those for which $p = 6k + 1$ and those for which $p = 6k + 5$.

One additional observation: if we take two numbers of the first category if $n_1 = 6k_1 + 1$ and $n_2 = 6k_2 + 1$, then their product

$$n_1 n_2 = (6k_1 + 1)(6k_2 + 1) = 6(6k_1 k_2 + k_1 + k_2) + 1 = 6k_3 + 1$$

will also be from the same category.

Now we assume that there are only finitely many primes p such that $p = 6k + 5$. Then there is the largest such prime. Let $p_1, p_2, \ldots, p_n, \ldots$ be the sequence of all primes in increasing order with p_n being the largest prime that gives remainder 5 on division by 6. Consider the number

$$N = p_1 p_2 \ldots p_n - 1.$$

Since $p_1 = 2$ and $p_2 = 3$, the product $p_1 p_2 \ldots p_n$ is divisible by 6. Hence N has remainder 5 on division by 6 and hence belongs to the second category. Let q be any prime that divides N. Obviously it is different from all of the $p_1, p_2, \ldots, p_n$. Since $q > p_n$, it must be of the type $q = 6k + 1$. Thus every prime that divides N has remainder 1 on division by 6, then, as we noted above, the same must be true for N, which contradicts to the fact that N has remainder 5 on division by 6.

8. There are many alternative proofs of the fact that the number of primes is infinite. Here is one of those. Assume on the contrary that there are only k primes $p_1, p_2, \ldots, p_k$. Given n, let us find an upper bound $f(n)$ for the number of products

$$p_1^{\alpha_1} p_2^{\alpha_2} \cdots p_k^{\alpha_k}$$

that do not exceed n by estimating the number of values that integers $\alpha_1, \alpha_2, \ldots, \alpha_k$ might assume. Since $n \geq p_i^{\alpha_i} \geq 2^{\alpha_i}$, we obtain $\alpha_i \leq \log_2 n$. Then the number of products which do not exceed n will be at most

$$f(n) = (\log_2 n + 1)^k.$$

It is easy to show that $f(n)$ grows slower than n for n sufficiently large. For example, we may use L'Hospital rule to show that

$$\lim_{n \to \infty} \frac{f(n)}{n} = 0.$$

This will be an absurdity since for large n there will be not enough prime factorisations for all positive integers between 1 and n.

Solutions to Exercises of Sect. 1.1.3

1. (a) Notice that since $\sqrt{210} < 17$, all composites below 210 divide 2, 3, 5, 7, 11 or 13. The primes to be found are 2, 3, 5, 7, 11, 13, 17, 19, 23, 29, 31, 37, 41, 43, 47, 53, 59, 61, 67, 71, 73, 79, 83,89, 97, 101, 103, 107, 109, 113, 127, 131, 137, 139, 149, 151, 157, 163, 167, 173, 179, 181, 191, 193, 197, 199. Hence $\pi(210) = 46$.
   (b) We have

$$\frac{210}{\ln 210} \approx 39,$$

   which is somewhat lower than 46. This shows that the approximation given by the prime number theorem is not very good for small values of n.
2. Straightforward.
3. (a) No, because $\sqrt{n} > 11093$ and n can be, for example, a square of a prime p such that $10000 < p \leq 11093$.
   (b) The number of possible prime divisors is approximately $x/ln(x)$ where $x = \sqrt{123123137}$, so approximately 1193 divisions are needed. The professor has already done $10000/\ln(10000) = 1085$ divisions; he needs to do another 108.
4. Since n is composite, $n = p_1 p_2 \ldots p_m$, where p_i is prime, for all $i = 1, 2, \ldots, m$, and we do not assume that all of them are different. We are given that $p_i > \lfloor \sqrt[3]{n} \rfloor$, and $m \geq 2$. Then, we have also $p_i > \sqrt[3]{n}$ because p_i is an integer. Suppose that $m \geq 3$. Then

$$n = p_1 p_2 \ldots p_m \geq p_1 p_2 p_3 > (\sqrt[3]{n})^3 = n,$$

   which is a contradiction. Thus $m = 2$ and p is a product of two primes.
5. Let $n > 6$ be an integer. If n is odd, then 2 and $n - 2$ are relatively prime (since $n - 2$ is odd) and $n = 2 + (n - 2)$ is a valid solution. More generally, if there is a prime p which is smaller than $n - 1$ and does not divide n, we can write $n = p + (n - p)$ and $\gcd((, p), n - p) = 1$.

Since $n > 6$ we may assume that $2|n$, $3|n$, $5|n$ and hence $30|n$. In particular, n is composite. Let q be the largest prime divisor of n. Then $n \geq 6q$ so $5 \leq q \leq n/6$. By Bertrand's postulate there is a prime p such that $q < p < 2q \leq n/3 < n$. Now $\gcd((, p), n) = 1$ and so $n = p + (n - p)$ is the solution (note that $n - p \geq n - n/3 > 1$).

6. The program does some kind of sieving but the result is very different from the result of the Sieve of Eratosthenes. It outputs all powers of 2 between 1 and 10^6, there are 20 such numbers in total.

Solutions to Exercises of Sect. 1.2.2

1. Firstly we represent the number as a product of primes:

$$2^2 \cdot 3^3 \cdot 4^4 \cdot 5^5 = 2^{10} \cdot 3^3 \cdot 5^5$$

and the number of divisors will be $(10 + 1)(3 + 1)(5 + 1) = 264$. Note that we cannot use the formula straight as 4 is not prime.

2. We factor this number with GAP:

```
gap> FactorsInt(123456789);
[ 3, 3, 3607, 3803 ]
```

so the prime factorisation is $123456789 = 3^2 \cdot 3607 \cdot 3803$. The number of divisors then will be $(2 + 1)(1 + 1)(1 + 1) = 12$.

3. The common divisors of 10650 and 6750 are the divisors of $\gcd(10650, 6750)$. So, let us calculate this number using the Euclidean algorithm. We will find:

$$10650 = 1 \cdot 6750 + 3900$$
$$6750 = 1 \cdot 3900 + 2850$$
$$3900 = 1 \cdot 2850 + 1050$$
$$2850 = 2 \cdot 1050 + 750$$
$$1050 = 1 \cdot 750 + 300$$
$$750 = 2 \cdot 300 + 150$$
$$300 = 2 \cdot 150$$

Hence

$$\gcd(10650, 6750) = 150 = 2 \cdot 3 \cdot 5^2.$$

Therefore the common divisors of 10650 and 6750 are the factors of 150, which are 1, 2, 3, 5, 6, 10, 15, 25, 30, 50, 75, 150.

4. (a) We have $\gcd(m, n) = 2^2 \cdot 5^4 \cdot 11^2$; $\mathrm{lcm}(m, n) = 2^4 \cdot 3^2 \cdot 5^7 \cdot 7^2 \cdot 11^3$.

(b) Using GAP we calculate

$$\gcd(m, n) \cdot \mathrm{lcm}(m, n) = 302500 \cdot 733713750000 = 221948409375000000.$$

Also

$$mn = 1361250000 \cdot 163047500 = 221948409375000000,$$

which is the same value.

5. The prime factorisation of 33 is $33 = 3 \cdot 11$. This number of divisors can occur when the number is equal to p^{32}, where p is prime, or when the number is $p^{10}q^2$, where p, q are primes. As $2^{32} > 10000$, the first possibility cannot occur. In the second, since $3^{10} > 10000$, the number can be only of the form $n = 2^{10}q^2$. The smallest unused prime is $q = 3$. This gives us the number $n = 10^2 \cdot 3^2 = 9216$. No other prime q works since $n = 10^2 \cdot 5^2 > 10000$. So the only such number is 9216.

6. Since the prime factorisation of 246 is $246 = 2 \cdot 3 \cdot 41$, the prime factorisation of 246^{246} will be

$$246^{246} = 2^{246} \cdot 3^{246} \cdot 41^{246}$$

and $d(246^{246}) = 247^3$. As $247 = 13 \cdot 19$ we have $247^3 = 13^3 \cdot 19^3$ and $d(d(246^{246})) = 4 \cdot 4 = 16$.

7. If d is a divisor of a and b, then $a = a'd$ and $b = b'd$ and $a - b = (a' - b')d$, whence d is a common divisor of a and $a - b$. If d is a divisor of a and $a - b$, then $a = a'd$ and $a - b = cd$. Then $b = a - (a - b) = (a' - c)d$ that is d is also a common divisor of a and b.

8. We use the previous exercise repeatedly. We have $\gcd(13n + 21, 8n + 13) = \gcd(8n + 13, 5n + 8) = \gcd(5n + 8, 3n + 5) = \gcd(3n + 5, 2n + 3) = \gcd(2n + 3, n + 2) = \gcd(n + 2, n + 1) = \gcd(n + 1, 1) = 1$.

9. (a) Suppose a^2 and $a + b$ have common prime divisor p. Then it is also a divisor of a and hence of $b = (a + b) - a$, contradiction.

   (b) As in Exercise 6 we notice that $\gcd(a, b) = \gcd(a, qa + b)$ for any integer q. Then, since $a^2 - b^2 = (a - b)(a + b)$ is divisible by $a + b$, we have

$$\gcd(a^2 + b^2, a + b) = \gcd((a^2 + b^2) + (a^2 - b^2), a + b) = \gcd(2a^2, a + b) = 2.$$

   This is because by (a) a^2 and $a + b$ do not have prime divisors in common. Since $a + b$ and $a^2 + b^2$ are not relatively prime, their greatest common divisor can be only 2. This can be realised taking two arbitrary odd relatively prime a and b, say $a = 25$ and $b = 49$.

10. Let $a > b$ and $a = qb + r$, where r is the remainder on division of a by b. Firstly, we note that

$$2^a - 1 = 2^{a-b}(2^b - 1) + 2^{a-b} - 1,$$

thus implying that

$$\gcd(2^a - 1, 2^b - 1) = \gcd(2^{a-b} - 1, 2^b - 1) = \ldots = \gcd(2^b - 1, 2^r - 1),$$

which in turn implies the statement (follow the Euclidean algorithm applied to a and b.)

11. Let F_i and F_j be two Fermat numbers with $i < j$. Then by Exercise 4 of Sect. 1.1.1 $F_0 F_1 \ldots F_{j-1} = F_j - 2$. Since the left-hand side is divisible by F_i, the only common divisor of F_i and F_j could be 2. However, these numbers are odd, hence coprime.

12. If there were only finite number k primes, then among any $k + 1$ Fermat numbers there will be two with a common prime factor. However this is not possible due to the previous exercise. Hence the number of primes is infinite.

Solutions to Exercises of Sect. 1.2.3

1. The Extended Euclidean algorithm gives

3773	1	0	
3596	0	1	1
177	1	−1	20
56	−20	21	3
9	61	−64	6
2	−386	405	4
1	1605	−1684	2

Looking at the last line we find that

$$\gcd(3773, 3596) = 1 = 3773 \cdot 1605 + (-1684) \cdot 3596.$$

So, $x = 1605$, $y = -1684$.

2. Perform the Extended Euclidean algorithm on 1995 and 1840 gives $\gcd(1995, 1840) = 5$ and hence $x = 95$ and $y = -103$ may be taken that will satisfy $1995x + 1840y = 5$.

Multiply now $1995x + 1840y = 5$ by (-2) to see that $z_0 = -2x = -190$ and $w_0 = -2y = 206$ satisfy $1995z_0 + 1840w_0 = -10$. Next, observe that $1995(-k \cdot 1840) + 1840(k \cdot 1995) = 0$, for any integer k. Sum the last two equations to obtain $1995(z_0 - 1840k) + 1840(w_0 + 1995k) = -10$, for any integer k. It is now easy to find two additional solutions, for example $z_1 = z_0 + 1840 = 1650$ and $w_1 = w_0 - 1995 = -1789$, or $z_2 = z_0 - 1840 = -2030$ and $w_2 = w_0 + 1995 = 2201$.

3. We are given that $N = kc + a$ and $N = td + b$ for some integers k and t. Subtracting the two equalities yields $0 = kc + a - td - b$. Therefore

$$a - b = kc - td.$$

Since the right-hand side is divisible by $\gcd(c, d)$, we see that $a - b$ is divisible by $\gcd(c, d)$ as well.

4. (a) The Extended Euclidean algorithm applied to 68 and 26 gives $2 = \gcd(68, 26) = 5 \cdot 68 + (-13) \cdot 26$. Multiplying both sides by $(35 - 9)/2 = 13$, we see that $35 - 9 = 13 \cdot 5 \cdot 68 - 13 \cdot 13 \cdot 26$. Hence, the number $x = 35 + 13 \cdot 13 \cdot 26 = 9 + 13 \cdot 5 \cdot 68 = 4429$ satisfies our congruences. (There are many other solutions, all of them are congruent modulo $884 = \operatorname{lcm}(26, 68)$, i.e., all these solutions are given by $4429 + 884 \cdot n, n \in Z$.)

   (b) The Extended Euclidean algorithm applied to 71 and 50 gives $1 = 27 \cdot 50 + (-19) \cdot 71$. Now, $15 = 19 - 4$ and the number $x' = 4 + 15 \cdot 27 \cdot 50 = 19 + 15 \cdot 19 \cdot 71 = 20254$ satisfies our congruences but is greater than 3550. But $x = x' \bmod 3550 = 2504$ is the unique solution of the two congruences which lies in the interval $[0, 3550)$.

5. (a) We know from Exercise 2 that $\gcd(1995, 1840) = 5$. If there were integers x and y satisfying $1840x + 1995y = 3$, then $3 = 5(368x + 399y)$ and 3 would be divisible by 5, a contradiction.

   (b) Let C be the set of integers c for which there exist integers x and y satisfying the equation $ax + by = c$, and let $d = \gcd(a, b)$. By the Extended Euclidean algorithm we know that there are some integers x_0, y_0, such that $ax_0 + by_0 = d$. Let k be an arbitrary integer. Then $a(kx_0) + b(ky_0) = kd$, showing that $kd \in C$, so C contains all multiples of $\gcd(a, b)$. Let us prove that C contains nothing else. Write $a = da'$ and $b = db'$, for some integers a' and b', and take an arbitrary $c \in C$. Then, for some integers x and y, we have $c = ax + by = d(a'x + b'y)$, showing that c is a multiple of d. Therefore, C is indeed the set of all multiples of $\gcd(m, n)$.

Solutions to Exercises of Sect. 1.3.1

1. By Fermat's little theorem $2^4 \equiv 1 \bmod 5$ so we need to find the remainder of 2^{2013} on division by 4. This remainder is obviously 0 so the remainder of $2^{2^{2013}}$ on division by 5 is 1.

2. By Fermat's little theorem we have $a^6 \equiv 1 \bmod 7$ for all $a \in \mathbb{Z}$, which are not divisible by 7. As $333 = 47 \cdot 7 + 4$ and $555 = 92 \cdot 6 + 3$,

$$333^{555} \equiv 4^3 \equiv 64 \equiv 1 \bmod 7.$$

As $555 = 79 \cdot 7 + 2$ and $333 = 55 \cdot 6 + 3$,

$$555^{333} \equiv 2^3 \equiv 1 \bmod 7.$$

Thus $333^{555} + 555^{333} \equiv 1 + 1 = 2 \bmod 7$, and the remainder is 2.

3. We compute $a^{n-1} \bmod n$ as follows:

```
gap> n:=1234567890987654321;
1234567890987654321
```

```
gap> a:=111111111;
111111111
gap> PowerMod(a,n-1,n);
385560404775530811
```

The result is not equal to 1, and this shows that by Fermat's little theorem n is not prime. Indeed, we see that n has four different prime factors:

```
gap> Factors(n);
[ 3, 3, 7, 19, 928163, 1111211111 ]
```

4. Let q be a prime divisor of $2^p - 1$. Then by Fermat's little theorem $q | 2^{q-1} - 1$ and by assumption $q | 2^p - 1$. By Exercise 10 of Sect. 1.2.2 this is possible only when $q - 1$ is a multiple of p that is $q - 1 = \ell p$. But since $q - 1$ is even we have also ℓ even, say $\ell = 2k$. Hence we obtain that $q = 2kp + 1$.

Solutions to Exercises of Sect. 1.3.2

1. (a) This follows from $m \mid (a - b)$ if and only if $dm \mid (da - db)$
   (b) Since $a \equiv b \bmod m$ means $m \mid (a - b)$, we see that for any divisor $d \mid m$ we have $d \mid (a - b)$ which is the same as $a \equiv b \bmod d$.
   (c) Indeed, $a \equiv b \bmod m_i$ is equivalent to $m_i \mid (a - b)$. This implies

$$\text{lcm}(m_1, m_2, \ldots, m_k) \mid (a - b),$$

   which means the equivalence holds also for the least common multiple of m_i's.
2. We have $72 \equiv -3 \bmod 25$, $47 \equiv -3 \bmod 25$ and $28 \equiv -3 \bmod 25$. Thus

$$72^{2n+2} - 47^{2n} + 28^{2n-1} \equiv (-3)^{2n+2} - (-3)^{2n} + 3^{2n-1} \bmod 25.$$

Since $2n + 2$ and $2n$ are even, the right-hand side will be

$$3^{2n+2} - 3^{2n} + 3^{2n-1} = 3^{2n}(27 - 3 + 1) = 25 \cdot 3^{2n},$$

which is obviously divisible by 25.
3. Using the prime factorisation of these numbers and the formula for $\phi(n)$ we compute:

$$\phi(125) = \phi(5^3) = 5^3 - 5^2 = 100,$$

$$\phi(180) = \phi(2^2 \cdot 3^2 \cdot 5) = 180 \left(\frac{1}{2}\right) \left(\frac{2}{3}\right) \left(\frac{4}{5}\right) = 48,$$

$$\phi(1001) = \phi(7 \cdot 11 \cdot 13) = 6 \cdot 10 \cdot 12 = 720.$$

4. We know that $n = 4386607 = pq$ for some prime integers p, q. In this case $\phi(n) = 4382136 = (p - 1)(q - 1)$. Thus $n - \phi(n) = 4471 = p + q - 1$, whence $p + q = 4472$. Solving the system of equations

$$p + q = 4472$$
$$pq = 4386607$$

we find that $p = 3019$ and $q = 1453$.

5. We have $\phi(m) = p_1 p_2 (p_1 - 1)(p_2 - 1) = 11424 = 2^5 \cdot 3 \cdot 7 \cdot 17$. Hence p_1 and p_2 can be only among the primes $2, 3, 7, 17$. By the trial and error method we find $p_1 = 7$, $p_2 = 17$ and $m = 14161$.

6. We have $\phi(3^x 5^y) = 3^{x-1} 5^{y-1} \cdot 2 \cdot 4 = 3^{x-1} 5^{y-1} \cdot 2^3$ and $600 = 2^3 \cdot 3 \cdot 5^2$. By uniqueness of prime factorisation, we have $x - 1 = 1$ and $y - 1 = 2$. Hence $x = 2$ and $y = 3$.

7. Let S be the set of positive integers a for which the congruence has a solution. We see that $243 = 3^5$ and $\phi(243) = 2 \cdot 3^4 = 162$. By Euler's theorem:

If $\gcd(x, n) = 1$, then $x^{\phi(n)} \equiv 1 \bmod n$.

Hence if $\gcd(x, 243) = 1$, then $x^{162} \equiv 1 \bmod 243$. Hence $1 \in S$. If $\gcd(x, n) > 1$, then $x = 3y$ and $x^{162} \equiv 0 \bmod 243$. Thus $S = \{0, 1\}$.

8. We are given that $n = pq$ where p and q are primes. Moreover, we know that $\phi(n) = \phi(p)\phi(q) = (p - 1)(q - 1) = pq - p - q + 1 = 3308580$, and therefore $p + q = n - 3308580 + 1$. We now determine p and q from the equations:

$$\begin{cases} pq = 3312913, \\ p + q = 4334. \end{cases}$$

This shows that p and q are the roots of the quadratic equation $x^2 - 4334x + 3312913 = 0$ in which roots are 3343 and 991. The result is $n = pq = 3343 \cdot 991$.

Solutions to Exercises of Sect. 1.4

1. By the distributive law (CR5) we have $a \cdot 0 + a \cdot 0 = a \cdot (0 + 0) = a \cdot 0$. Now subtracting $a \cdot 0$ on both sides we get $a \cdot 0 = 0$. We further argue as in Lemma 1.4.2.

2. (a) The invertible elements of $\mathbb{Z}_{16}$ are those elements that are relatively prime to $16 = 2^4$ (i.e., those which are odd). We have

$$1^2 = 7^2 = 9^2 = 15^2 = 1, \quad 3 \cdot 11 = 1, \quad 5 \cdot 13 = 1,$$

thus $1, 7, 9, 15$ are self-inverse, $3^{-1} = 11$, $11^{-1} = 3$, $5^{-1} = 13$ and $13^{-1} = 5$.

(b) The zero divisors of $\mathbb{Z}_{15}$ are those (nonzero) elements that are not relatively prime to $15 = 3 \cdot 5$ (i.e., multiples of 3 or 5). We have

$$\{3, 6, 9, 12\} \odot \{5, 10\} = 0.$$

That is, $a \odot b = 0$ whenever one of a and b is a multiple of 3 and another is a multiple of 5.

3. (a) Using the Euclidean algorithm, we find that $\gcd(111, 74) = 37$ and that $\gcd(111, 77) = 1$, so 77 is invertible and 74 is a zero divisor. Since $111 = 3 \cdot 37$, we have $74 \odot c = 0$ for any c that is a multiple of 3. From the Extended Euclidean algorithm $1 = 34 \cdot 111 - 49 \cdot 77$, hence $77^{-1} = -49 = 62$.

(b) We have

$$77 \odot x \oplus 21 = 10 \Rightarrow 77 \odot x = 10 \oplus (-21) = 10 \oplus (90) = 100.$$

Hence

$$x = (77^{-1}) \odot 100 = 62 \odot 100 = 95,$$

so that $x = 95$, while

$$74 \odot x \oplus 11 = 0 \Rightarrow 74 \odot x = -11 = 100,$$

and there are no solutions because $\{74 \odot x \mid x \in \mathbb{Z}_{111}\} = \{0, 37, 74\}$.

4. Since we will have only operations in $\mathbb{Z}_n$ for various n but not in $\mathbb{Z}$ we will write $+$ and $\cdot$ instead of $\oplus$ and $\odot$. Recall that a function from a set A to A itself is one-to-one, if no two (different) elements of A are mapped to the same element of A. For a finite set this is also equivalent to f being onto which can be also restated as the range of f being all of $\mathbb{Z}_{21}$.

(a) If a is a zero divisor in $\mathbb{Z}_{21}$, that is, if there is an element $d \neq 0$ in $\mathbb{Z}_{21}$, such that $ad = 0 \bmod 21$, then $f(d) = ad + b = b = f(0)$, and f is not one-to-one. On the other hand, if a is not a zero divisor, then $\gcd(a, 21) = 1$, and there exists (a unique) element $c \in \mathbb{Z}_{21}$, satisfying $ac = 1 \bmod 21$. But then $f(x_1) = f(x_2)$ implies $cf(x_1) = cf(x_2)$, or $c(ax_1 + b) = c(ax_2 + b)$, which reduces to $x_1 + cb = x_2 + cb$ and finally implies that $x_1 = x_2$, proving that f is one-to-one in this case. The set of pairs (a, b), for which the function f is one-to-one is therefore $\{(a, b) \mid a, b \in \mathbb{Z}_{21} \text{ and } \gcd(a, 21) = 1\}$.

(b) Since 7 is not relatively prime with 21 the function f is not one-to-one, and so the image of f is a proper subset of $\mathbb{Z}_{21}$. The expression $7x$, for $x \in \mathbb{Z}_{21}$, takes only three values in $\mathbb{Z}_{21}$, namely 0 if x is a multiple of 3, 7 if x is congruent to 1 modulo 3, and 14 if x is congruent to 2 modulo 3. The image of f is therefore $\{3, 10, 17\}$.

(c) Condition $f^{-1}(f(x)) = x$, for all $x \in \mathbb{Z}_{21}$, is equivalent to $c(ax + b) + d = x$, or $(ac)x + (cb + d) = x$. It is sufficient to take $ac = 1$ and $cb + d = 0$. We can find c by solving the equation $4c + 21y = 1$ using the Extended Euclidean algorithm, which gives us $c = -5$, $y = 1$, or better, $c = 16$, $y = -3$. Now, $d = -cb = -16 \cdot 15 = 12 \bmod 21$. So, $f^{-1}(x) = 16x + 12$.

5. Little Fermat theorem says that if p is prime and a is not divisible by p, then $a^{p-1} \equiv 1 \mod p$. Hence $x^{10} = 1$ in $\mathbb{Z}_{11}$. So $x^{102} = x^2$ in $\mathbb{Z}_{11}$. The equation $x^2 = 4$ has in $\mathbb{Z}_{11}$ two solutions: $x_1 = 2$ and $x_2 = -2 = 9$.

6. Since m is odd, $\gcd(m, 2) = 1$, whence $2^{\phi(m)} \equiv 1 \mod m$. Thus $2^{\phi(m)-1} \equiv 2^{-1} \mod m$ which is the inverse of 2 in $\mathbb{Z}_m$. Since m is odd, $m + 1$ is an even number and $(m + 1)/2$ is an integer. This number is the inverse of 2 in $\mathbb{Z}_m$ since $2 \odot (m + 1)/2 = 1$. Therefore $2^{\phi(m)-1} \equiv (m + 1)/2 \mod m$.

7. If $(p - 1)! \equiv -1 \mod p$, then $\gcd(j, p) = 1$ for all $j \in \mathbb{Z}_p^*$. Hence p is prime. If p is prime, then the equation $x^2 = 1$ in $\mathbb{Z}_p$ is equivalent to $(x - 1)(x + 1) = 0$, hence has only two solutions $x = \pm 1$ that is either $x = 1$ or $x = p - 1$. Then for every $j \in \{2, \ldots, p - 2\}$ we have $j^{-1} \neq j$. This means $2 \cdot 3 \cdot \ldots \cdot (p - 2) = 1$. Hence $(p - 1)! = p - 1 = -1$.

Solutions to Exercises of Sect. 1.5

1. $2002_{(10)} = 11111010010_{(2)}$; and $1100101_{(2)} = 2^6 + 2^5 + 2^2 + 1 = 99_{(10)}$.

2. (a) $2011_{(10)} = 11111011011_{(2)}$;
   (b) $101001000_{(2)} = 2^8 + 2^6 + 2^3 = 256 + 64 + 8 = 328_{(10)}$.

3. Observe first that the last three digits in the binary representation depend only on the remainder on division by 8. Namely if $a = a_n 2^n + \ldots + a_3 2^3 + a_2 2^2 + a_1 2 + a_0$ is the binary representation of a, then $a \equiv a_2 2^2 + a_1 2 + a_0 \mod 8$. Clearly $75^{1015} \equiv 3^{1015} \mod 8$. By Euler's theorem, $3^{\phi(8)} = 3^4 \equiv 1 \mod 8$. Therefore, $75^{1015} \equiv 3^{253 \cdot 4 + 3} \equiv 3^3 \equiv 3 \mod 8$. Since $3 = 11_{(2)}$, we see that the last three digits in the binary representation of 75^{1015} are 011.

4. We calculate as follows:

$$\underbrace{10\ldots01}_{n}{}_{(2)} \cdot \underbrace{10\ldots01}_{m}{}_{(2)} = (2^{n-1} + 1)(2^{m-1} + 1) = 2^{n+m-2} + 2^{n-1} + 2^{m-1} + 1.$$

Therefore, there are 4 nonzero digits if $m \neq n$, 3 nonzero digits if $m = n > 2$, and 2 nonzero digits if $m = n = 2$.

5. We are given that $n = a \cdot 7^3 + b \cdot 7^2 + c \cdot 7^1 + d$. Since $7 \equiv 1 \mod 6$, this means that $n \equiv a + b + c + d \mod 6$. Therefore $n \equiv 0 \mod 6$ if and only if $a + b + c + d \equiv 0 \mod 6$.

6. (a) $2A4F_{(16)} = 2 \cdot 16^3 + 10 \cdot 16^2 + 4 \cdot 16 + 15 = 10831$,
   (b) $1000 = 16 \cdot 62 + 8$, and $62 = 16 \cdot 3 + 14$, so

$$1000 = 62 \cdot 16 + 8 = (16 \cdot 3 + 14) \cdot 16 + 8 = 3 \cdot 16^2 + 14 \cdot 16 + 8 = 3E8_{(16)}.$$

11.2 Solutions to Exercises of Chap. 2

Solutions to Exercises of Sect. 2.1.1

1. To encrypt we will do the calculation $p_i \oplus k_i = c_i$ where p_i, k_i, c_i are the encodings of the ith positions of the plain text, the key and the cypher text, respectively.

i	1	2	3	4	5	6	7	8	9	10	11	12	13	14	15
Plaintext	B	U	Y	M	O	R	E	P	R	O	P	E	R	T	Y
p_i	1	20	24	12	14	17	4	16	17	14	16	4	17	19	24
Key	T	O	D	A	Y	I	W	I	L	L	G	O	O	N	C
k_i	19	14	3	0	24	8	22	8	11	11	6	14	14	13	2
$p_i + k_i = c_i$	20	8	1	12	12	25	0	24	2	25	22	18	5	6	0
Cyphertext	U	I	B	M	M	Z	A	Y	C	Z	W	S	F	G	A

So the cyphertext is **UIBMMZAYCZWSFGA**.

Conversely, to decrypt we add $(-k_i)$ to each side of the above to get $p_i = c_i + (-k_i)$.

i	1	2	3	4	5	6	7	8	9	10	11	12	13	14	15
Cyphertext	R	C	X	R	N	W	O	A	P	D	Y	W	C	A	U
c_i	17	2	23	17	13	22	14	0	15	3	24	22	2	0	20
Key	T	O	D	A	Y	I	W	I	L	L	G	O	O	N	C
$-k_i$	7	12	23	0	2	18	4	18	15	15	20	12	12	13	24
$c_i + (-k_i) = p_i$	24	14	20	17	15	14	18	18	4	18	18	8	14	13	18
Plaintext	Y	O	U	R	P	O	S	S	E	S	S	I	O	N	S

i	16	17	18	19	20	21	22	23	24	25	26	27	28	29
Cyphertext	E	R	K	Y	W	H	Z	R	G	S	X	Q	J	W
c_i	4	17	10	24	22	7	25	17	6	18	23	16	9	22
Key	E	A	G	A	I	N	I	N	T	O	L	I	F	E
$-k_i$	22	0	20	0	18	13	18	13	7	12	15	18	21	22
$c_i + (-k_i) = p_i$	0	17	4	24	14	20	17	4	13	4	12	8	4	18
Plaintext	A	R	E	Y	O	U	R	E	N	E	M	I	E	S

The plaintext is therefore **YOUR POSSESSIONS ARE YOUR ENEMIES**.

2. We will place the result in an array called random:

```
gap> random:=[1..20];;
gap> for i in [1..20] do
> random[i]:=Random([0..25]);
> od;
gap> random;
[ 24, 19, 16, 9, 1, 9, 24, 24, 15, 3, 12, 3, 10, 11, 21, 23, 19, 6, 19, 24 ]
```

3. The message as a numerical string will be [8, 7, 0, 21, 4, 13, 14, 19, 8, 12, 4, 19, 14, 7, 0, 19, 4].

```
gap>#Entering the key:
gap> k:=random;;
gap>#Entering the message:
gap> p:=[ 8, 7, 0, 21, 4, 13, 14, 19, 8, 12, 4, 19, 14, 7, 0, 19, 4 ];;
gap> c:=[1..Length(p)];
[ 1 .. 17 ]
gap> for i in [1..Length(p)] do
> c[i]:=(p[i]+k[i]) mod 26;
> od;
gap> c;
[ 6, 0, 16, 4, 5, 22, 12, 17, 23, 15, 16, 22, 24, 18, 21, 16, 23 ]
gap># which in letters will be GAQEFWMRXPQWYSVQX
gap> # Decoding back:
gap> q:=[1..Length(p)];;
gap> for i in [1..Length(p)] do
> q[i]:=(c[i]-k[i]) mod 26;
> od;
gap> p=q;
true
```

Solutions to Exercises of Sect. 2.1.2

1. $(13, 11)$ cannot be used as a key since 13 is not invertible in $\mathbb{Z}_{26}$ and the mapping $x \mapsto 13x + 11 \pmod{26}$ would not be one-to-one.
2. The cyphertext for **CRYPTO** will be **JSRWOL**. The inverse function for decrypting is

$$C \longrightarrow x \longrightarrow y = 19x + 13 \mod 26 \longrightarrow L$$

and the plaintext for **DRDOFP** is **SYSTEM**. We can calculate the latter using subprograms LtoN and NtoL:

```
gap> str := "DRDOFP"; ;
gap> outstr := "A";
gap> for i in [1..Length(str)] do
>    outstr[1] := NtoL( (19*LtoN( str[i] ) + 13) mod 26);
>    Print( outstr );
> od;
SYSTEM
```

3. Since the letter **F** was encrypted as **N**, the letter **K** was encrypted as **O**. Then for the encryption function $f(x) = ax + b \mod 26$ we will have $f(5) = 13$ and $f(10) = 14$. Solving the system of equations in $\mathbb{Z}_{26}$

$$5a + b = 13,$$
$$10a + b = 14$$

we find $a = 21$ and $b = 12$, hence the key is the pair $(21, 12)$.
With GAP this would be

```
gap> M:=[[5,1],[10,1]];
[ [ 5, 1 ], [ 10, 1 ] ]
gap> rhs:=[13,14];
[ 13, 14 ]
gap> [a,b]:=M^-1*rhs mod 26;
[ 21, 12 ]
```

4. A straightforward counting shows that the relative frequencies of the 26 letters in the cyphertext are as given in the table below.

letter	rel. freq.	letter	rel. freq.	letter	rel. freq.
a	0.049	j	0.076	s	0.017
b	0.052	k	0.045	t	0.000
c	0.135	l	0.076	u	0.062
d	0.000	m	0.031	v	0.007
e	0.000	n	0.031	w	0.007
f	0.000	o	0.035	x	0.101
g	0.000	p	0.093	y	0.021
h	0.007	q	0.017	z	0.000
i	0.101	r	0.066		

Since the most frequent letter in the cyphertext is c while in the English texts this role usually plays e, our guess is that the encryption function $f(x) = ax + b$ mod 26 maps the integer value of e, which is 4, to the integer value of c, which is 2. This gives the first equation:

$$4a + b = 2 \text{ mod } 26. \tag{11.1}$$

The second most frequent letter in English is t, while in our cyphertext the second place is shared by x and i. Suppose first that the letter t was encrypted to x. Then

$$19a + b = 23 \text{ mod } 26, \tag{11.2}$$

implying $15a = 21 \text{ mod } 26$, $a = 7 \cdot 21 = 17 \text{ mod } 26$ and $b = 2 - 4a = 12 \text{ mod } 26$. If the encryption function is $f(x) = ax + b$ mod 26, then the decryption function is $g(x) = cx - cb$ mod 26, where $ca = 1$ mod 26. In the case $a = 17$, $b = 12$, we get $c = 23$ and so $g(x) = 23x + 10$ mod 26. If we decrypt the cyphertext with this function we get

djree rctqk xmr ...

which is obviously not an English text. Our guess that t was encrypted to x must therefore be wrong. We get similar nonsense if we assume that t is encrypted to i. We can either proceed in this fashion until we get something meaningful, or observe that in our cyphertext the group of three letters ljc is very frequent. Since our guess is that c is in fact encrypted to e, it is very plausible that the group ljc

represents the word *the*. If this is the case, then t is encoded to l, which gives the equation

$$19a + b = 11 \text{ mod } 26, \tag{11.3}$$

This, together with (11.1), implies that $a = 11$ and $b = 10$. The decrypting function is then $g(x) = 19x + 18$. Decrypting the cyphertext with g gives the following plaintext:

> three rings for the elven kings under the sky seven for the dwarf lords in their halls of stone nine for mortal men doomed to die one for the dark lord on his dark throne in the land of mordor where the shadows lie one ring to rule them all one ring to find them one ring to bring them all and in the darkness bind them in the land of mordor where the shadows lie

Solutions to Exercises of Sect. 2.1.3

1. (a) Computing the determinants of these matrices we get

$$\det \begin{bmatrix} 1 & 12 \\ 12 & 1 \end{bmatrix} = 13, \quad \det \begin{bmatrix} 1 & 6 \\ 6 & 1 \end{bmatrix} = 17.$$

Since 13 is not relatively prime to 26 the first matrix is not invertible because its determinant is not invertible. Since $17^{-1} = 23$ exists, the second matrix is invertible with

$$\begin{bmatrix} 1 & 6 \\ 6 & 1 \end{bmatrix}^{-1} = 23 \begin{bmatrix} 1 & 20 \\ 20 & 1 \end{bmatrix} = \begin{bmatrix} 23 & 18 \\ 18 & 23 \end{bmatrix}.$$

   (b) We use the matrix

$$M = \begin{bmatrix} 1 & 6 \\ 6 & 1 \end{bmatrix}$$

for encryption. Encrypting **YEAR** we replace letters by the numbers

$$\textbf{YEAR} \rightarrow (24, 4, 0, 17)$$

then compute

$$M \begin{bmatrix} 24 \\ 4 \end{bmatrix} = \begin{bmatrix} 1 & 6 \\ 6 & 1 \end{bmatrix} \begin{bmatrix} 24 \\ 4 \end{bmatrix} = \begin{bmatrix} 22 \\ 18 \end{bmatrix}$$

and

$$M \begin{bmatrix} 0 \\ 17 \end{bmatrix} = \begin{bmatrix} 1 & 6 \\ 6 & 1 \end{bmatrix} \begin{bmatrix} 0 \\ 17 \end{bmatrix} = \begin{bmatrix} 24 \\ 17 \end{bmatrix}.$$

Now we can complete the encryption:

$$\textbf{YEAR} \rightarrow (24, 4, 0, 17) \rightarrow (22, 18, 24, 17) \rightarrow \textbf{WSYR}.$$

To decrypt **ROLK**, we represent letters by numbers

$$\mathbf{ROLK} \to (17, 14, 11, 10)$$

and then use the inverse matrix M^{-1} to compute

$$M^{-1} \begin{bmatrix} 17 \\ 14 \end{bmatrix} = \begin{bmatrix} 23 & 18 \\ 18 & 23 \end{bmatrix} \begin{bmatrix} 17 \\ 14 \end{bmatrix} = \begin{bmatrix} 19 \\ 4 \end{bmatrix}$$

and

$$M^{-1} \begin{bmatrix} 11 \\ 10 \end{bmatrix} = \begin{bmatrix} 23 & 18 \\ 18 & 23 \end{bmatrix} \begin{bmatrix} 11 \\ 10 \end{bmatrix} = \begin{bmatrix} 17 \\ 12 \end{bmatrix}.$$

Now we can complete decryption:

$$\mathbf{ROLK} \to (17, 14, 11, 10) \to (19, 4, 17, 12) \to \mathbf{TERM}.$$

2. Let $\mathbf{v} = (x, y)^T$ be the vector of numerical encodings for $\mathbf{X}$ and $\mathbf{Y}$, respectively. Then we know that $K(K\mathbf{v}) = \mathbf{v}$ that is $K^2\mathbf{v} = \mathbf{v}$. Of course, $\mathbf{v} = (0, 0)^T$ is one solution. If $\mathbf{v} \neq \mathbf{0}$, then it is an eigenvector of K^2 belonging to the eigenvalue 1. We have

$$K^2 - I = \begin{bmatrix} 5 & 4 \\ 4 & 5 \end{bmatrix} - \begin{bmatrix} 1 & 0 \\ 0 & 1 \end{bmatrix} = \begin{bmatrix} 4 & 4 \\ 4 & 4 \end{bmatrix}$$

The nullspace of this matrix is spanned by the vector $(1, -1)^T = (1, 25)^T$. The other eigenvectors will be $(2, -2)^T = (2, 24)^T$, etc., up to $(13, -13)^T = (13, 13)^T$. Together with $(0, 0)^T$ we will have 14 pairs:

$$\mathbf{XY} = \mathbf{AA}, \mathbf{AZ}, \mathbf{BY}, \mathbf{CX}, \dots, \mathbf{NN}.$$

3. Write the unknown secret key as

$$K = \begin{pmatrix} a & b \\ c & d \end{pmatrix}.$$

The first four letters of the ciphertext correspond to the vectors $(13, 22)$ and $(14, 11)$. The first four letters of the message correspond to $(3, 4)$ and $(0, 17)$. Hence, the secret key satisfies the equations

$$\begin{pmatrix} a & b \\ c & d \end{pmatrix} \begin{pmatrix} 3 \\ 4 \end{pmatrix} \equiv \begin{pmatrix} 13 \\ 22 \end{pmatrix}, \quad \begin{pmatrix} a & b \\ c & d \end{pmatrix} \begin{pmatrix} 0 \\ 17 \end{pmatrix} \equiv \begin{pmatrix} 14 \\ 11 \end{pmatrix} \quad \text{mod } 26.$$

Using the second equations first we find $17b \equiv 14 \pmod{26}$ and $17d \equiv 11 \pmod{26}$. Compute $17^{-1} \equiv -3 \equiv 23 \pmod{26}$ using the Extended Euclidean algorithm (or Gcdex in GAP) and hence determine that $b = 10$ and $d = 19$.

Now use the first equations to solve for a and c. One has $3a \equiv 13 - 4b$ (mod 26). Since $3^{-1} \equiv 9$ (mod 26), one determines that $a = 17$. Similarly, $3c \equiv 22 - 4d$ (mod 26) and so $c = 8$.

Now that one has the key, compute

$$K^{-1} = \begin{pmatrix} 5 & 22 \\ 2 & 25 \end{pmatrix}$$

and decrypt the ciphertext. The original message is

DEAR BOB I HAVE GOT IT X

(the letter X at the end was added to make the total number of letters in the message even).

4. Firstly we input of what is given: the key K and the cryptotext c:

```
gap> K:=[ [ 1, 2, 3, 4, 5 ], [ 9, 11, 18, 12, 4 ], [ 1, 2, 8, 23, 3 ],
  [ 7, 14, 21, 5, 1 ], [ 5, 20, 6, 5, 0 ] ];;
gap> c:= [ 24, 12, 9, 9, 4 ], [ 4, 25, 10, 4, 22 ], [ 7, 11, 16, 16, 8 ],
  [ 18, 3, 9, 24, 9 ], [ 2, 19, 24, 4, 20 ], [ 1, 24, 10, 5, 1 ],
  [ 22, 15, 1, 1, 4 ] ];;
gap>#Calculating the inverse of the key matrix:
gap> M:=K^-1 mod 26;
[ [ 21, 16, 22, 25, 11 ], [ 17, 9, 22, 21, 9 ], [ 6, 10, 23, 17, 20 ],
  [ 13, 14, 8, 11, 2 ], [ 22, 2, 3, 4, 17 ] ]
gap># Preparing the list for the plaintext:
gap> p:=[[],[],[],[],[],[],[],[],[],[],[],[]];;
gap>#Calculating the plaintext:
gap> for i in [1..12] do
> p[i]:=c[i]*M mod 26;
> od;
gap> p;
[ [ 12, 0, 19, 7, 4 ], [ 12, 0, 19, 8, 2 ], [ 8, 0, 13, 18, 0 ],
  [ 17, 4, 12, 0, 2 ], [ 7, 8, 13, 4, 18 ], [ 5, 14, 17, 2, 14 ],
  [ 13, 21, 4, 17, 19 ], [ 8, 13, 6, 2, 14 ], [ 5, 5, 4, 4, 8 ],
  [ 13, 19, 14, 19, 7 ], [ 4, 14, 17, 4, 12 ], [ 18, 25, 25, 25, 25 ] ]
gap># This reads: "Mathematicians are machines for conversion of coffee into
gap># theorems."
gap># This famous statement belongs to Paul Erd\"{o}s.
```

5. Let us start with some brief linear algebra preliminaries. Let R be a commutative ring and A be an $n \times n$ matrix with entries from R.

The (i, j) minor of A, denoted M_{ij}, is the determinant of the $(n-1) \times (n-1)$ matrix that results from deleting row i and column j of A. The cofactor matrix of A is the $n \times n$ matrix C whose (i, j) entry is the (i, j) cofactor $A_{ij} = (-1)^{i+j} M_{ij}$. Finally, the adjugate $\mathrm{adj}(A)$ of A is the transpose of C, that is, the $n \times n$ matrix whose (i, j) entry is the (j, i) cofactor of A.

The adjugate is defined so that the product of A and its adjugate yields a diagonal matrix whose diagonal entries are $\det(A)$:

$$A\,\mathrm{adj}(A) = \mathrm{adj}(A)A = \det(A)I_n,$$

where I_n is the identity matrix of order n. If $\det(A)$ is invertible, then $A^{-1} = \frac{1}{\det(A)} \mathrm{adj}(A)$ is the inverse of A.

On the other hand, if A is invertible and there exists a matrix B such that $AB = BA = I_n$, where I_n is the identity matrix of order n, then

$$\det(A)\det(B) = \det(AB) = \det(I_n) = 1$$

and $\det(A)$ is invertible.

Solutions to Exercises of Sect. 2.3.1

1. We have

$$\lim_{x\to\infty} \frac{(\log x)^2}{\sqrt{x}} = \lim_{x\to\infty} \frac{((\log x)^2)'}{(\sqrt{x})'} = 4\frac{\log x}{\sqrt{x}} \to 0$$

(as in Example 2.3.4).

2. Let us apply L'Hospital rule once:

$$\lim_{n\to\infty} \frac{n^{2007}}{2\sqrt{n}} = \lim_{x\to\infty} \frac{x^{2007}}{2\sqrt{x}} = \lim_{x\to\infty} \frac{(x^{2007})'}{(2\sqrt{x})'} = \lim_{x\to\infty} \frac{2007 \cdot x^{2006}}{\ln 2 \cdot 2^{\sqrt{x}} \cdot \frac{1}{2}x^{-1/2}} = k_1 \cdot \lim_{x\to\infty} \frac{x^{2006.5}}{2\sqrt{x}},$$

where $k_1 = \frac{4014}{\ln 2}$. If we continue applying L'Hospital rule 4014 times, we will obtain

$$\lim_{x\to\infty} \frac{x^{2007}}{2\sqrt{x}} = k_{4014} \cdot \lim_{x\to\infty} \frac{1}{2\sqrt{x}} = 0.$$

Hence $f(n) = o(g(n))$.

3. Straightforward.

4. Let $\chi(x) = \frac{x}{\ln x}$, so that we are trying to prove that $\psi(x) \sim \chi(x)$, or in other words, that

$$\lim_{x\to\infty} \frac{\chi(x)}{\psi(x)} = 1.$$

Since $\lim_{x\to\infty}\psi(x) = \lim_{x\to\infty}\chi(x) = \infty$, we can apply L'Hospital's rule to get

$$\lim_{x\to\infty} \frac{\chi(x)}{\psi(x)} = \lim_{x\to\infty} \frac{\chi'(x)}{\psi'(x)}$$

$$= \lim_{x\to\infty} \frac{\ln x - 1}{(\ln x)^2} \cdot \frac{1}{\frac{1}{\ln x}} = \lim_{x\to\infty} \left(1 - \frac{1}{\ln x}\right) = 1,$$

where we differentiated $\chi(x)$ using the quotient rule and $\psi(x)$ using the Fundamental Theorem of Calculus.

5. (a) We have $f(n) = o(g(n))$ since

$$\lim_{n \to \infty} \frac{(\ln n)^{1000}}{n^{10}} = 0.$$

This can be established by L'Hospital's rule:

$$\lim_{n \to \infty} \frac{(\ln n)^{1000}}{n^{10}} = \lim_{n \to \infty} \frac{1000(\ln n)^{999}}{10 n^{10}} = \ldots = \lim_{n \to \infty} \frac{(1000)!}{10^{1000} n^{10}} = 0.$$

Also $g(n) = o(h(n))$ since by L'Hospital's rule

$$\lim_{n \to \infty} \frac{n^{10}}{e^{n/3}} = \lim_{n \to \infty} \frac{10 n^9}{1/3 e^{n/3}} = \ldots = \lim_{n \to \infty} \frac{10!}{(1/3)^{10} e^{n/3}} = 0.$$

So the functions are already listed in the increasing order of magnitude.

(b) The function $f(n)$ is bounded from above and below: $|f(n)| \le e$ so $f(n) = O(1)$. We apply Stirling's formula to find

$$h(n) \sim \ln \sqrt{2\pi n} \left(\frac{n}{e} \right)^n = \frac{1}{2} (\ln(2\pi) + \ln n) + n(\ln n - 1) \sim n \ln n.$$

This is faster than constant but slower than n^2. Hence we have to order the functions as $f(n), h(n), g(n)$.

6. The algorithm is based on the observation that the equality

$$\left(\lfloor \sqrt[i]{n} \rfloor \right)^i = n$$

is true if and only if n is a ith power of an integer. Indeed, if n is not a perfect ith power, then $\sqrt[i]{n}$ is not an integer and $\lfloor \sqrt[i]{n} \rfloor < \sqrt[i]{n}$. Then $\left(\lfloor \sqrt[i]{n} \rfloor \right)^i < n$.
What is the maximal number i for which we shall try to check this equation. If $n = a^i$, then $i = \log_a n$ which reaches the maximal value at $\log_2 n$. Hence we should check the equation for $i = 2, 3, \ldots, \log_2 n$.
Thus the program performs $\log_2 n$ operations RootInt, hence its complexity is linear in the number of bits of n, which is the size of the input.

```
gap> i:=1;;
gap> ell:=LogInt(n,2);
2121
gap> while i<(ell+1) do
> if RootInt(n,i)^i=n then
> m:=RootInt(n,i);
> k:=i;
> fi;
> i:=i+1;
```

```
> od;
gap> m;
113
gap> k;
311
```

Hence $n = 113^{311}$.

7. (a) We have

$$\binom{n}{k} = \frac{n!}{k!(n-k)!} = \frac{1}{k!}n(n-1)\cdots(n-k+1) \sim \Theta(n^k)$$

so this growth is polynomial.

(b) The Stirling approximation asserts that $n! \sim \left(\frac{n}{e}\right)^n \sqrt{2\pi n}$. This gives

$$\binom{n}{\alpha n} \sim \frac{(n/e)^n}{(n\alpha/e)^{n\alpha}(n\beta/e)^{n\beta}} \cdot \frac{\sqrt{2\pi n}}{\sqrt{2\pi n\alpha}\sqrt{2\pi n\beta}},$$

where $\beta = 1 - \alpha$. After simplifications we get

$$\binom{n}{\alpha n} \sim \frac{1}{\alpha^{n\alpha}\beta^{n\beta}}\frac{1}{\sqrt{2\pi n\alpha\beta}}.$$

Upon taking logarithm to base 2, we get:

$$\frac{1}{n}\log_2\binom{n}{\alpha n} \sim -(\alpha\log_2\alpha + \beta\log_2\beta) - \log_2(2\pi n\alpha\beta)/(2n) = H(\alpha) + O\left(\frac{\ln n}{n}\right).$$

The function $H(\alpha) = -(\alpha\log_2\alpha + \beta\log_2\beta)$ is called the entropy function. We get as a result that

$$\binom{n}{\alpha n} \sim e^{nH(\alpha)},$$

which is exponential growth.

Solutions to Exercises of Sect. 2.3.2

1. (a) The exact number of bits required to input an integer N will be $\lceil\log_2 N\rceil$. We are interested in integers between 10^{99} and 10^{100}, so we have

$$\log_2 10^{99} = 99\log_2 10 \approx \log_2 10^{100} = 100\log_2 10 \approx 330.$$

(b) Using the repeated division algorithm, we have

$$1234567 = 100101101011010000111_2.$$

We can use GAP to establish the sequence of remainders:

```
gap> n:=1234567;
1234567
gap> base:=2;
2
gap> rems:=[];
[  ]
gap> pos:=1;
1
gap> while n>0 do;
> rems[pos]:=RemInt(n,base);
> n:=QuoInt(n,base);
> pos:=pos+1;
> od;
gap> n;
0
gap> rems;
[ 1, 1, 1, 0, 0, 0, 0, 1, 0, 1, 1, 0, 1, 0, 1, 1, 0, 1, 0, 0, 1 ]
```

They give us the digits but in the reverse order.

Since the binary representation is 21 digits long, we need to calculate $c^{2^0} = c, c^{2^1} = c^2, \ldots c^{2^{20}}$ first. This requires 20 multiplications. Since the binary representation contains 11 digits 1, we need to multiply 11 of these entries together to obtain c^n, which is a further 10 multiplications, for a total of 30.

2. Suppose that n items are being bubble sorted. We need to determine the greatest possible number of swaps that can occur. For each item x in the list, let S_x be the set of items in the list that are smaller than x, and ahead of x in the list. Each time a swap occurs where x is the larger item, the order of S_x decreases by one, and when S_x is empty no more swaps with x as the larger item can occur. While S_x is still non-empty, the list is not yet sorted. So $|S_x|$ is exactly the number of swaps that will occur with x as the larger item. Seeing that in every swap one of the two items involved is the larger, we simply need to add up these values for every item to determine the number of swaps. In other words, the total number of swaps to sort a list will be $\sum_x |S_x|$.

An obvious upper bound on $|S_x|$ is the number of items in the list that are smaller than x. The ith smallest item in the list can have at most $i - 1$ items that are smaller than it, so

$$\sum_x |S_x| \leq \sum_{i=1}^{n}(i - 1) = \frac{n(n - 1)}{2} = \Theta(n^2).$$

We also need to check that this can actually occur, and clearly it can, for example, the list $\{n, n - 1, \ldots, 2, 1\}$ will require this number of swaps.

3. In the worst-case scenario we might need $(\log_2 N)^3$ divisions. For large N, this is less than $\sqrt{N}/ \ln \sqrt{N}$ divisions required by the standard algorithm of factoring. Therefore some composite numbers together with primes might be declared to

be interesting. This algorithm has polynomial complexity. Indeed, since we may consider that $N \approx 2^n$, where n is the number of bits necessary to input N, the worst-case complexity function is $f(n) \approx (\log_2 2^n)^3 = n^3$. It is cubic.

4. (a) Obviously $f_n \geq f_{n-1}$. Hence $f_n = f_{n-1} + f_{n-2} \leq 2f_{n-1}$. We have $f_{n+5} = f_{n+4} + f_{n+3} = 2f_{n+3} + f_{n+2} = 3f_{n+2} + 2f_{n+1} = 5f_{n+1} + 3f_n = 8f_n + 5f_{n-1} > 8f_n + 4f_{n-1} \geq 8f_n + 2f_n = 10f_n$.

   (b) We may assume $a > b$. We use the Euclidean algorithm to find:

$$
\begin{aligned}
a &= q_1 b + r_1, & 0 < r_1 < b, \\
b &= q_2 r_1 + r_2, & 0 < r_2 < r_1, \\
r_1 &= q_3 r_2 + r_3, & 0 < r_3 < r_2, \\
&\ \ \vdots \\
r_{s-2} &= q_s r_{s-1} + r_s, & 0 < r_s < r_{s-1}, \\
r_{s-1} &= q_{s+1} r_s.
\end{aligned}
$$

where $r_s = \gcd(a, b)$. We may set $b = r_0$.

We have $r_s \geq 1 = f_2$ and $r_{s-1} \geq 2 = f_3$. By induction on i we prove that $r_{s-i} \geq f_{i+2}$ for all $i = 0, 1, \ldots, s$. Indeed, if the statement is true for all i such that $0 \leq i < k$, then $r_{s-k} = q_{s-k+2} r_{s-k+1} + r_{s-k+2} \geq r_{s-k+1} + r_{s-k+2} \geq f_{k+1} + f_k = f_{k+2}$, which proves the induction step. In particular, $b = r_0 \geq f_{s+2}$.

Suppose now that $s > 5k$, where k is the number of decimal digits in b. However, by (a) we get $b \geq f_{s+2} > f_{5k+2} > 10^k f_2 = 10^k$, which is a contradiction.

Solutions to Exercises of Sect. 2.4.1

1. As $e_1 = 2145$ and $\phi(n) = 11200$ are obviously not coprime (have a factor 5 in common), e_1 cannot be used in a public key. On the other hand, $e_2 = 3861$ is coprime with $\phi(n)$, and the Extended Euclidean algorithm gives us $1 = 1744 \cdot 11200 + (-5059) \cdot 3861$. So $d = 11200 - 5059 = 6141$. Checking with GAP:

```
gap> QuotientMod(1,3861,11200);
6141
```

2. We first need to calculate Bob's private key which is $e^{-1} \bmod \phi(n) = 113^{-1} \bmod 120 = 17$ and then calculate $97^{17} \bmod 143 = 15$. So the letter was "E".

3. Bob calculates $m^2, m^4, m^8, m^{16}, m^{32}$ by successive squaring. Then he multiplies $m^{32} \cdot m^8 \cdot m = m^{41}$ using in total seven multiplications.

5. (a) We calculate $\lfloor \sqrt{20687} \rfloor = 143$. Assuming that 20687 is a product of two three-digit primes, the smallest prime factor of 20687 should be one of these primes:

$$101, \ 103, \ 107, \ 109, \ 113, \ 127, \ 131, \ 137, \ 139,$$

Trying all of them we find that $20687 = 137 \cdot 151$. Thus $\phi(20687) = 136 \cdot 150 = 20400$. Now we may compute Alice's private key which is $d = 17179^{-1}$ mod 20400. We compute

$$
\begin{array}{rrr}
20400 & 1 & 0 \\
17179 & 0 & 1 \\
3221 & 1 & -1 \\
1074 & -5 & 6 \\
1073 & 11 & -13 \\
1 & -16 & 19
\end{array}
$$

Hence $d = 19$ (aren't we lucky that it is so small!). Thus the plaintext will be 353^{19} mod 20687. We note that $19 = 16 + 2 + 1 = (10011)_{(2)}$. Thus we have to compute 353^2 and 353^{16} and then 353^{19} operating in $\mathbb{Z}_{20687}$. We compute (fingers crossed) $353^2 = 487$, $353^4 = 487^2 = 9612$, $353^8 = 9612^2 = 2042$, $353^{16} = 2042^2 = 18618$. Hence $353^{19} = 18618 \odot 487 \odot 353 = 6060 \odot 353 = 8419$, which is the plaintext.

Checking with GAP:

```
gap> QuotientMod(1,17179,20400);
19
gap> PowerMod(353,19,20687);
8419
```

6. (a) The cyphertext Alice needs to send to Bob is $c = m^e$ mod $n = 183^{1003}$ mod 24613. Without GAP, this number can be efficiently calculated as follows: first, find the binary representation $e = 1111101011_{(2)}$ and construct the sequence (computed in Z_n)

$$
\begin{aligned}
m_0 = m = &\ 183, \\
m_1 = m_0^2 = &\ 8876, \\
m_2 = m_1^2 = m_0^{2^2} = &\ 21776, \\
m_3 = m_2^2 = m_0^{2^3} = &\ 118, \\
m_4 = m_3^2 = m_0^{2^4} = &\ 13924, \\
m_5 = m_4^2 = m_0^{2^5} = &\ 1175, \\
m_6 = m_5^2 = m_0^{2^6} = &\ 2297, \\
m_7 = m_6^2 = m_0^{2^7} = &\ 9027, \\
m_8 = m_7^2 = m_0^{2^8} = &\ 17699, \\
m_9 = m_8^2 = m_0^{2^9} = &\ 4950.
\end{aligned}
$$

Now,

$$
\begin{aligned}
c =&\ 183^{2^9+2^8+2^7+2^6+2^5+2^3+2+1} \bmod n = \\
&\ ((((((m_9 \odot m_8) \odot m_7) \odot m_6) \odot m_5) \odot m_3) \odot m_1) \odot m_0 = 20719.
\end{aligned}
$$

GAP, of course, simplifies calculations greatly:

```
gap> n:=24613; ; e:=1003;; m:=183;;
gap> PowerMod(m,e,n);
20719
```

(b) The private key d and the public key e satisfy the equation $ed = 1 \mod \phi(n)$, or equivalently, $ed + y\phi(n) = 1$. The Extended Euclidean algorithm gives a negative solution $d' = -533$, which is congruent to $d = d' + \phi(n) = 23767$ modulo $\phi(n)$.

```
gap> QuotientMod(1,e, 24300);
23767
```

(c). Bob can decrypt the cyphertext $c = 16935$ raising c to power d, $m = c^d \mod n = 16935^{23767}$. Applying the same procedure as in (a) we get $m = 135$. Alternatively we may use GAP as follows:

```
gap> PowerMod(16935,23767,n);
135
```

Solutions to Exercises of Sect. 2.4.2

1. The double encryption with e_1 and then with e_2 is the same as one encryption with $e = e_1 e_2$, since $c_2 \equiv c_1^{e_2} \equiv (m^{e_1})^{e_2} \equiv m^{e_1 e_2} \mod n$. As $\gcd(e_1 e_2, \phi(n)) = 1$, the product $e_1 e_2$ is another legitimate exponent. For decryption we can use exponent $d = d_1 d_2$, since $e_1 e_2 d_1 d_2 \equiv 1 \mod \phi(n)$ and $m \equiv (c_2^{d_2})^{d_1} \equiv c_2^{d_1 d_2} \mod n$. Thus double encryption is the same as a single encryption (with another exponent), and it does not increase security over single encryption.
2. Eve has to try to factorise n, and if it is successful, then calculate $\phi(n)$ and then Alice's private decryption exponent d.

```
gap> n:=30796045883;
30796045883
gap> e:=48611;
48611
gap> factors:=FactorsInt(n);
[ 163841, 187963 ]
gap> # So the factorisation was successful!
gap> phi:= (factors[1]-1)*(factors[2]-1);
30795694080
gap> d:=QuotientMod(1,e,phi);
20709535691
gap> # Eve inputs cryptotext in the list c
gap> c:=[ 5272281348, 21089283929, 3117723025, 26844144908, 22890519533,
 26945939925,  27395704341, 2253724391, 1481682985, 2163791130,
 13583590307, 5838404872, 12165330281, 28372578777, 7536755222 ];;
```

```
gap> # Now she decodes the crytpotext writing the output into the list m:
gap> m:=[0,0,0,0,0,0,0,0,0,0,0,0,0,0,0];;
gap> for i in [1..15] do
> m[i]:=PowerMod(c[i],d,n);
> od;
gap> m;
[ 2311301815, 2311301913, 2919293018, 1527311515, 2425162913, 1915241315,
  1124142431, 2312152830, 1815252835, 1929301815, 2731151524, 2516231130,
  1815231130, 1913292116, 1711312929 ]
gap> # This reads: "Mathematics is the queen of sciences and number theory
gap> # is the queen of mathematics KF GAUSS"
```

Solutions to Exercises of Sect. 2.4.3

1. (a) The test $(b, 91)$ reveals compositeness of 91 with probability 2/89 for the
   interval $b \in \{2, \ldots, 90\}$ as only $b = 7$ and $b = 13$ reveal that are divisors of
   91;
   (b) There are $n - \phi(n) - 1 = 18$ numbers b in $\{2, 3, \ldots, 90\}$ that are not relatively prime to $n = 91$ and will reveal compositeness of $n = 91$. The probability sought for is 18/89;
2. Since $\phi(91) = 72$, by Euler's theorem $5^{72} \equiv 1 \mod 91$. Hence $5^{90} \equiv 5^{18} \mod 91$
   and since $18 = 16 + 2 = (10010)_{(2)}$ we have to compute 5^2 and 5^{16}. We compute
   in $\mathbb{Z}_{91}$ as follows: $5^2 = 25$, $5^4 = 25^2 = 79$, $5^8 = 79^2 = 53$, $5^16 = 53^2 = 79$.
   Hence $5^{18} = 64$ and $5^{90} \not\equiv 1 \pmod{91}$. We know that 91 is composite by the
   third test.
3. We use Exercise 4 of Sect. 1.1 as follows:

$$2^{F_n} - 2 = 2\left(2^{F_n-1} - 1\right) = 2\left(2^{\left(2^{2^n}\right)} + 1 - 2\right) = 2(F_{2^n} - 2) = 2F_0 F_1 \ldots F_{2^n-1}.$$

   Since $n < 2^n - 1$, we get $F_n \mid 2^{F_n} - 2$ or $2^{F_n} \equiv 2 \mod F_n$. Hence, if F_n is not
   prime, it is a pseudoprime to base 2.
4. Here is a GAP program. It counts the number of integers i between 1 and n such
   that they are relatively prime to n and such that $i^{n-1} \equiv 1 \mod n$. If this number
   is equal to $\phi(n)$ and n is composite, the number n is a Carmichael number.

```
gap> n:=15841;;
gap> counter:=0;
0
gap> for i in [1..n] do
>      if GcdInt(i,n)=1 then
>              if PowerMod(i,n-1,n)=1 then
>              counter:=counter+1;
>              fi;
>      fi;
> od;
```

```
gap> counter=Phi(n);
true
gap> IsPrime(n);
false
```

We see that indeed $n = 15841$ is a Carmichael number.

5. We note that $561 = 3 \cdot 11 \cdot 17$. Let $\gcd(a, 561) = 1$. Then $\gcd(a, 3) = \gcd(a, 11) = \gcd(a, 17) = 1$. Hence by Fermat's little theorem we have $a^2 \equiv 1 \bmod 3$, $a^{10} \equiv 1 \bmod 11$ and $a^{16} \equiv 1 \bmod 17$. Thus $a^{560} = (a^2)^{280} \equiv 1 \bmod 3$, $a^{560} = (a^{10})^{56} \equiv 1 \bmod 11$, and $a^{560} = (a^{16})^{35} \equiv 1 \bmod 17$. By Chinese remainder theorem, these imply $a^{560} \equiv 1 \bmod 561$. This is true for all a relatively prime to 561. Hence 561 is a Carmichael number.

6. The output of the third pseudoprimality test is "inconclusive", because $7^{560} \equiv 1 \bmod 561$ Therefore 561 is pseudoprime to the base 7. Note that $561 = 3 \cdot 11 \cdot 17$ is not prime. Consider the following decomposition of $7^{560} - 1$:

$$7^{560} - 1 = (7^{35} - 1)(7^{35} + 1)(7^{70} + 1)(7^{140} + 1)(7^{280} + 1) \equiv 0 \quad \bmod 561.$$

Every expression in the brackets is not divisible by 561:

$$7^{35} - 1 \equiv 240 \quad \bmod 561, \; 7^{35} + 1 \equiv 242 \quad \bmod 561, \; 7^{70} + 1 \equiv 299 \quad \bmod 561,$$

$$7^{140} + 1 \equiv 167 \quad \bmod 561, \; 7^{280} + 1 \equiv 68 \quad \bmod 561.$$

So Miller–Rabin test will find 561 composite.

The product of expressions is divisible by 561, because of the presence of zero divisors 3, 11, 17 of $\mathbf{Z}_{561}$. More precisely, $7^{35} - 1$ is divisible by 3, $7^{35} + 1$ is divisible by 11 and $7^{280} - 1$ is divisible by 17.

7. We use the Rabin–Miller test with $b = 2$ to prove that $n = 294409$ is composite. Obviously $\gcd(n, 2) = 1$.

```
gap> n:=294409;
294409
gap> s:=0;;   t:=n-1;;
gap> while RemInt(t,2)=0 do
> s:=s+1;
> t:=t/2;
> od;
gap> s;
3
gap> t;
36801
gap> 2^s*t+1=n;
true
gap> # So our s is 3 and t is 36801.
gap> 2^t mod n;
```

```
512
gap> # i.e. b^t is not congruent to 1 and -1 mod n
gap> 2^(2*t) mod n;
262144
gap> # i.e. b^2t is not congruent to -1 mod n
gap> 2^(4*t) mod n;
1
gap> # b^4t is not congruent to -1 mod n
gap> Hence 2 is a Rabin-Miller witness that n is composite.
```

8. Suppose p is an odd prime and $n = p^k$, where $k > 1$. Then by Euler's theorem for any a relatively prime to p we have $a^{\phi(p^k)} = a^{p^k - p^{k-1}} \equiv 1 \bmod n$. If n is a Carmichael number, then also $a^{p^k-1} \equiv 1 \bmod n$. We have $\gcd(p^k - p^{k-1}, p^k - 1) = p - 1$. Hence we can find integers s and t such that $s(p^k - p^{k-1}) + t(p^k - 1) = p - 1$. Then we have

$$a^{p-1} = (a^{p^k - p^{k-1}})^s \cdot (a^{p^k - 1})^t \equiv 1 \bmod p^k.$$

As $k \geq 2$, this, in particular, implies

$$a^{p-1} \equiv 1 \bmod p^2.$$

This must be true for all a relatively prime to p^k and in particular for $a = p - 1$. But using binomial expansion we find that

$$(p-1)^{p-1} \equiv (p-1)p + 1 \equiv 1 - p \bmod p^2,$$

which is not 1 mod p^2, a contradiction. For $p = 2$ the argument is similar but easier. It is left to the reader.

Solutions to Exercises of Sect. 2.5

1. Here is the GAP program

```
gap> p:=100140889442062814140434711571;
100140889442062814140434711571
gap> g:=13;
13
gap> a:=123456789;
123456789
gap> # Alice received the following m from Bob:
gap> m:=926392043987322765326424904822;
926392043987322765326424904822
gap> # Alice has to send g^a to Bob:
gap> PowerMod(g,a,p);
```

```
49776677612066280125182950089
gap> # and take as the secret key m^a
gap> PowerMod(m,a,p);
16685041818541498009742672048
```

2. Firstly, lets find Bob's decryption exponent d_B. For this purpose compute $\phi(n_B) = (p_B - 1)(q_B - 1)$.

```
gap> pB:=8495789457893457345793;;
gap> qB:=9876345767834568934613;;
gap> phi:=(pB-1)*(qB-1);
83907354273436935926076409669190689472135 2704
```

We know that $e_B d_B \equiv 1 \mod \phi(n_B)$, so

```
gap> eB:=87697;;
gap> dB:=PowerMod(eB,-1,phi);
25995904256807890225566393955459263520507 1473
```

Bob needs to decrypt the message (m_1, s_1) using his private key d_B:

```
gap> m1:=119570441441889749705031896557386843883475475;;
gap> s1:=44368243049310248697807971950759679565772 9083;;
gap> nB:=pB*qB;;
gap> m:=PowerMod(m1,dB,nB);
 1234567890000000000987654321
gap> s:=PowerMod(s1,dB,nB);
12778075489862777682668011475893724302596 85176
```

Bob can verify that message is from Alice by computing $s^{e_A} \mod n_A$. If the message is from Alice, then the result will be m, which is indeed the case.

```
gap> nA:=17102470418361610970081806692519784151667 1277;;
gap> eA:=1571;;
gap> ms:=PowerMod(s,eA,nA);
1234567890000000000987654321
gap> ms=m;
 true
```

11.3   Solutions to Exercises of Chap. 3

Solutions to Exercises of Sect. 3.1.1

1. We have: .

   (a) $f \circ g(x) = \frac{1}{\sin x}$, and $g \circ f(x) = \sin \frac{1}{x}$;
   (b) $f \circ g(x) = \sqrt{e^x} = e^{x/2}$, and $g \circ f(x) = e^{\sqrt{x}}$.

2. Indeed, $R_\theta \circ R_{2\pi - \theta} = \mathrm{id}$ since this composition is a rotation through an angle of 2π.

3. We have $H \circ H = \mathrm{id}$.

4. Without loss of generality we may assume that our permutations fix elements $n - k + 1, n - k + 2, \ldots, n$. Any such permutation can be identified with permutations on the set $\{1, 2, \ldots, n-k\}$. Hence there are $(n - k)!$ of them.

5. We have

$$\sigma(1) = 5, \ \sigma(2) = 1, \ \sigma(3) = 6, \ \sigma(4) = 2, \ \sigma(5) = 7, \ \sigma(6) = 3, \ \sigma(7) = 8, \ \sigma(8) = 4,$$

   hence

$$\sigma = \begin{pmatrix} 1\,2\,3\,4\,5\,6\,7\,8 \\ 5\,1\,6\,2\,7\,3\,8\,4 \end{pmatrix}.$$

   The numbers in the last row are all different, hence this is a one-to-one mapping, hence a permutation.

6. Since for finite sets one-to-one implies onto, it is enough to prove that π is one-to-one. Suppose $\pi(k_1) = \pi(k_2)$. Then $3k_1 \equiv 3k_2 \bmod 13$, which implies $k_1 \equiv k_2 \bmod 13$ since 3 and 13 are coprime. Hence π is one-to-one and is a permutation.

7. Since $i^2 \equiv (13 - i)^2 \bmod 13$, the mapping is not one-to-one. We have $1^2 \bmod 13 = 1$, $2^2 \bmod 13 = 4$, $3^2 \bmod 13 = 9$, $4^2 \bmod 13 = 3$, $5^2 \bmod 13 = 12$, and $6^2 \bmod 13 = 10$. Therefore $2, 5, 6, 7, 8, 11$ are not in the range of τ, hence it is not onto.

8. We have

$$\rho = \begin{pmatrix} 1\,2\,3\,4\,5\,6 \\ 3\,4\,5\,6\,1\,2 \end{pmatrix}, \quad \rho^2 = \begin{pmatrix} 1\,2\,3\,4\,5\,6 \\ 5\,6\,1\,2\,3\,4 \end{pmatrix}, \quad \rho^3 = \mathrm{id}.$$

   Hence

$$\rho^{-1} = \rho^2 = \begin{pmatrix} 1\,2\,3\,4\,5\,6 \\ 5\,6\,1\,2\,3\,4 \end{pmatrix}.$$

   We also have $\tau^2 = \mathrm{id}$, hence $\tau^{-1} = \tau$.

9. $(\sigma\gamma)^{-1} = \begin{pmatrix} 1\,2\,3\,4\,5\,6\,7\,8\,9 \\ 9\,1\,3\,7\,6\,5\,8\,4\,2 \end{pmatrix}$. Calculating this with GAP:

```
gap> sigma:=PermList([2,4,5,6,1,9,8,3,7]);;
gap> gamma:=PermList([6,2,7,9,3,8,1,4,5]);
gap> mu:=sigma*gamma;;
gap> ListPerm(mu^-1);
[ 9, 1, 3, 7, 6, 5, 8, 4, 2 ]
```

thus confirming the result obtained.

10. We have to show that if f and g are permutations on $\{1, 2, \ldots, n\}$, then $f \circ g$ is also a permutation. It is enough to prove that it is one-to-one. Suppose not, then for two distinct elements $a, b \in \{1, 2, \ldots, n\}$ we have $f \circ g(a) = f \circ g(b)$. This means that $g(f(a)) = g(f(b))$. Since g is one-to-one, we conclude that $f(a) = f(b)$. However f is also one-to-one which implies $a = b$, a contradiction.

Solutions to Exercises of Sect. 3.1.3

1. Let $\pi = \sigma\tau$, where σ and τ are disjoint cycles. Suppose σ moves elements of the set I and τ moves elements of the set J. Since these cycles are disjoint, I and J have no elements in common. Let $K = \{1, 2, \ldots, n\} \setminus (I \cup J)$. Then $\pi(i) = \sigma(i)$ for $i \in I$, $\pi(j) = \tau(j)$ for $j \in J$ and $\pi(k) = k$ for $k \in K$. Exactly the same result we obtain for $\pi' = \tau\sigma$.

2. The calculation shows

$$\pi = \begin{pmatrix} 1\ 2\ 3\ \ 4\ \ 5\ 6\ 7\ 8\ 9\ 10\ 11\ 12 \\ 3\ 6\ 9\ 12\ 2\ 5\ 8\ 11\ 1\ \ 4\ \ 7\ \ 10 \end{pmatrix} = (1\ 3\ 9)(2\ 6\ 5)(4\ 12\ 10)(7\ 8\ 11).$$

This can be also done with GAP:

```
gap> s:=[1..12];;
gap> for i in [1..12] do
> s[i]:=3*s[i] mod 13;
> od;
gap> s;
[ 3, 6, 9, 12, 2, 5, 8, 11, 1, 4, 7, 10 ]
gap> PermList(s);
(1,3,9)(2,6,5)(4,12,10)(7,8,11)
```

3. $(1\ 4\ 3)(2\ 5)$.

4. $(\sigma\tau)^{-1} = \begin{pmatrix} 1\ 2\ 3\ 4\ 5\ 6\ 7\ 8\ 9 \\ 6\ 9\ 2\ 8\ 4\ 7\ 5\ 1\ 3 \end{pmatrix} = (1\ 6\ 7\ 5\ 4\ 8)(2\ 9\ 3).$

Solutions to Exercises of Sect. 3.1.4

1. (a) Since

$$\sigma = \begin{pmatrix} 1\ 2\ 3\ 4\ 5\ 6\ 7\ 8\ 9 \\ 5\ 3\ 6\ 7\ 1\ 2\ 8\ 9\ 4 \end{pmatrix} = (1\ 5)(2\ 3\ 6)(4\ 7\ 8\ 9),$$

the order of σ is $\mathrm{lcm}(2, 3, 4) = 12$.

(b) $\tau = (1\ 2)(2\ 3\ 4)(4\ 5\ 6\ 7)(7\ 8\ 9\ 10\ 11) = (1\ 3\ 5\ 6\ 8\ 9\ 10\ 11\ 7\ 4\ 2)$ so τ is actually a cycle of length 11 so its order is 11.

2. Let J be the Josephus permutation.

(a)
```
gap> j:=[ 3, 6, 9, 12, 15, 18, 21, 24, 27, 30, 33, 36, 39, 1, 5, 10, 14,
  19, 23, 28, 32, 37, 41, 7, 13, 20, 26, 34, 40, 8, 17, 29, 38, 11, 25, 2,
  22, 4, 35, 16, 31 ];;
gap> J:=PermList(j);
(1,3,9,27,26,20,28,34,11,33,38,4,12,36,2,6,18,19,23,41,31,17,14)(5,15)
(7,21,32,29,40,16,10,30,8,24)(13,39,35,25)(22,37)
```

(b) In which position did Josephus stand around the circle?

```
gap> 41^J;
31
```

So Josephus stood 31th.

(c) What is the order of the Josephus permutation?

```
gap> Order(J);
460
```

(d) Calculate σ^2 and σ^3.

```
gap> J^2;
(1,9,26,28,11,38,12,2,18,23,31,14,3,27,20,34,33,4,36,6,19,41,17)
(7,32,40,10,8)(13,35)(16,30,24,21,29)(25,39)
gap> J^3;
(1,27,28,33,12,6,23,17,3,26,34,38,36,18,41,14,9,20,11,4,2,19,31)(5,15)
(7,29,10,24,32,16,8,21,40,30)(13,25,35,39)(22,37)
```

3. The mapping $i \mapsto 13i \bmod 23$ is one-to-one mapping of S_{22} into itself since 13 and 23 are relatively prime. Now

```
gap> list:=[1..22];;
gap> for i in [1..22] do
> list[i]:=13*i mod 23;
> od;
gap> list;
[ 13, 3, 16, 6, 19, 9, 22, 12, 2, 15, 5, 18, 8, 21, 11, 1, 14, 4, 17, 7, 20, 10 ]
gap> PermList(list);
(1,13,8,12,18,4,6,9,2,3,16)(5,19,17,14,21,20,7,22,10,15,11).
```

The order of this permutation is $\operatorname{lcm}(11, 11) = 11$.

Solutions to Exercises of Sect. 3.1.5

1. The permutation corresponding to the shuffle will be

$$\begin{pmatrix} 1 & 2\ 3\ 4\ 5\ 6\ 7 & 8 & 9 & 10 & 11 & 12 & 13 & 14 & 15 \\ 11 & 2\ 3\ 4\ 5\ 6 & 12 & 13 & 14 & 15 & 1 & 7 & 8 & 9 & 10 \end{pmatrix} = (1\ 11)(7\ 12)(8\ 13)(9\ 14)(10\ 15).$$

The order of this permutation is 2 so repeating this shuffle twice will bring cards in the initial order.

2. We know that the interlacing shuffle is defined by the equation $\sigma(i) = 2i \bmod 105$. Thus we have

```
gap> lastrow:=[1..104];
[ 1 .. 104 ]
gap>  for i in [1..104] do
>  lastrow[i]:=2*i mod 105;
> od;
gap> s:=PermList(lastrow);
(1,2,4,8,16,32,64,23,46,92,79,53)(3,6,12,24,48,96,87,69,33,66,27,54)(5,10,20,
40,80,55)(7,14,28,56)(9,18,36,72,39,78,51,102,99,93,81,57)(11,22,44,88,71,37,
74,43,86,67,29,58)(13,26,52,104,103,101,97,89,73,41,82,59)(15,30,60)(17,34,68,
31,62,19,38,76,47,94,83,61)(21,42,84,63)(25,50,100,95,85,65)(35,70)(45,90,
75)(49,98,91,77)
gap> Order(s);
12
```

For this deck of cards this shuffle is very bad.

3. We assume, first, that each beetle has a number $1, 2, \ldots, n$ and each carries a little-coloured flag and all flags are of different colour. Suppose that when any two beetles meet, they exchange their flags. Now all flags move with the same constant speed without ever changing their directions, so after a certain time t they will occupy their initial positions. This means that the beetles will also occupy their initial positions but now in place of beetles $1, 2, \ldots, n$ we will find beetles $i_1, i_2, \ldots, i_n$. We will take time t as a unit of time. Hence, every unit time interval the beetles exchange places according to the permutation

$$\sigma = \begin{pmatrix} 1 & 2 & 3 & \ldots & n \\ i_1 & i_2 & i_3 & \ldots & i_n \end{pmatrix}.$$

When k units of time pass, they will exchange places according to the permutation σ^k. If σ is the product of m disjoint cycles of length $\ell_1, \ell_2, \ldots, \ell_m$, respectively, then σ^ℓ is the identity permutation for $\ell = \mathrm{lcm}(\ell_1, \ell_2, \ldots, \ell_m)$ being the order of σ. Hence after ℓ units of time all beetles will occupy their initial positions.

Solutions to Exercises of Sect. 3.1.6

1. Using equation (3.5) we get

$$(1\ 3\ 7)(5\ 8)(2\ 4\ 6\ 9) = (1\ 3)(1\ 7)(5\ 8)(2\ 4)(2\ 6)(2\ 9),$$

$$(1\ 3\ 7)(5\ 7\ 8)(2\ 3\ 4\ 6\ 9) = (1\ 3)(1\ 7)(5\ 7)(5\ 8)(2\ 3)(2\ 4)(2\ 6)(2\ 9).$$

Another representation can be obtained if we first represent the permutation as a product of disjoint cycles:

$$(1\ 3\ 7)(5\ 7\ 8)(2\ 3\ 4\ 6\ 9) = (1\ 4\ 6\ 9\ 2\ 3\ 8\ 5\ 7) = (1\ 4)(1\ 6)(1\ 9)(1\ 2)(1\ 3)(1\ 8)(1\ 5)(1\ 7).$$

2. It is odd. We can prove by induction that the product of an odd number of odd permutations is odd. Suppose this is true for any $2n - 1$ odd permutations. Consider the product $\Pi = \pi_1 \ldots \pi_{2n+1}$. We can write it as

$$\Pi = (\pi_1 \ldots \pi_{2n-1})(\pi_{2n}\pi_{2n+1}).$$

The induction hypothesis gives us that the first bracket is odd and the second by Theorem 3.1.7(ii) is even. Then Π is even by by Theorem 3.1.7(iii).

3. We must consider four cases, here we will consider only one: π is even and ρ is odd. By Theorem 3.1.7 (iv) ρ^{-1} is also odd. Then by Theorem 3.1.7 $\rho^{-1}\pi$ is odd and $\rho^{-1}\pi\rho$ is even. Hence π and $\rho^{-1}\pi\rho$ have the same parity. The other three cases are similar.

4. By the previous exercise $\pi^{-1}\rho^{-1}\pi$ has the same parity as ρ. Hence by Theorem 3.1.7 $\pi^{-1}\rho^{-1}\pi\rho$ is an even permutation.

5. If $n = 2k$, this permutation is a product

$$(1 \ n)(2 \ n - 1) \ldots (k \ k + 1).$$

If $n = 2k + 1$, then

$$(1 \ n)(2 \ n - 1) \ldots (k \ k + 2).$$

Hence the parity of this permutation is the parity of the number $\lfloor \frac{n}{2} \rfloor$.

Solutions to Exercises of Sect. 3.1.7

1. The two positions differ only by a switch of neighbouring squares 10 and 14. In this case the corresponding permutations will be of different parities, hence only one of them is realizable.

2. Calculating the corresponding permutations in GAP:

```
gap> first:=[1,3,2,4,6,5,7,8,9,13,15,11,14,10,12,16];
[ 1, 3, 2, 4, 6, 5, 7, 8, 9, 13, 15, 11, 14, 10, 12, 16 ]
gap> s:=PermList(first);
(2,3)(5,6)(10,13,14)(11,15,12)
```

This one is even, hence the first position is realisable.

```
gap> second:=[13,16,5,3,9,2,7,10,1,15,14,8,12,11,6,4];
[ 13, 16, 5, 3, 9, 2, 7, 10, 1, 15, 14, 8, 12, 11, 6, 4 ]
gap> t:=PermList(second);
(1,13,12,8,10,15,6,2,16,4,3,5,9)(11,14)
```

This one is odd, hence the second position is not realisable.

Solutions to Exercises of Sect. 3.2.1

1. It is a binary operation since, if a and b are two nonzero real numbers, $a \star b = a : b$ is also a nonzero real number. We have

$$(a \star b) \star c = \frac{a}{bc}, \qquad a \star (b \star c) = \frac{ac}{b},$$

which are different if $|c| \neq 1$. Hence this operation is not associative.

2. If e is a neutral element, then we must have $e \star a = a \star e = a$ for all $a \in \mathbb{R}_+$. This means $e^a = a^e = a$ for all a. Since for $e > 1$ the function e^x grows faster than x^e and for $e < 1$ the function e^x grows slower than x^e, the only option is $e = 1$ which is also impossible.

3. Obviously the identity element 1 is in $\mathbb{C}_n$. If z_1 and z_2 are any two roots of unity, that is $|z_1| = |z_2| = 1$, then $|z_1 z_2| = |z_1||z_2| = 1$, so their product also lies in $\mathbb{C}_n$. Hence multiplication is an algebraic operation on $\mathbb{C}_n$. Associative law for it follows from the properties of multiplication in $\mathbb{C}$. Also, if z is a root of unity, $|z^{-1}| = |z|^{-1} = 1$ and hence z^{-1} also belongs to $\mathbb{C}_n$. Hence $\mathbb{C}_n$ is a group relative to the operation of multiplication. (It is a subgroup of the multiplicative group $\mathbb{C}^*$ of $\mathbb{C}$). We know there are exactly n roots of degree n of unity:

$$\psi_i = \cos \frac{2i\pi}{n} + \sin \frac{2i\pi}{n}, \qquad i = 0, 1, 2, \ldots, n-1.$$

4. A matrix A is invertible if $\det(A) \neq 0$. If A, B are two invertible matrices, then $\det(AB) = \det(A)\det(B) \neq 0$ and hence AB is also invertible. The inverse of an invertible matrix is also invertible, and the identity matrix I_n is invertible. Thus $GL_n(\mathbb{R})$ contains the identity, inverses and its multiplication is associative. Hence $GL_n(\mathbb{R})$ is a group.

5. We have to use the associative law twice:

$$(g_1 g_2)(g_3 g_4) = (g_1 g_2)g_3)g_4 = (g_1(g_2 g_3))g_4.$$

The following are all possible arrangements of brackets on the product $g_1 g_2 g_3 g_4$:

$$g_1((g_2 g_3)g_4)) = g_1(g_2(g_3 g_4)) = (g_1 g_2)(g_3 g_4) = (g_1 g_2)g_3)g_4 = (g_1(g_2 g_3))g_4.$$

Let us now prove that all arrangements of brackets on $g_1 g_2 \ldots g_n$ will give us the same element as $(g_1 g_2 \ldots g_n)_r = (\ldots ((g_1 g_2)g_3) \ldots)g_n$. Suppose now that the bracket arrangement is arbitrary, this gives us the product uv, where u is $g_1 g_2 \ldots g_k$ with some arrangement of brackets and v is $g_{k+1} g_{k+2} \ldots g_n$ also with some arrangement of brackets. We have $|u| = k$ and $|v| = n - k$, where by $|w|$ we denote the number of group elements involved.

If $|v| = 1$, then the statement follows from the induction hypothesis. If $|v| > 1$, then we have $v = v_1 v_2$, where $|v_1| < |v|$ and $|v_2| < |v|$. Then we apply the associative law as follows:

$$uv = u(v_1 v_2) = (uv_1)v_2 = u'v',$$

where $|v'| < |v|$. By induction hypothesis $uv = (g_1 g_2 \ldots g_n)_r$ as required.

Solutions to Exercises of Sect. 3.2.2

1. Since in $\mathbb{Z}_n$

$$\text{ord}(i) = \frac{n}{\gcd(i,n)},$$

we calculate that the orders of the elements 5, 1331 and 594473 will be 16427202, 12342 and 7986, respectively. Indeed,

```
gap> n:=16427202;; i1:=5;; i2:=1331;; i3:=594473;;
gap> order1:=n/GcdInt(i1,n);order2:=n/GcdInt(i2,n);order3:=n/GcdInt(i3,n);
16427202
12342
7986
```

2. Using the formula as in the previous exercise we see that for having order 7 the element i must satisfy $\gcd(i, 84) = 84/7 = 12$. We have six such elements: 12, 24, 36, 48, 60, 72.

3. The order of $i \in \mathbb{Z}_n$ is calculated as

$$\text{ord}(i) = \frac{n}{\gcd(i,n)} = 87330619392.$$

Here is the calculation:

```
gap> n:=563744998038700032;; i:=41670852902912;;
gap> gcd:=GcdInt(n,i);
6455296
gap> order:=n/gcd;
87330619392
```

4. This is a group of order 4 but each nonzero element has order 2. Hence all cyclic subgroups have order 2 and the group is not cyclic.

5. We know that $\sigma_n(i) = 2i \bmod 2n + 1$ Suppose $\sigma_n^k = \text{id}$ for some k. Then $2^k i \equiv i \bmod 2n + 1$ for all i including those which are relatively prime to $2n + 1$. By Lemma 1.3.2(d) this is equivalent to $2^k \equiv 1 \bmod 2n + 1$. Hence the order of σ_n is equal to the order of 2 in $\mathbb{Z}_{2n+1}^*$.

Solutions to Exercises of Sect. 3.2.3

1. We have $gg^{-1} = e$, where e is the identity element of G. We know from the proof of Theorem 3.2.3 that $\sigma(e) = \epsilon$, where ϵ is the identity of H. Applying σ to $gg^{-1} = e$ and using the property that $\sigma(gh) = \sigma(g)\sigma(h)$ we obtain $\sigma(g)\sigma(g^{-1}) = \epsilon$. This means that $\sigma(g^{-1}) = \sigma(g)^{-1}$.

2. In Exercise 1 of Sect. 3.2.1 we defined roots of unity ψ_i. It is straightforward to check that

$$\psi_i \psi_j = \psi_{i \oplus j}.$$

This makes the mapping $i \mapsto \psi_i$ an isomorphism.

3. We define a mapping $\tau \colon \mathbb{C}^* \to G$ as

$$\tau(a + bi) = \begin{bmatrix} a & -b \\ b & a \end{bmatrix}.$$

This mapping is one-to-one and onto. Also let $z_1 = a + bi$ and $z_2 = c + di$. Then

$$z_1 z_2 = (a + bi)(c + di) = (ac - bd) + (ad + bc)i$$

and

$$\begin{bmatrix} a & -b \\ b & a \end{bmatrix} \begin{bmatrix} c & -d \\ d & c \end{bmatrix} = \begin{bmatrix} ac - bd & -ad - bc \\ ad + bc & ac - bd \end{bmatrix},$$

which means $\tau(z_1 z_2) = \tau(z_1)(z_2)$ and τ is an isomorphism.

4. Both 191 and 193 are primes, hence $|G_1| = 190$ and $|G_2| = 192$. The second number is not divisible by 19 and the first is. So G_2 has no elements of order 19 but G_1 has. Let g be a generator of G_1. Then element g^k will have order 19 if $\gcd(k, 190) = 10$. Hence elements g^k will have order 19 for $k = 1, \ldots, 18$, in total 18 such elements.

5. $\operatorname{ord}(2^{150}) = \dfrac{210}{\gcd(150, 210)} = \dfrac{210}{30} = 7.$

6. $\operatorname{ord}(264^{72}) = \dfrac{270}{\gcd(270, 72)} = \dfrac{270}{18} = 15.$

Solutions to Exercises of Sect. 3.2.4

1. If $\det(A) = \det(B) = 1$, then $\det(AB) = \det(A)\det(B) = 1$, also if $\det(A) = 1$, then $\det(A^{-1}) = \det(A)^{-1} = 1$. Finally we notice that $\det(I_n) = 1$. All requirements of a subgroup are satisfied: $SL_n(\mathbb{R})$ is closed under the multiplication, under inverses, and contains the identity element. Hence $SL_n(\mathbb{R})$ is a subgroup of $GL_n(\mathbb{R})$.

2. If m is a divisor of n, then $z^m = 1$ implies $z^n = 1$. Hence $\mathbb{C}_m \subseteq \mathbb{C}_n$. Since $\mathbb{C}_m$ is known to be a group on its own, it will be a subgroup of $\mathbb{C}_n$.

3. An element g^i is a generator of $< g >$ if $\operatorname{ord}(g^i) = n$. Since

$$\operatorname{ord}(g^i) = \frac{n}{\gcd(i, n)}$$

for g^i to be a generator, it is necessary and sufficient to have $\gcd(i, n) = 1$. We have exactly $\phi(n)$ such numbers i.

4. Suppose that for no element $1 \neq g \in G$ we have $g^2 = 1$. Then for no element $g \neq 1$ of G we have $g = g^{-1}$. Hence we can split the whole set $G \setminus \{1\}$ into disjoint pairs $\{g, g^{-1}\}$. Then $G \setminus \{1\}$ has an even number of elements and G has an odd number of elements. This is a contradiction.

5. Suppose G is a subgroup of $\mathbb{C}^*$ and $|G| = n$. Then by Corollary 3.2.3 $g^n = 1$ for every $g \in G$. This implies $G = \mathbb{C}_n$ since elements of G must then coincide with n roots of unity in $\mathbb{C}$.

Solutions to Exercises of Sect. 3.3.1

1. The discriminant of the cubic $X^3 + 4X + 11$ is zero, so the first equation does not define an elliptic curve, the discriminant of $X^3 + 6X + 11$ is 3, so the second equation does define an elliptic curve over $\mathbb{Z}_{13}$.
2. Compare coefficients of the polynomials in the right-hand side and in the left-hand side of the equation.
3. Direct calculation.
4. (a) Direct calculation.
   (b) With every point (x, y) there must be also the point $-(x, y) = (x, -y)$. This gives us another five points on E, namely $(1, 6), (2, 4), (3, 6), (4, 5), (6, 5)$. Also the point at infinity ∞.
   (c) $-(2, 3) = (2, 4), 2(4, 2) = (6, 5), (1, 1) + (3, 1) = (3, 6)$.
   (d) GAP shows that we found all points on E:

   ```
   gap> G:=EllipticCurveGroup(1,-1,7);
   EllipticCurveGroup(1,-1,7)
   gap> AsList(G);
   [ ( 1, 1 ), ( 1, 6 ), ( 2, 3 ), ( 2, 4 ), ( 3, 1 ), ( 3, 6 ), ( 4, 2 ),
   ( 4, 5 ), ( 6, 2 ), ( 6, 5 ), infinity ].
   ```

5. (a) Generating this elliptic curve with GAP:

   ```
   gap> G:=EllipticCurveGroup(5,1,13);
   EllipticCurveGroup(5,1,13)
   gap> AsList(G);
   [ ( 0, 1 ), ( 0, 12 ), ( 3, 2 ), ( 3, 11 ), ( 6, 0 ), ( 11, 3 ),
   ( 11, 10 ), infinity ]
   gap> Order(G);
   8
   ```

   (b) Since the order of E is 8, LagrangeÕs theorem tells us that the order of P is a factor of 8, i.e., 2 or 4 or 8. Then $2(0, 1) = (0, 1) + (0, 1) = (3, 11)$, $4(0, 1) = (3, 11) + (3, 11) = (6, 0)$, and $8(0, 1) = (6, 0) + (6, 0) = \infty$, so the order of P is 8 and the group G is a cyclic group with P as generator.

6.
   ```
   gap> p:=46301;;
   gap> G:=EllipticCurveGroup(7,11,p);
   EllipticCurveGroup(7,11,46301)
   gap> Order(G);
   46376
   gap> IsCyclic(G);
   true
   ```

Solutions to Exercises of Sect. 3.3.2

1. The complete addition table is

+	∞	(2,0)	(3,2)	(3,3)	(4,1)	(4,4)
∞	∞	(2,0)	(3,2)	(3,3)	(4,1)	(4,4)
(2,0)	(2,0)	∞	(4,1)	(4,4)	(3,2)	(3,3)
(3,2)	(3,2)	(4,1)	(3,3)	∞	(4,4)	(2,0)
(3,3)	(3,3)	(4,4)	∞	(3,2)	(2,0)	(4,1)
(4,1)	(4,1)	(3,2)	(4,4)	(2,0)	(3,3)	∞
(4,4)	(4,4)	(3,3)	(2,0)	(4,1)	∞	(3,2)

Firstly we note that $(4, 4) = -(4, 1)$ so they have the same order. We have

$$2(4, 1) = (3, 3), \ 3(4, 1) = (2, 0), \ 4(4, 1) = (3, 2), \ 5(4, 1) = (4, 4), \ 6(4, 1) = \infty$$

so the order of both $(4, 1)$ and $(4, 4)$ is 6.

2. In any field $x^2 = (-x)^2$, hence the following are all quadratic residues of $\mathbb{Z}_{17}$:

$$1^2 = 1, \ 2^2 = 4, \ 3^2 = 9, \ 4^2 = 16, \ 5^2 = 8, \ 6^2 = 2, \ 7^2 = 15, \ 8^2 = 13.$$

The answer is $\{1, 2, 4, 8, 9, 13, 15, 16\}$.

3. The number of points N on such elliptic curve by Hasse's theorem will be in the range

$$2012 - 2\sqrt{2011} \le N \le 2012 + 2\sqrt{2011}$$

As $\sqrt{2011} \approx 44.84417465$, we see that $1923 \le N \le 2101$.

4. We have three cases to consider:

   (a) The product of two quadratic residues is a quadratic residue. Indeed, if $g = g_1^2$ and $h = h_1^2$, then $gh = (g_1 h_1)^2$. The inverse of a quadratic residue is again a quadratic residue. Indeed, if $g = h^2$, then $g^{-1} = (h^{-1})^2$.

   (b) Suppose now that the product of a quadratic residue g and a quadratic non-residue h is a quadratic residue k. Then $h = g^{-1}k$ is a quadratic residue due to the two observations made above, a contradiction. Hence the product of a quadratic residue and a quadratic non-residue is a quadratic non-residue.

   (c) Let g be an arbitrary quadratic non-residue. Let us consider the function $\sigma: h \mapsto hg$ from $\mathbb{Z}_p^*$ to itself. Since g is invertible, this is a permutation of $\mathbb{Z}_p^*$. Let $q = (p - 1)/2$. Then Theorem 3.3.3 tells us that there are q quadratic residues, let us denote their set as R and q quadratic non-residues, let us denote their set N. Then we have shown that σ maps R into N. Since σ is invertible and R and N have the same cardinality, then σ maps R onto N. In such a case σ must map N onto R which means that the product of any quadratic non-residue and g is quadratic residue. Since g is arbitrary we have proved that the product of any two quadratic non-residues is a quadratic residue.

5. Since $\mathbb{Z}_p$ is a field, for every nonzero $c \in \mathbb{Z}_p$ the equation $cx = a$ has a solution c' in $\mathbb{Z}_p$ which is also nonzero. Also $c' \neq c$ since a is a non-residue. Thus all nonzero elements of $\mathbb{Z}_p$ are split into disjoint pairs $(c_1, c_1'), \ldots, (c_{(p-1)/2}, c_{(p-1)/2}')$ such that $c_i c_i' = a$. Then by Wilson's theorem

$$a^{\frac{p-1}{2}} = \prod_{i=1}^{(p-1)/2} c_i \cdot c_i' = (p-1)! = -1.$$

6. We generate numbers at random and then use Euler's criterion to check if it is a quadratic non-residue or not. In fact the first attempt gives us a non-residue,the second gives a quadratic residue:

```
gap> a:=Random([1..2^28]);
153521494
gap> PowerMod(a,(p-1)/2,p);
35933408596862283104196018859804366106538872695907983 6
# This is actually p-1 so 153521494 is a non-residue.
gap> a:=Random([1..2^28]);
199280309
gap> PowerMod(a,(p-1)/2,p);
1
# And this shows that 199280309 is a quadratic residue.
```

Solutions to Exercises of Sect. 3.3.3

1. $1729 = 11011000001_2$, so GAP will first perform 10 additions to calculate $2 \cdot P, 4 \cdot P, \ldots 1024 \cdot P$, and then a further 4 additions to compute the sum $P + 64 \cdot P + 128 \cdot P + 512 \cdot P + 1024 \cdot P$; 14 additions in total.

Solutions to Exercises of Sect. 3.4.1

1. We generate numbers at random and then use Euler's criterion to check if it is a quadratic non-residue or not. In fact the first attempt gives us a non-residue,the second gives a quadratic residue:

```
gap> a:=Random([1..2^28]);
153521494
gap> PowerMod(a,(p-1)/2,p);
35933408596862283104196018859804366106538872695907983 6
# This is actually p-1 so 153521494 is a non-residue.
gap> a:=Random([1..2^28]);
199280309
gap> PowerMod(a,(p-1)/2,p);
1
# And this shows that 199280309 is a quadratic residue.
```

```
gap> b:=RootMod(a,p);
286534778672701806664621728123564904392266164296221884
gap> a=b^2 mod p;
true
# So indeed b is a square root of a.
```

2. We make the following steps:

```
   CHRISTMAS --> [ CHR, IST, MAS ] --> [ 131828, 192930, 231129 ] -->

   [ (1318281, 15879309), (1929301, 3765260), (2311294, 6775980) ].
```

Here is the calculation:

```
gap> p:=17487707;;
gap> m:=[ 131828, 192930, 231129 ];;
gap> x1:=m[1]*10;
1318280
gap> f1:=(x1^3+123*x1+456) mod p;
7287640
gap> RootMod(f1,p);
fail
gap> x1:=x1+1;
1318281
gap> f1:=(x1^3+123*x1+456) mod p;
5117601
gap> RootMod(f1,p);
15879309
gap> x2:=m[2]*10;
1929300
gap> f2:=(x2^3+123*x2+456) mod p;
2898698
gap> RootMod(f2,p);
fail
gap> x2:=x2+1;
1929301
gap> f2:=(x2^3+123*x2+456) mod p;
3728942
gap> RootMod(f2,p);
3765260
gap> x3:=m[3]*10;
2311290
gap> f3:=(x3^3+123*x3+456) mod p;
14098022
gap> RootMod(f3,p);
```

```
fail
gap> x3:=x3+1;
2311291
gap> f3:=(x3^3+123*x3+456) mod p;
16049134
gap> RootMod(f3,p);
fail
gap> x3:=x3+1;
2311292
gap> f3:=(x3^3+123*x3+456) mod p;
14380285
gap> RootMod(f3,p);
fail
gap> x3:=x3+1;
2311293
gap> f3:=(x3^3+123*x3+456) mod p;
9091481
gap> RootMod(f3,p);
fail
gap> x3:=x3+1;
2311294
gap> f3:=(x3^3+123*x3+456) mod p;
182728
gap> RootMod(f3,p);
6775980
```

Solutions to Exercises of Sect. 3.4.2

1. I will first show how the message was encrypted and then show how to decrypt it.
   You have to do the opposite: first decrypt the message and then encrypt a message
   of your own.

```
gap> Read("elliptic.gd");
gap> Read("elliptic.gi");
gap> # Defining the curve:
EllipticCurveGroup(0,12345,95701)
gap> P:=Random(G);
( 91478, 65942 )
gap> # Encoding the message "I'm nobody. Who are you?"
gap> M:=[0,0,0,0,0,0,0,0,0,0,0,0];
[ 0, 0, 0, 0, 0, 0, 0, 0, 0, 0, 0, 0 ]
gap> M[1]:=EllipticCurvePoint(FamilyObj(P),[1942,37617]);
( 1942, 37617 )
gap> M[2]:=EllipticCurvePoint(FamilyObj(P),[2341,44089]);
( 2341, 44089 )
gap> M[3]:=EllipticCurvePoint(FamilyObj(P),[2425,89535]);
( 2425, 89535 )
gap> M[4]:=EllipticCurvePoint(FamilyObj(P),[1225,46279]);
( 1225, 46279 )
```

```
gap> M[5]:=EllipticCurvePoint(FamilyObj(P),[1435,60563]);
( 1435, 60563 )
gap> M[6]:=EllipticCurvePoint(FamilyObj(P),[43410,66195]);
( 43410, 66195 )
gap> M[7]:=EllipticCurvePoint(FamilyObj(P),[3318,58656]);
( 3318, 58656 )
gap> M[8]:=EllipticCurvePoint(FamilyObj(P),[25413,63045]);
( 25413, 63045 )
gap> M[9]:=EllipticCurvePoint(FamilyObj(P),[1128,14737]);
( 1128, 14737 )
gap> M[10]:=EllipticCurvePoint(FamilyObj(P),[1541,72018]);
( 1541, 72018 )
gap> M[11]:=EllipticCurvePoint(FamilyObj(P),[3525,29201]);
( 3525, 29201 )
gap> M[12]:=EllipticCurvePoint(FamilyObj(P),[3145,46983]);
( 3145, 46983 )
gap> M;
[ ( 1942, 37617 ), ( 2341, 44089 ), ( 2425, 89535 ), ( 1225, 46279 ),
  ( 1435, 60563 ), ( 43410, 66195 ), ( 3318, 58656 ), ( 25413, 63045 ),
  ( 1128, 14737 ), ( 1541, 72018 ), ( 3525, 29201 ), ( 3145, 46983 ) ]
gap> # In M[6] and M[8] we had to add an additional fifth digit in order
gap> # to get a point.

gap> # These are in the public domain:
gap> Q:=EllipticCurvePoint(FamilyObj(P),[88134,77186]);
( 88134, 77186 )
gap> QkA:=EllipticCurvePoint(FamilyObj(P),[27015, 92968]);
( 27015, 92968 )

gap> # All Bob needs for encryption is QkA which is in the public domain.
gap> C:=[0,0,0,0,0,0,0,0,0,0,0,0];
[ 0, 0, 0, 0, 0, 0, 0, 0, 0, 0, 0, 0 ]
gap> for i in [1..12] do
> C[i]:=[P,P];
> s:=Random([1..(p-1)]);
> C[i][1]:=Q^s;
> C[i][2]:=M[i]*(QkA)^s;
> od;
gap> C;
[ [ ( 87720, 6007 ), ( 59870, 82101 ) ], [ ( 34994, 7432 ), ( 36333, 86213 ) ],
  [ ( 50702, 2643 ), ( 33440, 56603 ) ], [ ( 34778, 12017 ), ( 81577, 501 ) ],
  [ ( 93385, 52237 ), ( 38536, 21346 ) ], [ ( 63482, 12110 ), ( 70599, 87781 ) ],
  [ ( 16312, 46508 ), ( 62735, 69061 ) ], [ ( 64937, 58445 ), ( 41541, 36985 ) ],
  [ ( 40290, 45534 ), ( 11077, 77207 ) ], [ ( 64001, 62429 ), ( 32755, 18973 ) ],
  [ ( 81332, 47042 ), ( 35413, 9688 ) ], [ ( 5345, 68939 ), ( 475, 53184 ) ] ]

gap> # Now Alice decrypts this message using her private key kA
gap> kA:=373;
373
gap> for i in [1..12] do
> M1[i]:=C[i][2]*((C[i][1])^kA)^-1;
> od;
gap> M1;
[ ( 1942, 37617 ), ( 2341, 44089 ), ( 2425, 89535 ), ( 1225, 46279 ),
  ( 1435, 60563 ), ( 43410, 66195 ), ( 3318, 58656 ), ( 25413, 63045 ),
  ( 1128, 14737 ), ( 1541, 72018 ), ( 3525, 29201 ), ( 3145, 46983 ) ]
gap> # Here we have to ignore any fifth digit in x-component which occurs.
gap> # Alice reads the message as "I'm nobody. Who are you?" which is the
gap> # first line of the following poem by Emily Dickinson:

I'm Nobody. Who are you?
Are you - Nobody - Too?
```

```
Then there's a pair of us?
Don't tell! They'd advertise - you know!

How dreary - to be - Somebody!
How public - like a Frog -
To tell one's name - the livelong June -
To an admiring Bog!
```

11.4   Solutions to Exercises of Chap. 4

Solutions to Exercises of Sect. 4.1.1

1. A negative of 1 does not exist in $\mathbb{Q}^+$ so F3 is violated.
2. The (multiplicative) inverse of 2 does not exist in $\mathbb{Z}$, hence F9 is violated.
3. To prove that $\mathbb{Q}(\sqrt{2})$ is a field we have to show that it contains 1, which is true since $1 = 1 + 0\sqrt{2}$, that $\mathbb{Q}(\sqrt{2})$ is closed under the multiplication, and that a^{-1} is in $\mathbb{Q}(\sqrt{2})$ whenever a is. Suppose $a = x + y\sqrt{2}$ and $b = x' + y'\sqrt{2}$. Then

$$ab = (x + y\sqrt{2})(x' + y'\sqrt{2}) = (xy + 2x'y') + (xy' + x'y)\sqrt{2}$$

and $\mathbb{Q}(\sqrt{2})$ is closed under the multiplication. To show that $a = x + y\sqrt{2}$ has an inverse in $\mathbb{Q}(\sqrt{2})$, we calculate the product

$$(x + y\sqrt{2})(x - y\sqrt{2}) = (x^2 - 2y^2)$$

and observe that it is in $\mathbb{Q}$. We also observe that for $a \neq 0$ we have $x^2 - 2y^2 \neq 0$, since $\sqrt{2}$ is irrational. Thus we have $aa^{-1} = 1$ for $a^{-1} = \frac{x}{x^2 - 2y^2} - \frac{y}{x^2 - 2y^2}\sqrt{2}$.
4. In $\mathbb{Q}(\sqrt{2})$ we will have $(2 - \sqrt{3})^{-1} = 2 + \sqrt{3}$. Further, $x = (2 + \sqrt{3})(1 + \sqrt{3}) = 5 + 3\sqrt{3}$.
5. All standard linear algebra techniques for finding such a solution works: you can find the solution by Gaussian elimination or by calculating the inverse of the matrix of this system of linear equations. Here we show how this can be solved with GAP:

```
gap> A:=[[3,1,4],[1,2,1],[4,1,4]];
[ [ 3, 1, 4 ], [ 1, 2, 1 ], [ 4, 1, 4 ] ]
gap> b:=[1,2,4];
[ 1, 2, 4 ]
gap> Determinant(A) mod 5;
3
# Hence the matrix is invertible.
gap> A^-1 mod 5;
[ [ 4, 0, 1 ], [ 0, 2, 2 ], [ 1, 2, 0 ] ]
# The solution can now be calculated as
```

```
gap> A^-1*b mod 5;
[ 3, 2, 0 ]
# Thus we have a unique solution x=3, y=2, z=0.
```

Solutions to Exercises of Sect. 4.1.2

4. We have to check, firstly, that S is an abelian group. We will use properties of the transpose. If A and B are symmetric, then $(A + B)^T = A^T + B^T = A + B$ and their sum is also symmetric. Hence the set of symmetric matrices is closed under the addition. It is also easy to see that the zero matrix is symmetric and, if A is symmetric, then $-A$ is also symmetric. Also we have to note that if A is symmetric, then λA, where λ is a scalar is also symmetric. The dimension of S will be $\frac{1}{2}n(n + 1)$. The basis or F can be taken consisting of all diagonal matrix units E_{ii} and of all matrices $E_{ij} + E_{ji}$ for $i \neq j$.

5. Let us prove first that $< V, \oplus >$ is an abelian group. Associative law for $\oplus$ follows from the associative law of multiplication. 1 is obviously the zero of this abelian group since $u \oplus 1 = u$. Also $u \oplus u^{-1} = 1$, hence $-u = u^{-1}$.
   Let us check further axioms. We have

$$1 \odot u = u^1 = u,$$
$$(ab) \odot u = u^{ab} = ((u^b)^a = a \odot (b \odot u),$$
$$(a + b) \odot u = u^{a+b} = u^a u^b = a \odot u \oplus b \odot u,$$
$$a \odot (u \oplus v) = (uv)^a = u^a v^a = a \odot u \oplus a \odot v.$$

Thus $< V, \oplus, \odot >$ is a vector space.

Solutions to Exercises of Sect. 4.1.3

1. We use GAP to factorise n_1 and n_2:

```
gap> n1:=449873499879757801;;
gap> n2:=449873475733618561;;
gap> FactorsInt(n1);
[ 670726099, 670726099 ]
gap> FactorsInt(n2);
[ 12347, 12347, 54323, 54323 ]
```

We see that $n_1 = 670726099^2$ is a power of a prime $p = 670726099$. Hence $\mathbb{Z}_p$ will be contained in $GF(n_1)$ and the dimension of $GF(n_1)$ over $\mathbb{Z}_p$ will be $\log_p(n_1) = 2$.
Also $n_2 = 12347^2 \cdot 54323^2$ is not a power of a prime. Hence $GF(n_2)$ does not exists.

2. Obviously the zero is a solution to this equation. Suppose $a \neq 0$. Then a belongs
   to the multiplicative group F^* of F which has $q - 1$ element. By Corollary 3.2.3
   $a^{q-1} = 1$ and, in particular, $a^q = a$.

Solutions to Exercises of Sect. 4.2.1

1. Let g, h, k be elements of G of orders $3, 5, 7$, respectively. Then by Corollary 4.2.1
   for the product $a = ghk$ we have $\text{ord}\,(a) = 3 \cdot 5 \cdot 7 = 105$. Since $\text{ord}\,(a) = |G|$,
   the group $G = < a >$ is cyclic.
2. Let g, h, k be elements of a finite abelian group G of orders $183618, 131726,$
   127308, respectively. We have to use g, h, k to construct an element x of G of
   order 1018264646281. The prime factorisations of the numbers involved are as
   follows:

```
gap> ordx:=1018264646281;
1018264646281
gap> FactorsInt(ordx);
[ 97, 97, 101, 101, 103, 103 ]
gap> ordg:=183618;
183618
gap> FactorsInt(ordg);
[ 2, 3, 3, 101, 101 ]
gap> ordh:=131726;
131726
gap> FactorsInt(ordh);
[ 2, 7, 97, 97 ]
gap> ordk:=127308;
127308
gap> FactorsInt(ordk);
[ 2, 2, 3, 103, 103 ]
```

As the order of the element x sought for is $97^2 \cdot 101^2 \cdot 103^2$ we need to construct
elements $x_1, x_2, x_3 \in G$ of orders $97^2, 101^2, 103^2$, respectively. Since orders of
g, h, k are $18 \cdot 101^2, 14 \cdot 97^2, 12 \cdot 103^2$, we can take $x_1 = h^{14}, x_2 = g^{18}, x_3 = k^{12}$. Then $x = x_1 x_2 x_3$.

Solutions to Exercises of Sect. 4.2.2

1. Let a be the generator of $\mathbb{Z}_p^*$. We factorise $p - 1$, which is the order of $\mathbb{Z}_p^*$ and
   both 11561 and 58380.

```
gap> p:=192837481;;
gap> FactorsInt(p-1);
[ 2, 2, 2, 3, 5, 11, 139, 1051 ]
gap> FactorsInt(11561);
[ 11, 1051 ]
```

```
gap> FactorsInt(58380);
[ 2, 2, 3, 5, 7, 139 ]
```

We see that 58380 does not divide $p - 1$ hence by Corollary 3.2.3 an element of this order cannot be in $\mathbb{Z}_p^*$.

Let us calculate $(p - 1)/11561 = 16680$. Then $\text{ord}(a^{16680}) = 11561$ by Lemma 4.2.4. So an element of order 11561 exists.

2. Let us divide n by m with remainder: $n = qm + r$ with $0 \le r < m$. Then

$$p^n - 1 = p^{n-m}(p^m - 1) + (p^{n-m} - 1),$$

that is $\gcd(p^n - 1, p^m - 1) = \gcd(p^{n-m} - 1, p^m - 1)$ from which we obtain $\gcd(p^n - 1, p^m - 1) = \gcd(p^r - 1, p^n - 1)$. This means that when we divide $p^n - 1$ by $p^m - 1$ the remainder will be $p^r - 1$. The statement follows from here.

3. Suppose now $GF(p^n)$ contains a subfield F of cardinality p^m. All elements of the first field satisfy the equation $x^{p^n-1} = 1$ and elements of the second satisfy $x^{p^m-1} = 1$. Let g be a primitive element of the smaller field F. Then $\text{ord}(g) = p^m - 1$. Since it lies in the multiplicative subgroup $GF(p^n)^*$ which cardinality is $p^n - 1$ the order of g must divide $p^n - 1$ (Corollary 3.2.3). Hence $p^m - 1$ divides $p^n - 1$ and $m \mid n$ by Exercise 14 of Section 1.2.2.

Suppose now that $F = GF(p^n)$ and $m \mid n$. Then $p^n - 1 = k(p^m - 1)$ and in the multiplicative group F^* there are elements of order $p^m - 1$. Indeed, by Lemma 4.2.4 if g is a primitive element of F^*, then $\text{ord}(g^k) = p^m - 1$. Let $h = g^k$. The subgroup $G = <h>$ then contains $p^m - 1$ elements. All elements of G and only they satisfy the equation $x^{p^m} - 1 = 0$ (indeed this equation may have no more than $p^m - 1$ roots). We need to show that G is a subfield. Since it closed under inverses and under the multiplication we only have to show that it is closed under the addition. Let $x, y \in G$. Then binomial theorem gives us $(x + y)^p = x^p + y^p$ and by induction $(x + y)^{p^m} = x^{p^m} + y^{p^m}$. Since $x^{p^m} = y^{p^m} = 1$, we have $(x + y)^{p^m} = 1$, i.e., $x + y \in G$. This finishes the proof.

Solutions to Exercises of Sect. 4.2.4

1. The number of primitive elements in the field $\mathbb{Z}_{1237}$ is the number of generators of its cyclic multiplicative group of order 1236, which is $\phi(1236) = \phi(2^2 \cdot 3 \cdot 103) = 1236 \cdot \frac{1}{2} \cdot \frac{2}{3} \cdot \frac{102}{103} = 408$.

2. (a) We have $2^8 \bmod 17 = 1$ and $3^8 \bmod 17 = 16 = -1$. Hence 2 is not primitive and 3 is a primitive element of F. Let us set $g = 3$.

   (b) Since $g = 3$ is a primitive element of $\mathbb{Z}_{17}$ all its powers of 3 in the following table are different:

n	1	2	3	4	5	6	7	8	9	10	11	12	13	14	15	16
3^n	3	9	10	13	5	15	11	16	14	8	7	4	12	2	6	1

   Therefore the table of logarithms to base 3 will be

n	1	2	3	4	5	6	7	8	9	10	11	12	13	14	15	16
$\log_3(n)$	16	14	1	12	5	15	11	10	2	3	7	13	4	9	6	8

(c) Let us note first that by the definition of the discrete log we have $g^{\log_g(x)} = x$. To prove

$$\log_g(ab) = \log_g(a) + \log_g(b) \mod q{-}1. \tag{11.4}$$

we take g, the primitive element to the power $k = \log_g(ab)$ and to the power $m = \log_g(a) + \log_g(b) \mod q{-}1$. Since all powers g^i are different for $i = 0, 1, \ldots, p - 2$ we get $g^k = g^m$ if and only if $k = m$ since $k, m \in \{0, 1, \ldots, q - 2\}$. Since $g^{q-1} = 1$ we have

$$g^m = g^{\log_g(a)+\log_g(b)} = g^{\log_g(a)} \cdot g^{\log_g(b)} = ab.$$

Also $g^k = g^{\log_g(ab)} = ab$, which proves (11.4).

Solutions to Exercises of Sect. 4.3

1. Bob's secret exponent k_B can be easily calculated. Indeed, since $2^5 = 32$ we have $k_B = 5$. Then the message M can be calculated as

$$M = C_2/C_1^{k_B} = C_2/C_1^5 = 42 \odot 30^{-5} = 42 \odot 23^5 = 12.$$

Hence Alice sent letter "B" to Bob.

2. All Bob needs to know is his private key $k_B = 5191$ and p. The calculation may be performed as follows:

```
gap> kB:=5191;
5191
gap> p:=123456789987654353003;
123456789987654353003
gap> c:= [ [ 83025882561049910713,  66740266984208729661 ],
           [ 117087132399404660932,  44242256035307267278 ],
           [ 67508282043396028407,  77559274822593376192 ],
           [ 60938739831689454113,  14528504156719159785 ],
           [ 50598400445619144427,  59498668430421643612 ],
           [ 92232942954165956522, 105988641027327945219 ],
           [ 97102226574752360229,  46166643538418294423 ] ]
gap> m:=[0,0,0,0,0,0,0];;
gap> for i in [1..7] do
> m[i]:=(c[i][2]*(PowerMod(c[i][1],kB,p))^-1) mod p;
> od;
gap> m;
[ 19244117112225192941, 16191522142944411631, 22224125164116222533,
  15282944412628192319, 30193215411522152315, 24302941141124131541,
  16252841182531282943 ]
gap> # Which reads: "In Galois fields, full of flowers, primitive elements
gap> # dance for hours."
```

11.5   Solutions to Exercises of Chap. 5

Solutions to Exercises of Sect. 5.1.1

1. We use long division:

$$
\begin{array}{r}
4x^2 + 2x \\
3x^2 + 2x + 1\overline{\smash{)}\,5x^4 + x^2 + 3x + 4} \\
\underline{5x^4 + x^3 + 4x^2 } \\
-x^3 - 3x^2 + 3x \\
\underline{-x^3 - 3x^2 - 5x } \\
x + 4
\end{array}
$$

We see that the quotient is $4x^2 + 2x$ and the remainder is $x + 4$, which means $5x^4 + x^2 + 3x + 4 = (3x^2 + 2x + 1)(4x^2 + 2x) + x + 4$.

2. We just evaluate $f(a)$ at each $a \in \mathbb{Z}_5$.

$$
\begin{array}{c|ccccc}
a & 0 & 1 & 2 & 3 & 4 \\
\hline
f(a) & 1 & 3 & 0 & 0 & 0
\end{array}
$$

So the roots of f are 2, 3 and 4, hence by Proposition 5.1.2 $f(x)$ has factors $x + 3$, $x + 2$ and $x + 1$. By long division (omitted), we find

$$
\begin{aligned}
f(x) &= (x + 1)(x^3 + x^2 + x + 1) \\
&= (x + 1)(x + 2)(x^2 + 4x + 3) \\
&= (x + 1)(x + 2)(x + 3)(x + 1) \\
&= (x + 1)^2(x + 2)(x + 3).
\end{aligned}
$$

Solutions to Exercises of Sect. 5.1.2

1. We use $k = 2$ and $\alpha_0 = 1$, $\alpha_1 = 2$, $\alpha_2 = 3$, $\beta_0 = \beta_1 = 1$ and $\beta_2 = 2$. The Lagrange interpolation formula then gives

$$
\begin{aligned}
f(x) &= \beta_0 \frac{(x - \alpha_1)(x - \alpha_2)}{(\alpha_0 - \alpha_1)(\alpha_0 - \alpha_2)} + \beta_1 \frac{(x - \alpha_0)(x - \alpha_2)}{(\alpha_1 - \alpha_0)(\alpha_1 \alpha_2)} + \beta_2 \frac{(x - \alpha_0)(x - \alpha_1)}{(\alpha_2 - \alpha_0)(\alpha_2 - \alpha_1)} \\
&= 1 \frac{(x - 2)(x - 3)}{(1 - 2)(1 - 3)} + 1 \frac{(x - 1)(x - 3)}{(2 - 1)(2 - 3)} + 2 \frac{(x - 1)(x - 2)}{(3 - 1)(3 - 2)} \\
&= 1 \frac{x^2 - 5x + 6}{6 \cdot 5} + 1 \frac{x^2 - 4x + 3}{1 \cdot 6} + 2 \frac{x^2 - 3x + 2}{2 \cdot 1} \\
&= \frac{1}{2}(x^2 + 2x + 6) + \frac{1}{6}(x^2 + 3x + 3) + \frac{2}{2}(x^2 + 4x + 2) \\
&= 4x^2 + 2x + 2.
\end{aligned}
$$

2. We need the constant term of the interpolation polynomial

$$f(x) = 3\frac{(x-3)(x-4)}{(1-3)(1-4)} + 2\frac{(x-1)(x-4)}{(3-1)(3-4)} + 1\frac{(x-1)(x-3)}{(4-1)(4-3)}$$

This will be

$$f(0) = 3\frac{(-3)(-4)}{(1-3)(1-4)} + 2\frac{(-1)(-4)}{(3-1)(3-4)} + 1\frac{(-1)(-3)}{(4-1)(4-3)} = 3.$$

3. We will use two methods of calculation.
   Method 1: is to use the formula for the Lagrange interpolation polynomial, $f(x)$ but evaluate just the constant term:

$$f(x) = 3\frac{(x-2)(x-3)(x-5)}{(1-2)(1-3)(1-5)} + 2\frac{(x-1)(x-3)(x-5)}{(2-1)(2-3)(2-5)}$$

$$+2\frac{(x-1)(x-2)(x-5)}{(3-1)(3-2)(3-5)} + \frac{(x-1)(x-2)(x-3)}{(5-1)(5-2)(5-3)}$$

The constant term of this polynomial is

$$\frac{3(-2)(-3)(-5)}{(-1)(-2)(-4)} + \frac{2(-1)(-3)(-5)}{1(-1)(-3)} + \frac{2(-1)(-2)(-5)}{2 \cdot 1 \cdot (-2)} + \frac{(-1)(-2)(-3)}{4 \cdot 3 \cdot 2} =$$

$$\frac{3}{4} - 3 - 5 - \frac{1}{4} = 6 - 3 - 5 - 2 = 6.$$

(Note that here $\frac{3}{4}$ means: find the inverse of 4 in $\mathbb{Z}_7$ and multiply the result by 3.)
Method 2: is to use linear algebra to determine the coefficients of the polynomial: we must find a_0, a_1, a_2, a_3 such that

$$\begin{bmatrix} 1 & 1 & 1^2 & 1^3 \\ 1 & 2 & 2^2 & 2^3 \\ 1 & 3 & 3^2 & 3^3 \\ 1 & 5 & 5^2 & 5^3 \end{bmatrix}\begin{bmatrix} a_0 \\ a_1 \\ a_2 \\ a_3 \end{bmatrix} = \begin{bmatrix} 1 & 1 & 1 & 1 \\ 1 & 2 & 4 & 1 \\ 1 & 3 & 2 & 6 \\ 1 & 5 & 4 & 6 \end{bmatrix}\begin{bmatrix} a_0 \\ a_1 \\ a_2 \\ a_3 \end{bmatrix} = \begin{bmatrix} 3 \\ 2 \\ 2 \\ 1 \end{bmatrix}$$

Form the augmented matrix and solve the system:

$$\begin{bmatrix} 1 & 1 & 1 & 1 & | & 3 \\ 1 & 2 & 4 & 1 & | & 2 \\ 1 & 3 & 2 & 6 & | & 2 \\ 1 & 5 & 4 & 6 & | & 1 \end{bmatrix} \longrightarrow \begin{bmatrix} 1 & 0 & 0 & 0 & | & 6 \\ 0 & 1 & 0 & 0 & | & 5 \\ 0 & 0 & 1 & 0 & | & 5 \\ 0 & 0 & 0 & 1 & | & 1 \end{bmatrix}$$

The constant term is 6, as before. We also have all the other coefficients of our polynomial: $f(x) = x^3 + 5x^2 + 5x + 6$ (but only the constant term of the polynomial was required).

4. This can be done by the following GAP command

```
gap> InterpolatedPolynomial( Integers, [ 1, 2, 3, 5 ], [ 5, 7, 0, 3 ] ) mod 13;
4*x_1^3+4*x_1^2+x_1+9
```

5. Suppose $|F| = q < \infty$ and $F = \{\alpha_1, \alpha_2, \ldots, \alpha_q\}$. Let $f(\alpha_i) = \beta_i$ for $i = 1, 2, \ldots, q$. Then Lagrange's interpolation formula gives us a polynomial $g(x)$ of degree at most $q - 1$ such that $g(\alpha_i) = \beta_i$.

Solutions to Exercises of Sect. 5.1.3

1. (a) True; (b) False.
2. There are nine quadratic polynomials in $\mathbb{Z}_3[7]$ but three of them with 0 constant term are divisible by x so we are left with

$$x^2 + 1, \ x^2 + 2, \ x^2 + x + 1, \ x^2 + x + 2, \ x^2 + 2x + 1, \ x^2 + 2x + 2.$$

We notice that $x^2 + 2x + 1 = (x + 1)^2$, $x^2 + x + 1 = (x + 2)^2$ and $x^2 + 2 = (x + 1)(x + 2)$. The remaining ones are

$$x^2 + 1, \ x^2 + x + 2, \ x^2 + 2x + 2 \tag{11.5}$$

are indeed irreducible since it is easy to check that they have no roots in $\mathbb{Z}_3$.
3. Checking irreducibility of degree 3 polynomials only require a search for roots. However, with a degree 4 polynomial f each irreducible monic quadratic polynomial must be checked as a potential factor of f. This entails compiling a list of all of the irreducible quadratics first. For larger fields this will be time consuming.
4. (i) Since $f(x)$ has no roots, it is irreducible.
   (ii) As $g(1) = 0$ and has a root in $\mathbb{Z}_3$. Hence $g(x)$ is reducible.
   (iii) We first determine that $h(x)$ has no roots. Then, we check each of the three monic reducible quadratics found in (11.5) as a potential factor by doing long division. Since none of these monic polynomials divide $h(x)$ (details are omitted), $h(x)$ is irreducible.
5. We need to check if $f(x)$ has divisors among irreducible polynomials of degree 1 and 2. As $f(0) = f(1) = 1 \neq 0$ it does not have linear factors. The only irreducible polynomial of degree 2 is $x^2 + x + 1$, so we have to try to divide by $x^2 + x + 1$. By the long division algorithm we get

$$f(x) = x^5 + x + 1 = (x^2 + x + 1)(x^3 + x^2 + 1).$$

Now the polynomial $x^3 + x^2 + 1$ is irreducible since it has no roots in $\mathbb{Z}_2$. Hence we have got the factorization sought for.

6. We need to check if $f(x)$ has divisors among irreducible polynomials of degree 1 and 2. As $f(0) = f(1) = 1 \neq 0$ it does not have linear factors. The only irreducible polynomial of degree 2 is $x^2 + x + 1$, so we have to try to divide by $x^2 + x + 1$. By the long division algorithm we get

$$f(x) = x^5 + x^2 + 1 = (x^2 + x + 1)(x^3 + x^2) + 1.$$

Thus $f(x) = x^5 + x^2 + 1$ is irreducible.

Solutions to Exercises of Sect. 5.1.4

1. Applying the Euclidean algorithm:

$$x^7 + 1 = (x^3 + x^2 + x + 1)(x^4 + x^3 + 1) + (x^2 + x)$$
$$x^3 + x^2 + x + 1 = (x^2 + x)x + (x + 1)$$
$$x^2 + x = (x + 1)x,$$

we get $\gcd(f, g)(x) = x + 1$. Now let us perform the Extended Euclidean algorithm:

$$
\begin{array}{c|c|c}
x^7 + 1 & 1 & 0 \\
x^3 + x^2 + x + 1 & 0 & 1 \\
x^2 + x & 1 & x^4 + x^3 + 1 \\
x + 1 & x & x^5 + x^4 + x + 1
\end{array}
$$

This yields the following:

$$x + 1 = x \cdot (x^7 + 1) + (x^5 + x^4 + x + 1) \cdot (x^3 + x^2 + x + 1).$$

2. Omitted.
3. (a) Straightforward calculation.
   (b) Suppose that a is a multiple root of $f(x)$. Then $f(x) = g(x)(x - a)^k$, where $k \geq 2$. By the product rule

$$f'(x) = g'(x)(x - a)^k + kg(x)(x - a)^{k-1} = [g'(x)(x - a)^{k-1} + kg(x)(x - a)^{k-2}](x - a)$$

   and a is also a root of the derivative. Hence it is also a root of $\gcd(f(x), f'(x))$.
   (c) The polynomial $f(x) = x^{p^n} - x$ does not have multiple roots in any field F of characteristic $p > 0$ since $f'(x) = -1$ and $f(x)$ is relatively prime to $f(x)$.

Solutions to Exercises of Sect. 5.2.1

1. We use the Extended Euclidean algorithm:

$$
\begin{array}{c|c|c}
x^5 + x^3 + 1 & 1 & 0 \\
x^3 + x^2 + x + 1 & 0 & 1 \\
x^2 & 1 & x^2 + x + 1 \\
x + 1 & x + 1 & x^3 \\
1 & x^2 & x^4 + x^3 + x^2 + x + 1
\end{array}
$$

Thus $(x^3 + x^2 + x + 1)^{-1} = x^4 + x^3 + x^2 + x + 1$.

2. Let $K = \mathbb{Z}_3[x]/(x^2 + 2x + 2)$.

   (a) To prove that K is a field, we need to show that $m(x) = x^2 + 2x + 2$ is irreducible over $\mathbb{Z}_3$. Since it is of degree 2, it is enough to show that it does not have roots in $\mathbb{Z}_3$. Indeed, $m(0) = m(1) = 2, m(2) = 1$ and no roots have been found.

   (b) The elements of K are all scalar and linear polynomials over $\mathbb{Z}_3$. That is

   $$K = \{0, 1, 2, x, x + 1, x + 2, 2x, 2x + 1, 2x + 2\}.$$

   (c) Let us calculate the powers of $2x + 1$ and form the "logarithm table"

2-tuple	polynomial	power of x	logarithm
00	0	0	$-\infty$
10	1	1	0
12	$1 + 2x$	$2x + 1$	1
22	$2 + 2x$	$(2x + 1)^2$	2
01	x	$(2x + 1)^3$	3
20	2	$(2x + 1)^4$	4
21	$2 + x$	$(2x + 1)^5$	5
11	$1 + x$	$(2x + 1)^6$	6
02	$2x$	$(2x + 1)^7$	7
	1	$(2x + 1)^8$	

   (d) Let $a = 2x + 1$. Now we can compute the following expression as follows:

   $$2x^7(x + 1)^{-5}(2x + 2) + (x + 2)^5 = a^4 \cdot (a^3)^7 \cdot (a^6)^{-5} \cdot a^2 + (a^5)^5 = a^5 + a = (2 + x) + (2x + 1) = 0.$$

   (e) There are $\phi(8) = 4$ primitive elements in this field. They are $a = 1 + 2x$, $a^3 = x, a^5 = 2 + x, a^7 = 2x$.

3. (a) Elements $Z(2^4)^5$ and $Z(2^4)^{10}$ are not listed in the form the other powers of $Z(2^4)$ did because $Z(2^4)^5 = Z(2^2)$ and $Z(2^4)^{10} = Z(2^2)^2$, i.e., they are elements of the subfield $GF(2^2)$.

(b) We generate $GF(2^4)$ as follows and denote for brevity $Z(2^4)^7$ as a:

```
gap> F:=GaloisField(2^4);
GF(2^4)
gap> e:=Elements(F);
[ 0*Z(2), Z(2)^0, Z(2^2), Z(2^2)^2, Z(2^4), Z(2^4)^2, Z(2^4)^3, Z(2^4)^4,
  Z(2^4)^6, Z(2^4)^7, Z(2^4)^8, Z(2^4)^9, Z(2^4)^11, Z(2^4)^12, Z(2^4)^13,
  Z(2^4)^14 ]
gap> a:=e[10];
Z(2^4)^7
```

We have to conduct an intelligent search for the polynomial. Firstly, it cannot have degree greater than 4 since 1, a, a^2, a^3, a^4 are already linearly dependent over $\mathbb{Z}_2$ being five vectors in a four-dimensional vector space. Since we are looking for a polynomial of minimal degree, it must be irreducible. Therefore the only polynomials we have to try are $x^2 + x + 1, x^3 + x + 1, x^3 + x^2 + 1$, $x^4 + x + 1$, $x^4 + x^3 + 1$, $x^4 + x^3 + x^2 + x + 1$. We substitute a into each of them one by one:

```
gap> a^2+a+1; a^3+a+1; a^3+a^2+1; a^4+a+1; a^4+a^3+1; a^4+a^3+a^2+a+1;
Z(2^4)^4
Z(2^2)
Z(2^4)^2
Z(2^2)^2
0*Z(2)
Z(2^4)
```

At this stage we see that $a = Z(2^4)^7$ is a root of $x^4 + x^3 + 1$. GAP can also do it for you with a command:

```
gap> MinimalPolynomial(GF(2),a);
x_1^4+x_1^3+Z(2)^0
```

Solutions to Exercises of Sect. 5.2.2

1. As $|GF(16)| = |GF(4)|^2$ the dimension is 2.
2. Since this field was studied in the lectures, it is easy for us to compute what we want:

(a) $\alpha = 1 + x + x^2$;
   We calculate the coordinate tuples of the following powers of α:

$$
\begin{aligned}
\alpha^0 &= (1 + x + x^2)^0 = 1 & &\to 1000 \\
\alpha^1 &= (1 + x + x^2)^1 = 1 + x + x^3 & &\to 1110 \\
\alpha^2 &= (1 + x + x^2)^2 = x + x^2 & &\to 0110 \\
\alpha^3 &= (1 + x + x^2)^3 = 1 & &\to 1000 \\
\alpha^4 &= (1 + x + x^2)^4 = 1 + x + x^2 & &\to 1110
\end{aligned}
$$

These five will be already linearly dependent, so we do not have to compute any further powers. Now we use Linear Dependency Relationship algorithm to find linear dependency between these tuples. We place them as columns in a matrix and take it to the row reduced echelon form

$$\begin{pmatrix} 1\ 1\ 0\ 1\ 1 \\ 0\ 1\ 1\ 0\ 1 \\ 0\ 1\ 1\ 0\ 1 \\ 0\ 0\ 0\ 0\ 0 \end{pmatrix} \xrightarrow{rref} \begin{pmatrix} 1\ 0\ 1\ 1\ 1 \\ 0\ 1\ 1\ 0\ 1 \\ 0\ 0\ 0\ 0\ 0 \\ 0\ 0\ 0\ 0\ 0 \end{pmatrix},$$

from which it follows that $1, \alpha$ are linearly independent hence no annihilating polynomials of degree ≤ 2 and that $\alpha^2 = 1 + \alpha$ (clearly seen without any row reduction), whence the minimal annihilating polynomial will be $f(t) = t^2 + t + 1$.

(b) Let now $\alpha = 1 + x$. We calculate the coordinate tuples of the first five powers of α:

$$\begin{aligned} \alpha^0 &= (1+x)^0 = 1 & &\to 1000 \\ \alpha^0 &= (1+x)^1 = 1+x & &\to 1100 \\ \alpha^1 &= (1+x)^2 = 1+x^2 & &\to 1010 \\ \alpha^2 &= (1+x)^3 = 1+x+x^2+x^3 & &\to 1111 \\ \alpha^3 &= (1+x)^4 = x & &\to 0100 \end{aligned}$$

These five will be already linearly dependent, so we do not have to compute any further powers. Now we use Linear Dependency Relationship algorithm to find linear dependency between these tuples. We place them as columns in a matrix and take it to the row reduced echelon form

$$\begin{pmatrix} 1\ 1\ 1\ 1\ 0 \\ 0\ 1\ 0\ 1\ 1 \\ 0\ 0\ 1\ 1\ 0 \\ 0\ 0\ 0\ 1\ 0 \end{pmatrix} \xrightarrow{rref} \begin{pmatrix} 1\ 0\ 0\ 0\ 1 \\ 0\ 1\ 0\ 0\ 1 \\ 0\ 0\ 1\ 0\ 0 \\ 0\ 0\ 0\ 1\ 0 \end{pmatrix}$$

from which it follows that $1, \alpha, \alpha^2, \alpha^3$ are linearly independent hence no annihilating polynomials of degree ≤ 4 and that $\alpha^4 = 1 + \alpha$, whence the minimal annihilating polynomial will be $f(t) = t^4 + t + 1$.

3. (a) The table can be calculated as follows:

4-tuple	polynomial	power of x	logarithm
0000	0	0	$-\infty$
1000	1	1	0
0100	x	x	1
0010	x^2	x^2	2
0001	x^3	x^3	3
1001	$1 + x^3$	x^4	4
1101	$1 + x + x^3$	x^5	5
1111	$1 + x + x^2 + x^3$	x^6	6
1110	$1 + x + x^2$	x^7	7
0111	$x + x^2 + x^3$	x^8	8
1010	$1 + x^2$	x^9	9
0101	$x + x^3$	x^{10}	10
1011	$1 + x^2 + x^3$	x^{11}	11
1100	$1 + x$	x^{12}	12
0110	$x + x^2$	x^{13}	13
0011	$x^2 + x^3$	x^{14}	14

We see from it that x is a primitive element of K since all powers of x are different and represent every element of K. We denote $\alpha = x$.

(b) What is the minimal irreducible polynomial $m_1(t)$ of α? Let $p(t) = t^4 + t^3 + 1$. Then $m_1(x) = x^4 + x^3 + 1 = 0$ in K, i.e., x is a root of $p(t) = t^4 + t^3 + 1$. On the other hand, from the table we see that $1, x, x^2$ and x^3 are linearly independent, hence 4 is the minimal degree of an anihilating polynomial. Hence $m_1(t) = p(t) = t^4 + t^3 + 1$.

To compute the minimal annihilating polynomial for $\beta = x^3$, we use the table to find $\beta^2 = x^6 = 1 + x + x^2 + x^3$, $\beta^3 = x^9 = 1 + x^2$, $\beta^4 = x^{12} = 1 + x$. These elements must be already linearly dependent (as any other five vectors in a four- dimensional space) and we use Linear Dependency Relationship algorithm to find that dependency:

$$
\begin{pmatrix} 1\,0\,1\,1\,1 \\ 0\,0\,1\,0\,1 \\ 0\,0\,1\,1\,0 \\ 0\,1\,1\,0\,0 \end{pmatrix}
\longrightarrow
\begin{pmatrix} 1\,0\,1\,1\,1 \\ 0\,1\,1\,0\,0 \\ 0\,0\,1\,1\,0 \\ 0\,0\,1\,0\,1 \end{pmatrix}
\longrightarrow
\begin{pmatrix} 1\,0\,0\,0\,1 \\ 0\,1\,1\,0\,0 \\ 0\,0\,1\,1\,0 \\ 0\,0\,0\,1\,1 \end{pmatrix}
\longrightarrow
$$

$$
\begin{pmatrix} 1\,0\,0\,0\,1 \\ 0\,1\,1\,0\,0 \\ 0\,0\,1\,0\,1 \\ 0\,0\,0\,1\,1 \end{pmatrix}
\longrightarrow
\begin{pmatrix} 1\,0\,0\,0\,1 \\ 0\,1\,0\,0\,1 \\ 0\,0\,1\,0\,1 \\ 0\,0\,0\,1\,1 \end{pmatrix}
$$

and we see that $\beta^4 = 1 + \beta + \beta^2 + \beta^3$, while $1, \beta, \beta^2, \beta^3$ are linearly independent. Thus, $m_3(t) = t^4 + t^3 + t^2 + t + 1$ is the minimal annihilating polynomial for β.

To find the minimal annihilating polynomial $m_5(t) \in \mathbb{Z}_2[t]$ of the element $\gamma = x^5$ we calculate the coordinate tuples of the following powers of γ:

$$
\begin{aligned}
\gamma^0 &= x^0 = 1 & &\rightarrow 1000 \\
\gamma^1 &= x^5 = x + x^3 & &\rightarrow 1101 \\
\gamma^2 &= x^{10} = x + x^3 & &\rightarrow 0101
\end{aligned}
$$

These first three powers are already linearly dependent, so we don't have to compute any more powers. We even do not need to use the Linear Dependency Relationship Algorithm to find a linear dependency between these tuples. It is obvious that $\gamma^2 = 1 + \gamma$, whence the minimal annihilating polynomial will be $f(t) = t^2 + t + 1$ (because there can be no annihilating polynomials of degree 1 as $x \notin \mathbb{Z}_2$).

(c) We can now calculate using the table:

$$(x^{100} + x + 1)(x^3 + x^2 + x + 1)^{15} = (x^3 + 1)(x^3 + x^2 + x + 1)^{15} = x^3 + 1.$$

Thus

$$(x^{100} + x + 1)(x^3 + x^2 + x + 1)^{15} + x^3 + x + 1 = x.$$

Solutions to Exercises of Sect. 5.3.2

1. Suppose f is not one-to-one and $f(a) = f(b)$ for distinct $a, b \in \mathbb{Z}_{2^n}$ that is

$$a(2a + 1) - b(2b + 1) \equiv 0 \bmod 2^n.$$

This is equivalent to

$$[2(a + b) + 1](a - b) \equiv 0 \bmod 2^n.$$

Since $2(a + b) + 1$ is odd and relatively prime to 2^n, by Lemma 1.3.2(d) we have $(a - b) \equiv 0 \bmod 2^n$ which is impossible since both a and b satisfy $0 \le a < 2^n$ and $0 \le b < 2^n$.

2. If we could represent f as a polynomial in $\mathbb{Z}_4[x]$, then we would have

$$1 = f(3) \equiv f(1) = 0 \quad \bmod 2,$$

which is a contradiction.

11.6   Solutions to Exercises of Chap. 6

Solutions to Exercises of Sect. 6.1.1
1. $\{3, 4\}, \{1, 2, 3\}, \{1, 3, 4\}, \{2, 3, 4\}, \{1, 2, 3, 4\}$.
2. $\{1, 2\}, \{3, 4, 5\}, \{1, 3, 4\}, \{1, 3, 5\}, \{1, 4, 5\}, \{2, 3, 4\}, \{2, 3, 5\}, \{2, 4, 5\}$.
3. In a minimal authorised coalition all permanent members must be present and any four non-permanent members. Hence there are $\binom{10}{4} = 210$ minimal authorised coalitions.
4. We will prove only (a) since (b) is similar. All we need to show is the monotone property. Suppose X, Y are both subsets of U, $X \subseteq Y$, and $X \in \Gamma_1 + \Gamma_2$. Then $X \cap U_1 \in \Gamma_1$ or $X \cap U_2 \in \Gamma_2$ and we suppose that the former is true. But then $Y \cap U_1 \supseteq X \cap U_1 \in \Gamma_1$ and $Y \cap U_1 \in \Gamma_1$ due to the fact that Γ_1 is monotonic. Hence $Y \in \Gamma_1 + \Gamma_2$.

Solutions to Exercises of Sect. 6.1.2
1. (a) We are looking for the polynomial of degree ≤ 2 such that $f(1) = f(2) = 4$, and $f(4) = 0$. Lagrange interpolation formulae gives us

$$f(x) = 4\frac{(x-2)(x-4)}{(1-2)(1-4)} + 4\frac{(x-1)(x-4)}{(2-1)(2-4)} = 4\frac{(x-2)(x-4)}{3} + 4\frac{(x-1)(x-4)}{5} =$$

$$4\left(5(x-2)(x-4) + 3(x-1)(x-4)\right) = 6(x-2)(x-4) + 5(x-1)(x-4)$$

$$= 4(x+1)(x+3) = 4x^2 + 2x + 5$$

(b) As $f(3) = 5$, the remaining card will be $\boxed{\begin{array}{c} 3 \\ 5 \end{array}}$

2. GAP helps us to find the interpolation polynomial. Then we calculate the constant term substituting 0 into it:

```
gap> f:=InterpolatedPolynomial(GF(31),[1,5,7],[16,7,22]);
Z(31)^29*x_1^2+Z(31)^4*x_1+Z(31)^28
gap> Int(Value(f,0));
7
```

So the secret is 7.
3. We do the interpolation as follows:

```
gap> f:=InterpolatedPolynomial(GF(97),[1,2,4,6],[56,40,22,34]);
Z(97)^46*x^3+Z(97)^58*x^2+Z(97)^77*x+Z(97)^87
gap> # Calculating the secret:
gap> Int(Value(f,0));
55
gap> # Calculating the share of the third and the fifth board member:
```

```
gap> Int(Value(f,3));
96
gap> Int(Value(f,5));
4
```

Hence the secret is 55, and the cards of the two remaining board members are

likely to be
$\dfrac{3}{96}$
$\dfrac{5}{4}$

Solutions to Exercises of Sect. 6.2.1

1. (a) $S_0 = \{0, 1\}$ and $S_1 = \cdots = S_6 = \{0, 1, 2\}$.
   (b) The coalition $\{1, 2\}$ is authorised since $s_0 = s_0'$ whenever $s_1 = s_1'$ and $s_2 = s_2'$.
       Let us pay attention to the first and the seventh rows of the distribution table.
       The shares of participants are the same: 0, 1, 2, but the secrets are different.
       Thus this coalition would not know which row the dealer had chosen and
       would not know the secret.
   (c) The secret recovery function will be

s_1 s_2	$f_{\{1,2\}}(s_1, s_2)$
0 0	0
1 0	1
0 1	1
1 1	0
0 2	1
2 0	1
1 2	1
2 1	1
2 2	0

Solutions to Exercises of Sect. 6.2.2

1. Let $h_0, h_1, \ldots, h_4$ be the rows of H. We note that $h_3 = 3h_1$ and $h_4 = 2h_2$. Only
   one from each pair can be in a minimal authorised coalition. The coalition $\{1, 2\}$
   is authorised since $h_0 = \frac{1}{2}h_1 + \frac{1}{4}h_2$. Similarly, the coalitions $\{1, 4\}, \{2, 3\}, \{3, 4\}$
   are authorised. These four are minimal. A coalition of size 3 cannot be minimal
   authorised since it contains either h_1 and h_3 or h_2 and h_4.
2. Let us denote $h_0, h_1, \ldots, h_6$ the rows of H. Any three rows among h_0, h_1, h_2, h_3
   are linearly independent (the corresponding determinant is the Vandermonde
   determinant). The implications of this comment are the following: the coalition
   $\{1, 2, 3\}$ is authorised and no subset of it is. Hence this is a minimal authorised
   coalition. Let us consider the determinant

$$\begin{vmatrix} 1 & a & a^2 \\ 1 & b & b^2 \\ 0 & 0 & c \end{vmatrix} = (b - a)c.$$

If $b \neq a$ and $c \neq 0$ it is nonzero. The implications are that the coalitions $\{1, 2, 4\}$, $\{1, 2, 5\}$, $\{1, 2, 6\}$, $\{1, 3, 4\}$, $\{1, 3, 5\}$, $\{1, 3, 6\}$, $\{2, 3, 4\}$, $\{2, 3, 5\}$, $\{2, 3, 6\}$ are minimal authorised. It remains to note that no coalitions containing two of the users 4, 5, 6 are minimal authorised: if it is, then one of these users can be removed without making coalition losing (indeed their respective rows are one multiple of another). Therefore the minimal authorised coalitions listed so far are all minimal authorised coalitions.

3. See next problem which is more general.
4. Let $\{1, 2, \ldots, n\}$ be a set of users. For a linear secret sharing scheme with matrix H with rows $\mathbf{h}_0, \mathbf{h}_1, \ldots, \mathbf{h}_n$ a coalition $\{i_1, i_2, \ldots, i_k\}$ is authorised if $\mathbf{h}_0$—which is normally taken to be $(1, 0, \ldots, 0)$—is in the span of $\mathbf{h}_{i_1}, \ldots, \mathbf{h}_{i_k}$.
   It is immediate that $\{1, 2\}$ and $\{3, 4, 5\}$ are authorized for any distinct nonzero a_1, a_2, a_3, a_4, a_5. Indeed, the determinant $\begin{vmatrix} 1 & a_1 \\ 1 & a_2 \end{vmatrix} \neq 0$, hence $(1, 0) = x_1(1, a_1) + x_2(1, a_2)$. But then $(1, 0, 0) = x_1(1, a_1, 0) + x_2(1, a_2, 0)$ too. Also

$$\begin{vmatrix} 1 & a_3 & a_3^2 \\ 1 & a_4 & a_4^2 \\ 1 & a_5 & a_5^2 \end{vmatrix} \neq 0.$$

as this is the Vandermonde determinant. So the row $(1, 0, 0)$ can be expressed as a linear combination of the rows of this matrix. Let us show that $\{1, 2\}$ and $\{3, 4, 5\}$ are minimal authorized coalitions. While it is obvious that the first is minimal, it is not so clear for the second. To prove this, it would be sufficient to show that

$$D = \begin{vmatrix} 1 & 0 & 0 \\ 1 & a_i & a_i^2 \\ 1 & a_j & a_j^2 \end{vmatrix} \neq 0.$$

for $i, j \in \{3, 4, 5\}$. Expanding D using cofactors of the first row we will get

$$D = \begin{vmatrix} a_i & a_i^2 \\ a_j & a_j^2 \end{vmatrix} = a_i a_j \begin{vmatrix} 1 & a_i \\ 1 & a_j \end{vmatrix} = a_i a_j (a_j - a_i) \neq 0.$$

Coalitions $\{i, j, k\}$, where $i \in \{1, 2\}$ and $j, k \in \{3, 4, 5\}$ may or may not be authorized depending on the values a_1, a_2, a_3, a_4, a_5. To find out the exact condition when $\{i, j, k\}$ is authorized, let us consider the determinant

$$\begin{vmatrix} 1 & a & 0 \\ 1 & b & b^2 \\ 1 & c & c^2 \end{vmatrix} = \begin{vmatrix} b & b^2 \\ c & c^2 \end{vmatrix} - a \begin{vmatrix} 1 & b^2 \\ 1 & c^2 \end{vmatrix} = bc(c - b) - a(c^2 - b^2).$$

Thus this determinant is zero if and only if $a = \dfrac{bc}{b + c}$.

Now let us consider the coalition $\{i, j, k\}$, where $i \in \{1, 2\}$ and $j, k \in \{3, 4, 5\}$. If $\mathbf{h}_i \in \mathrm{Span}\{\mathbf{h}_j, \mathbf{h}_k\}$, then we know that this coalition is not authorized since $\mathbf{h}_0$

as we know is not in Span$\{\mathbf{h}_j, \mathbf{h}_k\}$. On the other hand, if $\mathbf{h}_i \notin \text{Span}\{\mathbf{h}_j, \mathbf{h}_k\}$, then $\{\mathbf{h}_i, \mathbf{h}_j, \mathbf{h}_k\}$ form a basis of $\mathbb{R}^3$, $\mathbf{h}_0$ is in the span of this set and coalition $\{i, j, k\}$ is authorized. So coalition $\{i, j, k\}$ is authorized if and only if $a_i \neq \dfrac{a_j a_k}{a_j + a_k}$.

5. (a) Let us try A first. We will apply Linear Dependency Relationship algorithm. Consider the matrix

$$H' = [\mathbf{h}_1^T\ \mathbf{h}_2^T\ \mathbf{h}_3^T\ \mathbf{e}_1^T] = \begin{bmatrix} 1 & 1 & 11 & 1 \\ 2 & 3 & 5 & 0 \\ 3 & 3 & 2 & 0 \\ 0 & 0 & 0 & 0 \end{bmatrix}.$$

We row reduce it to the reduced echelon form:

$$H' \to \begin{bmatrix} 1 & 1 & 11 & 1 \\ 0 & 1 & 14 & 29 \\ 0 & 0 & 0 & 1 \\ 0 & 0 & 0 & 0 \end{bmatrix}.$$

We may stop row reducing here. As the last column contains a pivot, $\mathbf{e}_1$ is not a linear combination of $\mathbf{h}_1, \mathbf{h}_2, \mathbf{h}_3$. So A is not authorized.

Let us now try B. Consider the matrix

$$H' = [\mathbf{h}_1^T\ \mathbf{h}_4^T\ \mathbf{h}_5^T\ \mathbf{e}_1^T] = \begin{bmatrix} 1 & 0 & 0 & 1 \\ 2 & 1 & 6 & 0 \\ 3 & 1 & 1 & 0 \\ 0 & 2 & 1 & 0 \end{bmatrix}.$$

We row reduce it to the reduced echelon form:

$$H' \to \begin{bmatrix} 1 & 0 & 0 & 1 \\ 0 & 1 & 6 & 29 \\ 0 & 1 & 1 & 28 \\ 0 & 2 & 1 & 0 \end{bmatrix} \to \begin{bmatrix} 1 & 0 & 0 & 1 \\ 0 & 1 & 6 & 29 \\ 0 & 0 & 26 & 30 \\ 0 & 0 & 20 & 4 \end{bmatrix} \to \begin{bmatrix} 1 & 0 & 0 & 1 \\ 0 & 1 & 6 & 29 \\ 0 & 0 & 1 & 25 \\ 0 & 0 & 20 & 4 \end{bmatrix} \to \begin{bmatrix} 1 & 0 & 0 & 1 \\ 0 & 1 & 6 & 29 \\ 0 & 0 & 1 & 25 \\ 0 & 0 & 0 & 0 \end{bmatrix}$$

$$\to \begin{bmatrix} 1 & 0 & 0 & 1 \\ 0 & 1 & 0 & 3 \\ 0 & 0 & 1 & 25 \\ 0 & 0 & 0 & 0 \end{bmatrix}.$$

This means that $\mathbf{e}_1 = \mathbf{h}_1 + 3\mathbf{h}_4 - 6\mathbf{h}_5$. So B is authorized.

(b) The secret now can be calculated as $s_0 = s_1 + 3s_4 - 6s_5 = 29$.

6. Suppose that

$$\sum_{j=1}^{r} \lambda_j \mathbf{h}'_j = \mathbf{0} \text{ but } \sum_{j=1}^{r} \lambda_j c_j = c \neq 0.$$

In such a case we will have

$$c^{-1} \sum_{j=1}^{r} \lambda_j \mathbf{h}_j = \mathbf{e}_1,$$

which means that the coalition $\{i_1, i_2, \ldots, i_r\}$ is authorised.

7. Let $|U| = p$ and $|V| = q$ and M and N be $(p + 1) \times k$ and $(q + 1) \times r$ matrices, respectively, with rows $\mathbf{m}_0, \mathbf{m_1}, \ldots, \mathbf{m_p}$ and $\mathbf{n}_0, \mathbf{n}_1, \ldots, \mathbf{n}_q$, where $\mathbf{m}_0$ and $\mathbf{n}_0$ be the target vectors $(1, 0, \ldots, 0)$ of dimensions p and q, respectively. We represent the matrix M as $M = (M_1, M_{-1})$, where M_1 is the first column of M and M_{-1} consists of all other columns. Similarly, $N = (N_1, N_{-1})$.

(a) Let us construct the following $(p + q + 1) \times (k + r - 1)$ matrix for the sum $\Gamma_S = \Gamma_M + \Gamma_N$:

$$S = \begin{bmatrix} 1 & \mathbf{0} & \mathbf{0} \\ M_1 & M_{-1} & \mathbf{0} \\ N_1 & \mathbf{0} & N_{-1} \end{bmatrix},$$

Obviously, all coalitions authorised in Γ_M or in Γ_N will be authorised in Γ_S. Suppose now that two coalitions $\{i_1, i_2, \ldots, i_s\} \subseteq U$ and $\{j_1, j_2, \ldots, j_t\} \subseteq V$ are non-authorised in Γ_M and in Γ_N, respectively, but $\{i_1, i_2, \ldots, i_s\} \cup \{j_1, j_2, \ldots, j_t\}$ is authorised in Γ_S. Then for some scalars $\beta_1, \beta_2, \ldots, \beta_s$ and $\delta_1, \delta_2, \ldots, \delta_t$ we have

$$\sum_{i=1}^{s} \beta_i \mathbf{m}_i + \sum_{j=1}^{t} \delta_j \mathbf{n}_j = \mathbf{t},$$

where $\mathbf{t}$ is the target vector which is the top row of matrix S. Let $\mathbf{m}_i = (c_i, \mathbf{m}'_i)$ and $\mathbf{n}_j = (d_j, \mathbf{n}'_j)$. Then

$$\sum_{i=1}^{s} \beta_i \mathbf{m}'_i = \sum_{j=1}^{t} \delta_j \mathbf{n}'_j = \mathbf{0}$$

and by the previous exercise

$$\sum_{i=1}^{s} \beta_i c_i = \sum_{j=1}^{t} \delta_j d_j = 0,$$

from which

$$\sum_{i=1}^{s} \beta_i \mathbf{m}_i = \sum_{j=1}^{t} \delta_j \mathbf{n}_j = \mathbf{0},$$

a contradiction.

(b) The matrix P such that $\Gamma_P = \Gamma_M \times \Gamma_N$ can be constructed as follows:

$$P = \begin{bmatrix} 1 & 0 & 0 \\ M & M_1 & 0 \\ 0 & N_1 & N_{-1} \end{bmatrix}.$$

We leave the proof to the reader.

8. Suppose that we have an $5 \times n$ matrix H with rows $\mathbf{h}_0, \mathbf{h}_1, \dots, \mathbf{h}_4$ such that $\mathbf{h}_0 = (1, 0, \dots, 0)$ and for $\Gamma = \Gamma_H$ we have $\Gamma_{min} = \{\{1, 2\}, \{2, 3\}, \{3, 4\}\}$. As $\{1, 2\}$ is authorized, we have coefficients $a, b, c, d, e, f \in F$ such that $a\mathbf{h}_1 + b\mathbf{h}_2 = c\mathbf{h}_2 + d\mathbf{h}_3 = e\mathbf{h}_3 + f\mathbf{h}_4 = \mathbf{h}_0$. Assume first that $b \neq c$. Then $\frac{ac}{b}\mathbf{h}_1 - d\mathbf{h}_3 = (\frac{c}{b} - 1)\mathbf{h}_0$ and $\mathbf{h}_0 \in \mathrm{Span}\{\mathbf{h}_1, \mathbf{h}_3\}$. This is not the case since $\{1, 3\}$ is not authorized. Then $b = c$ and $\mathbf{h}_3 = \frac{a}{d}\mathbf{h}_1$. But then $\frac{ae}{d}\mathbf{h}_1 + f\mathbf{h}_4 = \mathbf{h}_0$ and $\{1, 4\}$ is authorized which is again not the case. Hence it is impossible to find such matrix H.

9. Let H be an $(n + 1) \times k$ matrix over a field F with rows $\mathbf{h}_0, \mathbf{h}_1, \dots, \mathbf{h}_n$ and Γ_H be the access structure given by Γ_{min}. Since coalitions $\{1, i\}$ are authorised and $\{1\}$ is not, this means that $\mathbf{h}_i \in \mathrm{span}\{\mathbf{h}_0, \mathbf{h}_1\}$. It cannot happen that the $\mathrm{span}\{\mathbf{h}_2, \dots, \mathbf{h}_n\}$ is one dimensional since in such a case all singletons $\{2\}, \dots, \{n\}$ would be authorised. If there are two vectors, say $\mathbf{h}_2$ and $\mathbf{h}_3$ that are not multiples of each other, then $\mathrm{span}\{\mathbf{h}_2, \mathbf{h}_3\} = \mathrm{span}\{\mathbf{h}_0, \mathbf{h}_1\}$ and, in particular, $\mathbf{h}_0 \in \mathrm{span}\{\mathbf{h}_2, \mathbf{h}_3\}$, so $\{2, 3\}$ is authorised. This contradiction proves the statement.

Solutions to Exercises of Sect. 6.2.3

1. (i) If C is authorised, then their shares are compatible with only one value of a secret, hence $\#\mathcal{T}_{C'} = \#\mathcal{T}_C$.
   (ii) If C is not authorised, then, since the scheme is perfect, their shares are compatible with any of the q secrets, hence $\#\mathcal{T}_{C'} = q \cdot \#\mathcal{T}_C$.
3. (i) Follow the argument in Example 6.2.6.
   (ii) Let A be a maximal authorised coalition and $A' = A \cup \{0\}$. Then arguing as in Example 6.2.6 we can prove that $\#\mathcal{T}_{A'} = q^2$. Suppose $A \neq U$, then, due to connectedness, we will have $i \in A$ and $j \notin A$ such that $\{i, j\}$ is authorised. Let us fix a share s_i. Then as in the proof of Theorem 6.2.3 we obtain a one-to-one correspondence between S_0 and S_j. Hence the secret s and s_i uniquely determine s_j. This implies that $\#\mathcal{T}_{A' \cup \{j\}} = q^2$ which contradicts to maximality of A.
   (iii) Due to (ii) we have $\#\mathcal{T}_{\{i, j\}} = q$ or $\#\mathcal{T}_{\{i, j\}} = q^2$. Let us prove that the second option cannot happen. Take any two shares s_i and s_j. Since $\{i, j\}$ is not authorised, we have at least q rows in $\mathcal{T}$ containing s_i and s_j (as there must be such row for every secret s). This implies $\#\mathcal{T}_{\{i, j\}} = q$.
   (iv) This follows from (iv) since $\#\mathcal{T}_{\{i, j\}} = \#\mathcal{T}_{\{j, k\}} = q$ implies $\#\mathcal{T}_{\{i, k\}} = q$.
   (v) From (iv) we deduce that the relation

$$i \equiv j \iff \{i, j\} \notin \Gamma_{min}$$

is an equivalence relation. This implies the statement of the theorem.

11.7   Solutions to Exercises of Chap. 7

Solutions to Exercises of Sect. 7.1.1

1. (a) $wt(\mathbf{u}) = 5$ and $wt(\mathbf{v}) = 4$. Also $d(\mathbf{u}, \mathbf{v}) = 3$.
   (b) The error vector $\mathbf{e} = \mathbf{y} - \mathbf{x} = (0\,0\,1\,0\,0\,0\,1\,0\,0\,0)$. Two mistakes have occurred.
2. Firstly, this is the word $\mathbf{x} = (1\,0\,1\,0)$ itself. Then comes four vectors

$$(0\,0\,1\,0), \quad (1\,1\,1\,0), \quad (1\,0\,0\,0), \quad (1\,0\,1\,1),$$

   whose distance from $\mathbf{x}$ is 1. Then comes six vectors

$$(0\,1\,1\,0), \ (0\,0\,0\,0), \ (0\,0\,1\,1), \ (1\,1\,0\,0), \ (1\,1\,1\,1), \ (1\,0\,0\,1)$$

   whose distance from $\mathbf{x}$ is 2.
3. $|B_3(\mathbf{x})| = \binom{7}{0} + \binom{7}{1} + \binom{7}{2} + \binom{7}{3} = 1 + 7 + 21 + 35 = 64$.
4. The cardinality of $B_k(\mathbf{x})$ does not depend on $\mathbf{x}$ since it consists of all vectors $\mathbf{x} + \mathbf{e}$, where $wt(\mathbf{e}) \le k$ and the number of the latter does not depend on $\mathbf{x}$.

Solutions to Exercises of Sect. 7.1.2

1. This condition is that in every set $\{e_i, e_{i+4}, e_{i+8}\}$, for $i = 1, 2, 3, 4$, there are at most one 1. An error vector of weight 4 that will be corrected is, for example, $\mathbf{e} = (1, 1, 1, 1, 0, 0, 0, 0, 0, 0, 0, 0)$.
2. If one mistake happened, say in the ith row and jth column, then $e_i = f_j = 1$, while $e_s = 0$ for $s \ne i$ and $f_t = 0$ for all $t \ne j$. This will allow us to locate the exact position of the mistake. If three mistakes happen, then either in one of the columns or in one of the rows there will be a single 1. This means that at least one symbol among $e_1, e_2, \ldots, e_{m_1}, f_1, f_2, \ldots, f_{m_2}$ will be equal to 1. This will show that at least one mistake took place.

Solutions to Exercises of Sect. 7.1.3

1. Let us prove this by induction. For this it is sufficient to prove that, if H is an Hadamard $n \times n$ matrix, then $H' = \begin{bmatrix} H & H \\ H & -H \end{bmatrix}$ is a $2n \times 2n$ Hadamard matrix. Since H' is a ± 1 matrix, it is sufficient to prove that, given that the system of rows $\{\mathbf{h}_1, \mathbf{h}_2, \ldots, \mathbf{h}_n\}$ of H is orthogonal, then the system of rows $\{\mathbf{h}'_1, \ldots, \mathbf{h}'_{2n}\}$ of H' is also orthogonal. We have $\mathbf{h}'_i = (\mathbf{h}_i, \mathbf{h}_i)$ for $i = 1, \ldots, n$ and $\mathbf{h}'_i = (\mathbf{h}_i, -\mathbf{h}_i)$ for $i = n + 1, \ldots, 2n$. Now $\mathbf{h}'_i \cdot \mathbf{h}'_j = 2(\mathbf{h}_i \cdot \mathbf{h}_j) = 0$ for $i, j \in \{1, \ldots, n\}$ or $i, j \in \{n + 1, \ldots, 2n\}$ with $i \ne j$. Also $\mathbf{h}'_i \cdot \mathbf{h}'_j = \mathbf{h}_i \cdot \mathbf{h}_j - \mathbf{h}_i \cdot \mathbf{h}_j = 0$ for all $i \in \{1, \ldots, n\}$ and $j \in \{n + 1, \ldots, 2n\}$. This proves the statement.

2. Let $h_1, h_2, \ldots, h_n$ be the system of rows of H. It is, as we know, orthogonal. Consider the vector $-h_i$ for some $i = 1, \ldots, n$. It will remain orthogonal to all vectors h_j for $j \neq i$, hence it will have $n/2$ agreements with h_i and $n/2$ disagreements. After the change of -1s to 0s, the Hamming distance between these vectors will be $n/2$. Also h_i and $-h_i$ will have n disagreements and after the change of -1s to 0s the distance between these vectors will be n. Finally, the distance between $-h_i$ and $-h_j$, with $i \neq j$, after the change will be $n/2$ since $-h_i$ and $-h_j$ are orthogonal.

3. Let $\mathbf{a} = (a_1, a_2, a_3)$ and $\mathbf{b} = (b_1, b_2, b_3)$. Then $\mathbf{a} + \mathbf{b} = (a_1 + b_1, a_2 + b_2, a_3 + b_3)$. We have

$$E_1(\mathbf{a} + \mathbf{b}) = (a_1 + b_1, a_2 + b_2, a_3 + b_3, (a_1 + b_1) + (a_2 + b_2), (a_2 + b_2) + (a_3 + b_3), (a_1 + b_1) + (a_3 + b_3), 0) =$$

$$(a_1, a_2, a_3, a_1 + a_2, a_2 + a_3, a_1 + a_3, 0) + (b_1, b_2, b_3, b_1 + b_2, b_2 + b_3, b_1 + b_3, 0) = E_1(\mathbf{a}) + E_1(\mathbf{b})$$

so E_1 is linear. E_2 is not linear since $E_2(\mathbf{0}) \neq \mathbf{0}$.

4. Let C be a binary linear code. Suppose not all codewords have even Hamming weight and there is a codeword $\mathbf{c}_0$ whose Hamming weight is odd. Let $E \subseteq C$ be the set of all codewords of even Hamming weight. Then the set

$$E + \mathbf{c}_0 = \{\mathbf{c} + \mathbf{c}_0 \mid \mathbf{c} \in E\}$$

consists of codewords of odd length and its cardinality is the same as the cardinality of E. Let us show that every codeword with an odd Hamming weight is in $E + \mathbf{c}_0$. Let $\mathbf{d}$ be such a codeword. Then $\mathbf{d} + \mathbf{c}_0$ is also a codeword and has an even Hamming weight, that is, $\mathbf{d} + \mathbf{c}_0 = \mathbf{c} \in E$. But then $\mathbf{d} = \mathbf{c} + \mathbf{c}_0 \in E + \mathbf{c}_0$.

Solutions to Exercises of Sect. 7.1.4

1. We get

$$E(\mathbf{a}) = (a_1, a_2, a_3, a_1 + a_2 + a_4, a_2 + a_3, a_1 + a_3 + a_4, a_4) =$$
$$a_1(1, 0, 0, 1, 0, 1, 0) + a_2(0, 1, 0, 1, 1, 0, 0) + a_3(0, 0, 1, 0, 1, 1, 0) + a_4(0, 0, 0, 1, 0, 1, 1) =$$

$$\begin{bmatrix} 1 & 0 & 0 & 1 & 0 & 1 & 0 \\ 0 & 1 & 0 & 1 & 1 & 0 & 0 \\ 0 & 0 & 1 & 0 & 1 & 1 & 0 \\ (0 & 0 & 0 & 1 & 0 & 1 & 1 \end{bmatrix} \begin{bmatrix} a_1 \\ a_2 \\ a_3 \\ a_4 \end{bmatrix} = G\mathbf{a}.$$

2. Straightforward.

3. It is known from linear algebra that elementary row operations performed on G do not change the row space of G, which is exactly the set of codewords.

4. (a) $(1\ 1\ 1)G = (1\ 0\ 0\ \ldots)$, hence it is not systematic as the first three coordinates do not represent the message.

(b) We row reduce G as follows using only elementary row operations:

$$G = \begin{bmatrix} 1\,0\,1\,0\,1\,0 \\ 1\,1\,0\,0\,1\,1 \\ 1\,1\,1\,0\,0\,0 \end{bmatrix} \rightarrow \begin{bmatrix} 1\,0\,1\,0\,1\,0 \\ 0\,1\,1\,0\,0\,1 \\ 0\,1\,0\,0\,1\,0 \end{bmatrix} \rightarrow \begin{bmatrix} 1\,0\,1\,0\,1\,0 \\ 0\,1\,1\,0\,0\,1 \\ 0\,0\,1\,0\,1\,1 \end{bmatrix}$$

$$\rightarrow \begin{bmatrix} 1\,0\,0\,0\,0\,1 \\ 0\,1\,0\,0\,1\,0 \\ 0\,0\,1\,0\,1\,1 \end{bmatrix}.$$

The latter is the generator matrix of a systematic code C_2. The codewords are the rows of the following matrix:

$$\begin{bmatrix} 0\,0\,0\,0\,0\,0 \\ 1\,0\,0\,0\,0\,1 \\ 0\,1\,0\,0\,1\,0 \\ 0\,0\,1\,0\,1\,1 \\ 1\,1\,0\,0\,1\,1 \\ 1\,0\,1\,0\,1\,0 \\ 0\,1\,1\,0\,0\,1 \\ 1\,1\,1\,0\,0\,0 \end{bmatrix}.$$

The minimum distance is the minimum weight, which is 2.

Solutions to Exercises of Sect. 7.1.5

1. Let us row reduce A to its row reduced echelon form:

$$A = \begin{bmatrix} 1\,2\,1\,2\,1 \\ 1\,2\,1\,0\,2 \\ 2\,1\,0\,1\,0 \end{bmatrix} \longrightarrow \begin{bmatrix} 1\,2\,0\,0\,1 \\ 0\,0\,1\,0\,1 \\ 0\,0\,0\,1\,1 \end{bmatrix}.$$

The latest matrix is already in the reduced row echelon form with columns 1,3 and 4 being pivotal.

(a) The equation $A\mathbf{x} = \mathbf{0}$ is equivalent to the system

$$x_1 = x_2 + 2x_5$$
$$x_3 = 2x_5$$
$$x_4 = 2x_5$$

with x_2, x_5 being independent variables and x_1, x_3, x_4 being dependent. One of the possible bases for $NS(A)$ is

$$\mathbf{f}_1 = \begin{bmatrix} 1 \\ 1 \\ 0 \\ 0 \\ 0 \end{bmatrix}, \qquad \mathbf{f}_2 = \begin{bmatrix} 2 \\ 0 \\ 2 \\ 2 \\ 1 \end{bmatrix}.$$

(b) Apart from these two, there are seven other vectors in $NS(A)$, namely

$$\mathbf{0} = (0\,0\,0\,0\,0)^T,$$
$$2\mathbf{f}_1 = (2\,2\,0\,0\,0)^T,$$
$$2\mathbf{f}_2 = (1\,0\,1\,1\,2)^T,$$
$$\mathbf{f}_1 + \mathbf{f}_2 = (0\,1\,2\,2\,1)^T,$$
$$\mathbf{f}_1 + 2\mathbf{f}_2 = (2\,1\,1\,1\,2)^T,$$
$$2\mathbf{f}_1 + \mathbf{f}_2 = (1\,2\,2\,2\,1)^T,$$
$$2\mathbf{f}_1 + 2\mathbf{f}_2 = (0\,2\,1\,1\,2)^T.$$

(c) We have $\mathrm{wt}(\mathbf{f}_1) = \mathrm{wt}(2\mathbf{f}_1) = 2$. These are the two vectors which have the minimum Hamming weight, which is 2.

2. (a) Let us row reduce the parity check matrix to the form $(A \mid I_4)$:

$$H = \begin{bmatrix} 0\,0\,1 & 1\,1\,0\,1 \\ 0\,1\,0 & 1\,0\,1\,1 \\ 1\,0\,0 & 0\,1\,1\,1 \\ 1\,1\,1 & 1\,1\,1\,0 \end{bmatrix} \longrightarrow \begin{bmatrix} 0\,0\,1 & 1\,0\,0\,0 \\ 1\,1\,1 & 0\,1\,0\,0 \\ 1\,0\,0 & 0\,0\,1\,0 \\ 1\,1\,1 & 0\,0\,0\,1 \end{bmatrix}.$$

Hence $G = (I_3 \mid A^T)$ and $\mathcal{C}$ has three information symbols.

$$G = \begin{bmatrix} 1\,0\,0 & 0\,1\,1\,1 \\ 0\,1\,0 & 0\,1\,0\,1 \\ 0\,0\,1 & 1\,1\,0\,1 \end{bmatrix}.$$

(b) Yes, because all columns of H are different.

(c) No, it will not because $\mathbf{h}_2 + \mathbf{h}_5 = \mathbf{h}_7$, hence if two mistakes will occur in the second and fifth positions, they will give the same syndrome as one mistake in seventh position. As the decoding is maximum likelyhood decoding, the decoder will decide that a single error in the seventh position has occurred and that will result in decoding error. This is not the only example. There are other double mistakes that will not be corrected.

(d) Yes, because all columns of H are different and hence $\mathbf{h}_i + \mathbf{h}_j \neq \mathbf{0}$ for all $i \neq j$. Thus a syndrome of a double mistake is never $(0\,0\,0\,0)^T$.

(e) $(1\,1\,1)G = (1\,1\,1\,1\,1\,1\,1)$.

(f) $H\mathbf{y}_1^T = (0\,1\,0\,1)^T = \mathbf{h}_2$, hence we assume one mistake occurred in the second position and decode to $(1\,0\,0)$. Also $H\mathbf{y}_2^T = (1\,1\,1\,1)^T$. This syndrome is not equal to any of the columns and also not even to the sum of two columns. The easiest way to see this is to notice that to have the last coordinate 1 one should take the last column and any other column. Hence we have more than two mistakes! We have $\mathbf{h}_1 + \mathbf{h}_2 + \mathbf{h}_3 = (1\,1\,1\,1)^T$ but also $\mathbf{h}_3 + \mathbf{h}_4 + \mathbf{h}_5 = (1\,1\,1\,1)^T$, hence the decoding will not be unique here. The decoder may well report a decoding failure, depending on how it was programmed.

(g) The syndrome $Hz^T = (1\ 0\ 1\ 0)^T$ is different from all columns of H, hence such a syndrome cannot be a result of one mistake. As $(1\ 1\ 1\ 0)^T = \mathbf{h}_1 + \mathbf{h}_3$, this syndrome may appear as a result of two mistakes in positions 1 and 3.

Solutions to Exercises of Sect. 7.1.6

1. (a) To encode the vector $\mathbf{u} = (1\ 1\ 0\ 1)$, we compute

$$\mathbf{u}G = (1\ 1\ 0\ 1)\begin{bmatrix} 1\ 0\ 0\ 0\ 0\ 1\ 1 \\ 0\ 1\ 0\ 0\ 1\ 0\ 1 \\ 0\ 0\ 1\ 0\ 1\ 1\ 0 \\ 0\ 0\ 0\ 1\ 1\ 1\ 1 \end{bmatrix} = (1\ 1\ 0\ 1\ 0\ 0\ 1).$$

(b) Suppose that the vector $\mathbf{v} = (1\ 0\ 0\ 1\ 1\ 0\ 0)$ was received. Computing the syndrome

$$Hv^T = \begin{bmatrix} 0\ 0\ 0\ 1\ 1\ 1\ 1 \\ 0\ 1\ 1\ 0\ 0\ 1\ 1 \\ 1\ 0\ 1\ 0\ 1\ 0\ 1 \end{bmatrix}\begin{bmatrix} 1 \\ 0 \\ 0 \\ 1 \\ 1 \\ 0 \\ 0 \end{bmatrix} = \begin{bmatrix} 0 \\ 0 \\ 0 \end{bmatrix},$$

hence $\mathbf{v}$ is a codevector. We assume that no mistakes happen and decode it to $(1\ 0\ 0\ 1)$. Suppose that the vector $\mathbf{w} = (1\ 1\ 1\ 1\ 0\ 0\ 0)$ was received. Computing the syndrome

$$Hw^T = \begin{bmatrix} 0\ 0\ 0\ 1\ 1\ 1\ 1 \\ 0\ 1\ 1\ 0\ 0\ 1\ 1 \\ 1\ 0\ 1\ 0\ 1\ 0\ 1 \end{bmatrix}\begin{bmatrix} 1 \\ 1 \\ 1 \\ 1 \\ 0 \\ 0 \\ 0 \end{bmatrix} = \begin{bmatrix} 1 \\ 0 \\ 0 \end{bmatrix} = \mathbf{h}_4,$$

which is the fourth column of H. Thus we immediately know that a mistake occurred in the fourth position. Then we decode $\mathbf{w}$ by correcting the fourth coordinate and cutting off the three last check symbols to get $(1\ 1\ 1\ 0)$.

(c) We saw in (a) that $\mathbf{u} = (1\ 1\ 0\ 1)$ will be encoded to $\mathbf{z} = (1\ 1\ 0\ 1\ 0\ 0\ 1)$. This vector will be decoded back to $\mathbf{u}$. Moreover, if one mistake happens during the transmission, then we receive a vector $\mathbf{z} + \mathbf{e}$, with $\mathrm{wt}(\mathbf{e}) = 1$, then again $\mathbf{z} + \mathbf{e}$ will be decoded to $\mathbf{u}$ since the code corrects all single mistakes. No $\mathbf{z} + \mathbf{e}$, when $\mathrm{wt}(\mathbf{e}) > 1$, will be corrected to $\mathbf{u}$, since the Hamming code does not correct any combination of two or more mistakes. There are seven possibilities to choose $\mathbf{e}$, hence apart from $\mathbf{z}$, the only vectors which are decoded to $\mathbf{u}$ are

$$(0\ 1\ 0\ 1\ 0\ 0\ 1),\ (1\ 0\ 0\ 1\ 0\ 0\ 1),\ (1\ 1\ 1\ 1\ 0\ 0\ 1),\ (1\ 1\ 0\ 0\ 0\ 0\ 1),$$

$$(1\ 1\ 0\ 1\ 1\ 0\ 1),\ (1\ 1\ 0\ 1\ 0\ 1\ 1),\ (1\ 1\ 0\ 1\ 0\ 0\ 0).$$

2. Such a code contains $2^k - k - 1$ information symbols, hence 2^{2^k-k-1} codewords. The number of points in a ball of radius 1 is $2^k - 1 + 1 = 2^k$. So if we surround each codeword with a ball of radius 1, they will collectively contain $2^{2^k-k-1} \cdot 2^k = 2^{2^k-1}$ codewords and these are all vectors in $\mathbb{Z}^{2^k-1}$. This means that all single mistakes will be corrected but no more mistakes will be.

Solutions to Exercises of Sect. 7.1.7

1. (a) We find $\mathbf{a}(x) = 1 + x^2 + x^4$ and

$$\mathbf{b}(x) = \mathbf{a}(x)g(x) = (1 + x^2 + x^4)(1 + x + x^3) = 1 + x^2 + x^4 + x^7$$

Hence $\mathbf{b} = (1\ 0\ 1\ 0\ 1\ 0\ 0\ 1)$.

(b) The matrix G has five rows and eight columns:

$$G = \begin{bmatrix} 1\ 1\ 0\ 1\ 0\ 0\ 0\ 0 \\ 0\ 1\ 1\ 0\ 1\ 0\ 0\ 0 \\ 0\ 0\ 1\ 1\ 0\ 1\ 0\ 0 \\ 0\ 0\ 0\ 1\ 1\ 0\ 1\ 0 \\ 0\ 0\ 0\ 0\ 1\ 1\ 0\ 1 \end{bmatrix}.$$

(c) Row reducing, we find a matrix G' for a systematic code with the same minimum distance:

$$G \rightarrow G' = \begin{bmatrix} 1\ 0\ 0\ 0\ 0\ 0\ 0\ 1 \\ 0\ 1\ 0\ 0\ 0\ 1\ 1\ 0 \\ 0\ 0\ 1\ 0\ 0\ 0\ 1\ 1 \\ 0\ 0\ 0\ 1\ 0\ 1\ 1\ 1 \\ 0\ 0\ 0\ 0\ 1\ 1\ 0\ 1 \end{bmatrix}.$$

From this we form the parity check matrix

$$H' = \begin{bmatrix} 0\ 1\ 0\ 1\ 1\ 1\ 0\ 0 \\ 0\ 1\ 1\ 1\ 0\ 0\ 1\ 0 \\ 1\ 0\ 1\ 1\ 1\ 0\ 0\ 1 \end{bmatrix}.$$

Solutions to Exercises of Sect. 7.1.8

1. (a) This binary BCH code C has parameters $n = 15$ (length of codewords) and $d = 7$ (minimum distance). We have to choose a primitive element α of K, then the generating polynomial of the code C will be given by a formula

$$g(t) = \operatorname{lcm}\left[(m_1(t), m_2(t), m_3(t), m_4(t), m_5(t), m_6(t)\right].$$

where $m_i(t)$ is the minimal annihilating polynomial of α^i. Since every element has the same minimal annihilating polynomial as its square, we have $m_1(t) = m_2(t) = m_4(t)$ and $m_3(t) = m_6(t)$, hence

$$g(t) = (m_1(t)m_3(t)m_5(t).$$

The polynomial $m(x) = 1 + x^3 + x^4$ does not have roots in $\mathbb{Z}_2$ as $m(0) = m(1) = 1$. It is not divisible by the only irreducible polynomial of degree 2, namely $x^2 + x + 1$. Indeed,

$$x^4 + x^3 + 1 = (x^2 + 1)(x^2 + x + 1) + x.$$

This means that $m(x)$ is irreducible in $\mathbb{Z}_2[x]$. If it were reducible, then it could be factorised into a product of two polynomials, one of degree 1 and one of degree 3 or else it can be $(x^2 + x + 1)^2$. The latter, as we saw, is not true. The former is not possible since a linear factor means that $m(x)$ has a root in $\mathbb{Z}_2$. We know that the ring $K = \mathbb{Z}_2[x]/(m(x))$ is a field if $m(x)$ is irreducible. Since $m(x)$ is irreducible, K is a field. We see from it that x is a primitive element of K since all powers of x are different and represent every element of K. We take $\alpha = x$.

What is the minimal irreducible polynomial $m_1(t)$ of α? Let $p(t) = t^4 + t^3 + 1$. Then $m_1(x) = x^4 + x^3 + 1 = 0$ in K, i.e., x is a root of $p(t) = t^4 + t^3 + 1$. On the other hand, from the table we see that $1, x, x^2$ and x^3 are linearly independent, hence 4 is the minimal degree of an anihilating polynomial. Hence $m_1(t) = p(t) = t^4 + t^3 + 1$.

4-tuple	polynomial	power of x	logarithm
0000	0	0	$-\infty$
1000	1	1	0
0100	x	x	1
0010	x^2	x^2	2
0001	x^3	x^3	3
1001	$1 + x^3$	x^4	4
1101	$1 + x + x^3$	x^5	5
1111	$1 + x + x^2 + x^3$	x^6	6
1110	$1 + x + x^2$	x^7	7
0111	$x + x^2 + x^3$	x^8	8
1010	$1 + x^2$	x^9	9
0101	$x + x^3$	x^{10}	10
1011	$1 + x^2 + x^3$	x^{11}	11
1100	$1 + x$	x^{12}	12
0110	$x + x^2$	x^{13}	13
0011	$x^2 + x^3$	x^{14}	14

To compute the minimal annihilating polynomial for $\beta = x^3$, we use the table to find $\beta^2 = x^6 = 1 + x + x^2 + x^3$, $\beta^3 = x^9 = 1 + x^2$, $\beta^4 = x^{12} = 1 + x$. These elements must be already linearly dependent (as any other five vectors in a four- dimensional space), and we use Linear Dependency Relationship algorithm to find that dependency:

$$
\begin{bmatrix} 1\,0\,1\,1\,1 \\ 0\,0\,1\,0\,1 \\ 0\,0\,1\,1\,0 \\ 0\,1\,1\,0\,0 \end{bmatrix}
\rightarrow
\begin{bmatrix} 1\,0\,1\,1\,1 \\ 0\,1\,1\,0\,0 \\ 0\,0\,1\,1\,0 \\ 0\,0\,1\,0\,1 \end{bmatrix}
\rightarrow
\begin{bmatrix} 1\,0\,0\,0\,1 \\ 0\,1\,1\,0\,0 \\ 0\,0\,1\,1\,0 \\ 0\,0\,0\,1\,1 \end{bmatrix}
\rightarrow
$$

$$
\begin{bmatrix} 1\,0\,0\,0\,1 \\ 0\,1\,1\,0\,0 \\ 0\,0\,1\,0\,1 \\ 0\,0\,0\,1\,1 \end{bmatrix}
\rightarrow
\begin{bmatrix} 1\,0\,0\,0\,1 \\ 0\,1\,0\,0\,1 \\ 0\,0\,1\,0\,1 \\ 0\,0\,0\,1\,1 \end{bmatrix}
$$

and we see that $\beta^4 = 1 + \beta + \beta^2 + \beta^3$, while $1, \beta, \beta^2, \beta^3$ are linearly independent. Thus, $m_3(t) = t^3 + t^2 + t + 1$ is the minimal annihilating polynomial for β.

To find the minimal annihilating polynomial $m_5(t) \in \mathbb{Z}_2[t]$ of the element $\gamma = x^5$, we calculate the coordinate tuples of the following powers of γ:

$$
\begin{aligned}
\gamma^0 &= x^0 = 1 & &\rightarrow 1000 \\
\gamma^1 &= x^5 = x + x^3 & &\rightarrow 1101 \\
\gamma^2 &= x^{10} = x + x^3 & &\rightarrow 0101
\end{aligned}
$$

These first three powers are already linearly dependent, so we do not have to compute any more powers. We even do not need to use the Linear Dependency

Relationship algorithm to find a linear dependency between these tuples. It is obvious that $\gamma^2 = 1 + \gamma$, whence the minimal annihilating polynomial will be $f(t) = t^2 + t + 1$ (because there can be no annihilating polynomials of degree 1 as $x \notin \mathbb{Z}_2$). We can now calculate

$$g(t) = (m_1(t)m_3(t)m_5(t) = (t^4 + t^3 + 1)(t^4 + t^3 + t^2 + t + 1)(t^2 + t + 1) =$$

$$t^{10} + t^9 + t^8 + t^6 + t^5 + t^2 + 1.$$

(b) The number of information symbols is $m = n - \deg(g) = 15 - 10 = 5$.
(c) This must be 5×15 matrix. Here it is

$$G = \begin{bmatrix} 1 & 0 & 1 & 0 & 0 & 1 & 1 & 0 & 1 & 1 & 1 & 0 & 0 & 0 & 0 \\ 0 & 1 & 0 & 1 & 0 & 0 & 1 & 1 & 0 & 1 & 1 & 1 & 0 & 0 & 0 \\ 0 & 0 & 1 & 0 & 1 & 0 & 0 & 1 & 1 & 0 & 1 & 1 & 1 & 0 & 0 \\ 0 & 0 & 0 & 1 & 0 & 1 & 0 & 0 & 1 & 1 & 0 & 1 & 1 & 1 & 0 \\ 0 & 0 & 0 & 0 & 1 & 0 & 1 & 0 & 0 & 1 & 1 & 0 & 1 & 1 & 1 \end{bmatrix}.$$

(d) $(1\ 1\ 1\ 1\ 1)G = (1\ 1\ 0\ 0\ 0\ 1\ 0\ 1\ 0\ 0\ 1\ 1\ 0\ 1)$.
2. We need to construct the field $GF(2^8)$, find a primitive element α of this field and then calculate the generator polynomial as

$$g(x) = \text{lcm}(m_1(x), \ldots, m_6(x)) = m_1(x)m_3(x)m_5(x),$$

where $m_i(x)$ is the minimal annihilator polynomial of α^i. Here is the calculation:

```
gap> F:=GF(256);
GF(2^8)
gap> elts:=Elements(F);
[ 0*Z(2), Z(2)^0, Z(2^2), Z(2^2)^2, Z(2^4), Z(2^4)^2, Z(2^4)^3, Z(2^4)^4,
Z(2^4)^6, Z(2^4)^7, Z(2^4)^8, Z(2^4)^9, Z(2^4)^11, Z(2^4)^12, Z(2^4)^13,
Z(2^4)^14, Z(2^8), Z(2^8)^2, Z(2^8)^3, ...

gap> a:=elts[17];
Z(2^8)
gap> m1:=MinimalPolynomial(GF(2),a);
x_1^8+x_1^4+x_1^3+x_1^2+Z(2)^0
gap> m3:=MinimalPolynomial(GF(2),a^3);
x_1^8+x_1^6+x_1^5+x_1^4+x_1^2+x_1+Z(2)^0
gap> m5:=MinimalPolynomial(GF(2),a^5);
x_1^8+x_1^7+x_1^6+x_1^5+x_1^4+x_1+Z(2)^0

gap> g(x)=m1*m3*m5;
x_1^24+x_1^23+x_1^21+x_1^20+x_1^19+x_1^17+x_1^16+x_1^15+x_1^13+x_1^8+x_1^7+
x_1^5+x_1^4+x_1^2+Z(2)^0
```

Hence the generator polynomial $g(x)$ of this code will be

$$g(x) = x^{24} + x^{23} + x^{21} + x^{20} + x^{19} + x^{17} + x^{16} + x^{15} + x^{13} + x^8 + x^7 + x^5 + x^4 + x^2 + 1.$$

Solutions to Exercises of Sect. 7.2.1

1. No, the second column is negative of the last column. By Theorem 7.2.3 this code
   does not correct all single errors.

   (a) All columns are different and no one is the negative of another one. By Theo-
       rem 7.2.3 this code corrects all single errors.
   (b) Let us row reduce

$$H_2 = \begin{bmatrix} 1\,2\,1\,2\,1\,1 \\ 1\,2\,1\,0\,2\,2 \\ 2\,1\,0\,1\,0\,1 \end{bmatrix} \longrightarrow \begin{bmatrix} 1\,2\,1\,2\,1\,1 \\ 1\,2\,1\,0\,2\,2 \\ 0\,0\,1\,0\,1\,2 \end{bmatrix} \longrightarrow$$

$$\begin{bmatrix} 2\,1\,2\,1\,2\,2 \\ 1\,2\,1\,0\,2\,2 \\ 0\,0\,1\,0\,1\,2 \end{bmatrix} \longrightarrow \begin{bmatrix} 1\,2\,1\,1\,0\,0 \\ 1\,2\,1\,0\,2\,2 \\ 0\,0\,1\,0\,1\,2 \end{bmatrix} \longrightarrow \begin{bmatrix} 1\,2\,1\,1\,0\,0 \\ 2\,1\,2\,0\,1\,1 \\ 0\,0\,1\,0\,1\,2 \end{bmatrix} \longrightarrow$$

$$\begin{bmatrix} 1\,2\,1\,1\,0\,0 \\ 2\,1\,2\,0\,1\,1 \\ 1\,2\,2\,0\,0\,1 \end{bmatrix} \longrightarrow \begin{bmatrix} 1\,2\,1 & 1\,0\,0 \\ 1\,2\,0 & 0\,1\,0 \\ 1\,2\,2 & 0\,0\,1 \end{bmatrix} = (A \mid I_3).$$

   Hence the generator matrix for this code will be

$$G = (I_3 \mid -A^T) = \begin{bmatrix} 1\,0\,0 & 2\,2\,2 \\ 0\,1\,0 & 1\,1\,1 \\ 0\,0\,1 & 2\,0\,1 \end{bmatrix}.$$

   (c) Decoding **y** we calculate

$$H\mathbf{y}^T = \begin{bmatrix} 2 \\ 2 \\ 0 \end{bmatrix} = 2\mathbf{h}_3.$$

   Hence the error vector was (0 0 2 0 0 0) and the codevector sent was (0 2 0 2 2 2).
   Hence the message was (0 2 0).

Solutions to Exercises of Sect. 7.2.2

1. (a) The powers of α in the following table generate all of F^*, so α is a generator
       of F.

i	α^i
0	$x^0 = 1$
1	$x^1 = x$
2	$x^2 = (x^2 + 2x + 2) + x + 1 = x + 1$
3	$x^3 = x(x+1) = x^2 + x = (x^2 + 2x + 2) + 2x + 1 = 2x + 1$
4	$x^4 = x(2x+1) = 3x^2 + x = 2(x^2 + 2x + 2) + 2 = 2$
5	$x^5 = 2x$
6	$x^6 = 2x^2 = 2(x^2 + 2x + 2) + 2x + 2 = 2x + 2$
7	$x^7 = x(2x+2) = 2(x^2 + 2x + 2) + x + 2 = x + 2$
8	$x^8 = x(x+2) = x^2 + 2x = (x^2 + 2x + 2) + 1 = 1$

(b) First, we note that these quadratic polynomials are irreducible (it is straight-forward to check that they have no roots) and monic.

Evaluating, we find that $m_1(\alpha) = \alpha^2 + 2\alpha + 2 = (x + 1) + 2x + 2 = 3x + 3 = 0$. So $m_1(x)$ is an annihilating polynomial of α. Since it is irreducible and monic, it is the minimal annihilating polynomial of α.

Similarly $m_2(\alpha^2) = \alpha^4 + 1 = 2 + 1 = 3 = 0$, so $m_2(x)$ is the minimal annihilating polynomial of α^2.

Finally $\qquad m_3(\alpha^3) = \alpha^6 + 2\alpha^3 + 2 = (2x + 2) + 2(2x + 1) + 2 = 6x + 6 = 0$ is the minimal annihilating polynomial of α^3.

(c) Note that $m_1(x) = m_3(x)$.
$g(x) = \text{lcm}(m_1(x), m_2(x), m_3(x)) = m_1(x)m_2(x) = (x^2 + 2x + 2)(x^2 + 1) = x^4 + 2x^3 + 2x + 2$.

(d) s $m = n - deg(g)$, so there are $8 - 4 = 4$ information symbols.

(e) The generator matrix will be

$$G = \begin{bmatrix} 2 & 2 & 0 & 2 & 1 & 0 & 0 & 0 \\ 0 & 2 & 2 & 0 & 2 & 1 & 0 & 0 \\ 0 & 0 & 2 & 2 & 0 & 2 & 1 & 0 \\ 0 & 0 & 0 & 2 & 2 & 0 & 2 & 1 \end{bmatrix}.$$

Solutions to Exercises of Sect. 7.3.1

1. Let $X = \{\mathbf{x}_1, \mathbf{x}_2, \mathbf{x}_3\}$, where

$$\mathbf{x}_1 = (1\ 1\ 1\ 0\ 0\ 0\ 2\ 2\ 2),$$
$$\mathbf{x}_2 = (1\ 1\ 2\ 2\ 0\ 0\ 1\ 1\ 2),$$
$$\mathbf{x}_3 = (1\ 2\ 2\ 0\ 2\ 0\ 1\ 2\ 0).$$

(a) $P_1(X) = \{1\}$, $P_2(X) = \{1, 2\}$, $P_3(X) = \{1, 2\}$.

(b) There are $2^7 = 128$ elements in the envelope desc(X).

(c) We use the "majority" rule to obtain $(1\ 1\ 2\ 0\ 0\ 0\ 1\ 2\ 2)$. This vector can be produced by any pair of vectors in X.

2. For example,

$$\mathbf{c}_1 = (1\ 1), \quad \mathbf{c}_2 = (2\ 1), \quad \mathbf{c}_3 = (1\ 2), \quad \mathbf{c}_4 = (2\ 2).$$

3. In the ith coordinate of a descendant we can find any element from $P_i(X)$ and they can be chosen independently of each other. Hence the total number of descendants is $m_1 \cdot \ldots \cdot m_n$.

Solutions to Exercises of Sect. 7.3.2

The code $\mathcal{C} \subset \{1, 2, 3\}^6$ consists of six codewords:

$$\mathbf{c}_1 = (1\ 1\ 1\ 1\ 1\ 1),$$
$$\mathbf{c}_2 = (2\ 2\ 2\ 2\ 2\ 2),$$
$$\mathbf{c}_3 = (3\ 3\ 3\ 3\ 3\ 3),$$
$$\mathbf{c}_4 = (1\ 2\ 3\ 1\ 2\ 3),$$
$$\mathbf{c}_5 = (2\ 3\ 1\ 2\ 3\ 1),$$
$$\mathbf{c}_6 = (3\ 1\ 2\ 3\ 1\ 2).$$

1. By inspection we see that $d_{\min}(\mathcal{C}) = 4$.
2. To prove that it is 2-frameproof we use Theorem 7.3.1. Indeed, $d_{\min}(\mathcal{C}) = 4 > 3 = 6(1 - 1/2)$.

Solutions to Exercises of Sect. 7.3.3

1. Let us choose a set $X = \{\mathbf{x}_1, \mathbf{x}_2, \ldots, \mathbf{x}_{w+1}\}$ consisting of any $w + 1$ codewords from $\mathcal{C}$. Let $1 \leq i \leq n$ and let us consider the projection $P_i(X)$. Since $q < w + 1$ at least one element of $P_i(X)$ is the ith coordinate of at least two vectors in X. Let us call this element c_i. Then a vector $\mathbf{c} = (c_1, c_2, \ldots, c_n)$ belongs to any $\mathrm{desc}(X')$, where X' is a subset of X of cardinality w. Hence $\mathcal{C}$ does not have the w-IPP property.
2. Using Reed–Solomon code $\mathcal{C}$ over $\mathbb{Z}_{17}$ of length 16 with the minimum distance 13, we can show that there exists a fingerprinting code with the identifiable parent property of order 2 containing 83521 codewords. Such a Reed–Solomon code, if we take 3 as the primitive element, will have a generating polynomial

$$g(x) = (x - 3)(x - 9)(x - 10)(x - 13)(x - 5)(x - 15)(x - 11)(x - 16)(x - 14)(x - 8)(x - 7)(x - 4).$$

of degree 12. Being of length 16, it will have distance 13 and have 12 check symbols and four information symbols, hence $7^4 = 83521$ codewords. Since

$$d_{\min} = 13 > 12 = 16(1 - 1/2^2),$$

it will have the identifiable parent property of order 2 by Theorem 7.3.3.

11.8   Solutions to Exercises of Chap. 8

Solutions to Exercises of Sect. 8.1.2

1. We have three words of length 2 and two words of length 3. Since

$$\frac{1}{2^2} + \frac{1}{2^2} + \frac{1}{2^2} + \frac{1}{2^3} + \frac{1}{2^3} = 1$$

such a code exists due to Kraft's inequality. The tree would be, for example,

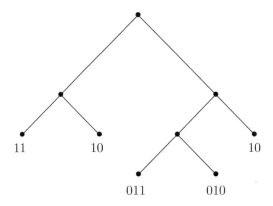

2. For example, $\{11, 10, 011, 010, 0011, 0010, 0001, 0000\}$.
3. (a) We calculate

$$\frac{1}{2^2} + \frac{1}{2^3} + \frac{1}{2^3} + \frac{1}{2^3} + \frac{1}{2^3} + \frac{1}{2^3} + \frac{1}{2^4} + \frac{1}{2^5} + \frac{1}{2^2} + \frac{1}{2^5} = \frac{1}{4} + 5 \cdot \frac{1}{8} + \frac{1}{16} + 2 \cdot \frac{1}{32} = 1.$$

   Hence such a prefix code does exist.
   (b) Straightforward.
   (c) Example of such a code:

$$\{11, 100, 101, 011, 010, 001, 0001, 00001, 00000\}.$$

Solutions to Exercises of Sect. 8.1.3

1. Let the alphabet consist of a, b, c, d with frequencies $0.1, 0.2, 0.3$ and 0.4, respectively. After joining a and b into a single vertex u the frequencies of u, c, d will be $0.3, 0.3$ and 0.4, respectively. At this point there are two choices for d to partner with: either with c or with subtree combining a and b. There will be two different Huffman's trees as a result.
2. Omitted.
3. The average bit number for the given code is

$$(0.26 + 0.24) \times 2 + (0.14 + 0.13 + 0.12 + 0.11) \times 3 = 2.5.$$

The code is clearly not a Huffman code since the level 4 leaf does not have a sibling. But to decide on its optimality we have to construct the Huffman code. It will be slightly different:

Source symbol	Probability	Codeword
x_1	0.26	00
x_2	0.24	01
x_3	0.14	100
x_4	0.13	101
x_5	0.12	110
x_6	0.11	111

but with the same average bit per symbol. The original code is not Huffman's code since the two least frequent symbols x_5 and x_6 in Huffman's code must be siblings.

4. Since the distribution is uniform, the Huffman tree will consist of word lengths of $\lceil \log_2 100 \rceil = 7$ and $\lfloor \log_2 100 \rfloor = 6$. As $2^6 < 100$ we should have codewords of length 7. Suppose in the contrary we have a leaf of depth 5 corresponding to symbol x. Then we can give to this leaf two children converting it into an internal node and make both children leaves and assigning x to one of them. This will leave us with an unassigned node of level 6 to which we move one of the words that were previously assigned a code of length 7. At this point our changes are neutral (unless x was at level 4 or higher). That word x had a sibling y also of depth 7, which will now become the only child of a certain node of depth 6. In such case that node of degree 6 can be deleted and y will move to level 6. Then average bit length of the code will be improved, contradiction.
   There are 64 nodes of depth 6, of which $(64 - k)$ will be leaf nodes; and there are k internal nodes of depth 6 which will be parents of $2k$ leaf nodes of depth 7. Since the total number of leaf nodes is 100, we have $(64 - k) + 2k = 100$, from which $k = 36$. So there are $64 - 36 = 28$ codewords of length 6 and $2 \cdot 36 = 72$ codewords of length 7.

5. This proves that for every full tree the Kraft's inequality becomes an equality. Let $m = \max_{i=1}^{q} m_i$ be the maximal length of a codeword, and S_i be the number of codewords of length i. Arguing as in the proof of Kraft's inequality we will obtain that

$$2^m - S_1 \cdot 2^{m-1} - S_2 \cdot 2^{m-2} - \ldots - S_{m-1} \cdot 2 - S_m > 0,$$

which means that at the last mth level not all words are prohibited and we can add another one and connect it with a path to the root. The internal vertex to which this path is connected had therefore only one child. This means that the tree was not full, contradiction.

6. Huffman's tree is full, and the result follows from the previous exercise.

Solutions to Exercises of Sect. 8.2.1

1. $wt(\mathbf{u}) = 4$ and $wt(\mathbf{v}) = 5$.
2. There are $\binom{7}{3} = 35$ vectors of Hamming weight 3 in $\mathbb{Z}_2^7$ and $\binom{7}{5} = 21$ vectors of Hamming weight 5.

Solutions to Exercises of Sect. 8.2.2

1. 5 bits.
2. The magician has five cards with numbers on them and no two numbers coincide. He can put them in increasing order, decreasing order or something in between. There are $5! = 120$ ways to arrange the cards, and each arrangement can be associated with an integer between 0 and 100. The magician and the assistant must agree in advance on the way arrangements of cards are associated with the numbers.

Solutions to Exercises of Sect. 8.2.3

1. $a \prec c \prec b$.
2. The number of vectors of weight k such that $3 < k \leq 5$ in $\mathbb{Z}^{10}$ is equal to

$$\binom{10}{4} + \binom{10}{5} = 210 + 252 = 462.$$

3. We have $n_1 = 3, n_2 = 4, n_3 = 6, n_4 = 7$, hence

$$N(\mathbf{w}) = \binom{7-3}{4} + \binom{7-4}{3} + \binom{7-6}{2} + \binom{7-7}{1} = 2.$$

4. (a) The maximum weight is 15 and $\lceil \log_2 15 \rceil = 4$. This is the length of the prefix.
   (b) As $wt(\mathbf{x}) = 3$, the prefix will be 0011.
   (c) The length of the suffix will be $\lceil \log_2 \binom{15}{3} \rceil = 9$.
   (d) As $n_1 = 5, n_2 = 7, n_3 = 13$, the number of $\mathbf{x}$ in the orbit is

$$N(\mathbf{x}) = \binom{15-5}{3} + \binom{15-7}{2} + \binom{15-13}{1} = 150 = 10010110_{(2)}.$$

   As $N(\mathbf{x})$ has 8 binary digits we have to put an additional zero in front of its 8 digits immediately after the prefix, i.e., the suffix would be 010010110.
   (e) Hence $\mathbf{x}$ will be encoded into 0011010010110.

Solutions to Exercises of Sect. 8.2.4

1. We separate the first four bits 0010 of $\psi(\mathbf{y})$, this is the prefix and we discover that $wt(\mathbf{y}) = 2$. From the remaining part we see that the number $N(\mathbf{y})$ of $\mathbf{y}$ in the orbit is $11110_{(2)} = 30$. From Pascal's triangle we solve the equation

$$\binom{15-n_1}{2} + \binom{15-n_2}{1} = N(\mathbf{y}) = 30$$

by finding

$$\binom{15-n_1}{2} = 28 = \binom{8}{2}, \qquad \binom{15-n_2}{1} = 2 = \binom{2}{1}.$$

Hence $n_1 = 7$ and $n_2 = 13$, and $\mathbf{y} = 000000100000100$.

Since our codelength is very short, no real compression occurs in this case.

2. We will make use of the Pascal's triangle

$$
\begin{array}{ccccccccccccccc}
 & & & & & & & 1 & & & & & & & \\
 & & & & & & 1 & & 1 & & & & & & \\
 & & & & & 1 & & 2 & & 1 & & & & & \\
 & & & & 1 & & 3 & & 3 & & 1 & & & & \\
 & & & 1 & & 4 & & 6 & & 4 & & 1 & & & \\
 & & 1 & & 5 & & 10 & & 10 & & 5 & & 1 & & \\
 & 1 & & 6 & & 15 & & 20 & & 15 & & 6 & & 1 & \\
1 & & 7 & & 21 & & 35 & & 35 & & 21 & & 7 & & 1 \\
\end{array}
$$

We see that $\binom{7}{4} = 35 \le 43 < 56 = \binom{8}{4}$, hence $x_1 = 7$. Then $43 - 35 = 8$ and 4 is maximal such that $\binom{4}{3} = 4 \le 8$, hence $x_2 = 4$. Now $8 - 4 = 4$ and 3 is maximal such that $\binom{3}{2} = 3 \le 4$. Hence $x_3 = 3$. Now $4 - 3 = 1$ and 1 is maximal such that $\binom{1}{1} \le 1$, hence $x_4 = 1$.

3. (a) The maximum weight is 15 and $\lceil \log_2 16 \rceil = 4$. This is the length of the prefix.
   (b) The length of the suffix will be $\lceil \log_2 \binom{15}{3} \rceil = 9$.
   (c) As $wt(\mathbf{x}) = 3$, the prefix will be 0011. As $n_1 = 4$, $n_2 = 9$, $n_3 = 12$, the number of $\mathbf{x}$ in the orbit is

$$N(\mathbf{x}) = \binom{15-4}{3} + \binom{15-9}{2} + \binom{15-12}{1} = 165 + 15 + 3 = 183 = 10110111_{(2)}.$$

As $N(\mathbf{x})$ has 8 binary digits we have to put an additional zero in front of its 8 digits immediately after the prefix. Hence $\mathbf{x}$ will be encoded into $\psi(\mathbf{x}) = 0011010110111$.

(e) We separate the first four bits 0010 of $\psi(\mathbf{y})$, this is the prefix and we discover that $wt(\mathbf{y}) = 2$. From the remaining part we see that the number $N(\mathbf{y})$ of $\mathbf{y}$ in the orbit is $11110_{(2)} = 30$. From Pascal's triangle we solve the equation

$$\binom{15-n_1}{2} + \binom{15-n_2}{1} = N(\mathbf{y}) = 30$$

by finding

$$\binom{15-n_1}{2} = 28 = \binom{8}{2}, \qquad \binom{15-n_2}{1} = 2 = \binom{2}{1}.$$

Hence $n_1 = 7$ and $n_2 = 13$, and $\mathbf{y} = 000000100000100$.

Literature

1. Agrawal, M., Kayal, N., and Saxena, N. Primes is in P. Dept of Computer Science and Engineering, Indian Institute of Technology, Kanpur, India, 6 August, 2002.
2. Alford, W.R., Granville, A., and Pomerance, C. There are infinitely many Carmichael numbers. Ann. Math. **140** (1994), 703–722.
3. Atkins, D., Graff, M., Lenstra, A.K., and Leyland, P.C. The magic words are squeamish ossifrage. ASIACRYPT-94, Lecture Notes in Computer Science, 917, Springer, 1995.
4. Boneh, D and Shaw, J. Collusion-Secure Fingerprinting for Digital Data. IEEE Transactions on Information Theory, 44(5), 1897–1905, 1998.
5. Brickell, E.F. and Davenport, D.M. (1991) On the classification of ideal secret sharing schemes. Journal of Cryptology, 4: 123–134.
6. Daniel Shanks. Five Number Theoretic Algorithms. Proceedings of the Second Manitoba Conference on Numerical Mathematics. pp. 51–70, 1973.
7. Desmedt, Y. G. (1988) Society and group oriented cryptography: a new concept. In Advances in Cryptology, Proceedings of Crypto 1987 (Lecture Notes in Computer Science 293) C. Pomerance, Ed. Springer–Verlag pp. 120–127.
8. Desmedt, Y. G. (1994) Threshold cryptography. European Transactions on Telecommunications 5: 449–457.
9. Diffie, W., and Hellman, M., 1976, New directions in cryptography, IEEE Transactions on Information Theory IT–22, 644–654.
10. Fitingof B.M. (1966) Optimal Encoding under an Unknown or Changing Statistics. Problems of Information Transmission, Vol. 2(2), 3–11.
11. Koblitz, N. Algebraic Aspects of Cryptography. Springer, 1998
12. Koblitz, N. Elliptic Curve Cryptosystems. Mathematics of Computation. 48: pp. 203–209.
13. Kolmogorov A.N. (1965) Three Approaches to the Definition of the Concept "the Quantity of Information". Problems of Information Transmission, Vol. 1(1), 3–11.
14. Lenstra, A.K., Lenstra, H.W., Manasse, M.S., Pollard, J.M. The number field sieve. Proc. 22nd Ann. ACM Symposium on Theory of Computing. Baltimore, May 14-16, 1990. pp 564–572.
15. Macwilliams, F.J., and Sloane, N.J.A. The Theory of Error-Correcting Codes. North-Holland, 1977.
16. Miller, V. Uses of Elliptic Curves in Cryptography. Advances in Cryptology - Crypto'85. pp. 417–426. 1986.
17. Peterson W.W., and Weldon, E.J. Error-Correcting Codes. 2nd ed. MIT Press, 1972.

© Springer Nature Switzerland AG 2020
A. Slinko, *Algebra for Applications*, Springer Undergraduate Mathematics Series,
https://doi.org/10.1007/978-3-030-44074-9

18. Rivest, R.L., Shamir, A. and Adelman, L. A method for obtaining digital signatures and public key cryptosystems. Commun. ACM, **21(2)** (1978), 120–126.
19. Ross, K. and Wright, K. Discrete Mathematics, Prentice Hall. 1999.
20. Schwenk, Jörg, and Klaus Huber. (1998) Public key encryption and digital signatures based on permutation polynomials. Electronics Letters 34.8: 759–760.
21. Shamir, A. (1979) How to share a secret. Communications of the ACM 22 : 612–613.
22. Song Y. Y. Primality testing and integer factorization in public key cryptography. Kluwer, 2004.
23. Staddon, J. N., Stinson, D. R. and Wei, R. Combinatorial properties of frameproof and traceability codes. IEEE Transactions on Information Theory. 47(3), 1042–1049, 2001.
24. Stinson, D.R. (1992) An explication of secret sharing schemes. Designs, Codes and Cryptography, 2:357–390.
25. Trappe, W., and Washington, L.C. Introduction to Cryptography with Coding Theory. Prentice Hall. 2002.
26. Williams, H.C. Primality testing on a computer. Ars Combinatoria, **5** (1978), 127–185.

Index

A

Access structure, 172, 180
  k-out-of-n, 172, 182
  linear, 182
  threshold, 172
Addition of points on elliptic curve, 115
Advanced Encryption Standard (AES), 86, 162
Algorithm
  Agrawal–Kayal–Saxena (AKS) for primality
    testing, 73
  of decoding of Fitingof's code, 251
  division of polynomials, 149
  Double and Add, 123
  Euclidean, 17, 19, 60, 62, 67
  Euclidean for polynomials, 157, 273
  Extended Euclidean, 19, 22, 32
  Extended Euclidean for polynomials, 157,
    158, 161, 273
  Lagrange's interpolation, 168
  linear dependency relationship, 165, 220,
    275
  of Schwenk and Huber, 169
  of secret recovery, 177
  Square and Multiply, 59, 62, 71, 120, 123
  Tonelli–Shanks, 120
  Trial Division, 10–12, 61
All-or-nothing property, 189
Asymptotically equal, 54
Authorised coalition, 172
  minimal, 172

B

Ball, 194
Basis of a vector space, 135

Bertrand's Postulate, 12
Binary (m, n)-code, 195
Binary entropy function, 254
Binary tree
  rooted, 237
Bit of information, 37, 235, 245

C

Certification authority, 65
Characteristic of a field, 137
Chinese remainder theorem, 21, 26, 66, 169,
    262
Cipher
  Atbash, 42
  Caesar, 42
  permutation, 84, 85
  product, 85
  substitution, 46
Ciphertext, 43, 64, 169
Code
  BCH, 217
  binary, 237
  compression, 244
  equivalent, 205
    Fitingof's, 249
  frameproof, 230
  Hamming, 212, 214
  Huffman's, 240
  Huffman's optimal, 241
  (m, n), 201
  optimal, 241
  parity check, 195
  perfect, 214
  polynomial, 215

© Springer Nature Switzerland AG 2020
A. Slinko, *Algebra for Applications*, Springer Undergraduate Mathematics Series,
https://doi.org/10.1007/978-3-030-44074-9

prefix, 237
Reed-Solomon (RS), 224
systematic, 204
triple repetition, 196
w-frameproof, 230
with identifiable parent property, 231
w-traceable, 233
Codevector, 195
Codeword, 195
Commutative group, 30
Complexity
average case, 54
worst case, 54
Complexity of
Euclidean algorithm, 59, 60, 63
Square and Multiply, 59
Trial Division, 61
Composition
of mappings, 79
of permutations, 81
Cryptology
public key, 51
secret key, 42
Cryptosystem
affine, 46
based on a permutation polynomial, 168
Elgamal, 144, 145
public key, 52, 62, 74
RSA, 62, 66, 67, 75, 79, 168, 190
secret key, 43
Cycle permutation, 86
Cycles
disjoint, 87

D

Data Encryption Standard (DES), 85
Decomposition into disjoint cycles, 87
Degree of a polynomial, 148
Descendant, 229
Diffie–Hellman
problem, 74
for elliptic curves, 126
secret key exchange, 74, 76, 126, 144
Digital signature, 74
Digits, 34
Dimension of a vector space, 136
Discrete logarithm, 143, 272
problem, 143
Discriminant of the cubic, 113
Disjoint cycles, 87, 267
Distribution table, 176
Divisible, 4
Division of polynomials with remainder, 149
Division with remainder, 4

Divisor, 4
Dummy, 172, 186

E
Elliptic curve, 113, 268
Encryption, 168
Error vector, 193
Euler's criterion, 120, 142
Euler's ϕ-function, 26, 63
Euler's theorem, 25, 27, 28, 38, 66
Euler's totient function, 26, 261
Exponent
decryption, 190
encryption, 190
Extended Euclidean algorithm, 63

F
Factorisation of integers, 57, 61, 68
Factorisation of polynomials, 156
Fermat numbers, 7
Fermat primes, 7
Fermat's little theorem, 24, 25, 27, 66 71, 72,
120
Field, 33, 130
extension, 164
finite, 159, 271
finite-dimensional, 134
Galois, 137, 162
$GF(p^n)$, 162
$\mathbb{Z}_p$, 129
15-puzzle, 97
Fingerprinting (watermarking) code, 228
Fitingof's theorem, 250
Function $\pi(x)$, 11, 57
Fundamental theorem of arithmetic, 5

G
Generator matrix, 203, 208, 209
Generator of
the cyclic group, 103
Generator polynomial, 215
Greatest common divisor, 15, 261
for polynomials, 157, 159
Group, 79, 83, 101
abelian, 101
alternating of degree n, 96
of an elliptic curve, 268
commutative, 83, 101
cyclic, 103
multiplicative of a field, 138, 145
symmetric
of degree n, 83, 96, 268
Growth

exponential, 57
factorial, 54
measure of, 56
polynomial, 57

H

Hadamard matrix, 199
Hamming distance, 193
Hamming weight, 193, 244
Hasse's theorem, 121
Hill's cryptosystem, 47, 50

I

Information ratio of the secret sharing scheme, 180
Information relative to a partition, 245
Integers, 4
    composite, 68
    modulo m, 29
    modulo n, 102
    positive, 1
Inverse of
    an element of the group, 101
    a permutation, 83
Invertible element, 31
Isomorphism, 106

K

Known plaintext attack, 50
Kraft's Inequality, 238

L

Lagrange's interpolation polynomial, 173
Lagrange's theorem, 110, 111, 139
Least common multiple, 15
    for polynomials, 158
Length of the secret, 180
L'Hospital's rule, 55, 57
Linearly dependent set of vectors, 134

M

Maximum likelihood decoding, 196
Mersenne primes, 8
Minimum distance, 198, 201
Minimum weight, 200
Mono-alphabet scheme, 42
Multiple, 15
Multiplicative inverse, 31
Multiplicity of a root of a polynomial, 151
Mutual inverses, 80

N

Natural numbers, 1

Notation
    big-Oh, 54
    big-Theta, 54
    little-oh, 54
Number
    Carmichael, 71, 73, 74
    composite, 5
    Fibonacci, 62
    floating-point real, 262
    prime, 5–12, 63, 64, 66, 69, 258
Number of positive divisors, 15
Numbers
    congruent modulo m, 23
    coprime, 17
    relatively prime, 17
    twin prime, 260

O

One-time pad, 43, 45, 46
One-way function, 52, 63
Order
    lexicographic, 248
Order of
    an element, 111, 138–141
    an element multiplicative, 262
    an element of the group, 104
    a permutation, 88, 268
    the group, 102

P

Parity check matrix, 207–209
Parity of permutation, 95
Permutation, 79, 81, 267
    even, 94
    interlacing shuffle, 91
    Josephus, 90
    odd, 94
    of degree n, 81, 83
    random, 268
Permutation group, 87
Permutations
    of degree n, 81
Plaintext, 43, 64, 168
Polynomials, 147
    annihilating, 164, 165
        minimal, 164, 165, 274
    coprime, 158
    irreducible, 154–156
    Lagrange's interpolation, 179
    modulo $m(x)$, 159, 161
    monic, 154
    permutation, 167–169
    reducible, 154
    relatively prime, 158

Positional notation of numbers, 34
Prefix of the codeword, 250
Prime factorisation, 5, 6, 8, 9, 11, 14–16, 27, 66, 260
Prime Number Theorem, The, 11
Primitive element, 161
Primitive element of a field, 141, 143
Primitive root modulo m, 263
Principle
  of mathematical induction, 2
  of strong mathematical induction, 2
  of the least integer, 1
Private exponent in RSA, 64
Private key, 53, 63
Private multiplier, 126
Pseudoprimality test, 69
  good, 69, 71
  Rabin–Miller, 71
Public exponent in RSA, 64
Public key, 53, 63, 66
Public Key Infrastructure (PKI), 65

Q
Quadratic non-residue, 119
Quadratic residue, 119, 124, 142, 143, 263
Quotient, 4, 261

R
Remainder, 4, 261
Representation of a number
  binary, 37, 58
  decimal, 34
  to base b, 35
Restricted envelope, 229
Ring
  commutative, 31
  of integers modulo n, 29
  of polynomials, 148, 272
Root of a polynomial, 151

S
Secret recovery function, 177
Secret sharing scheme, 172
  generalised linear, 188
  ideal, 187

  linear, 181
  perfect, 179, 180
  Shamir's, 174, 175
  threshold, 174
Sieve of Eratosthenes, 10
Singleton bound, 225
Smallest prime divisor, 10
Span, 134
Steps of an algorithm, 53
Stirling's formula, 54, 278
Subfield, 130, 164
Subgroup, 109, 110
Substitution methods, 42
Suffix of the codeword, 250
Symbols
  check, 204
  information, 204
Symmetric channel, 192
Symmetric group, 83
Syndrome, 208

T
Threshold cryptography, 190
Transposition, 93
Trapdoor, 52
  function, 52
Tree
  binary, 237
  binary rooted, 237
  fbinary
    ull, 241

U
Universal encoding, 243

V
Vandermonde determinant, 182, 276
Vector space, 132

W
Wilson's Theorem, 34

Z
Zero divisor, 32

Printed in the United States
By Bookmasters